STEEL CARRIAGE BY SEA

FOURTH EDITION

Arthur Sparks FNI is a qualified Master Mariner and served 17 years at sea in the British Merchant Service. After leaving the sea he spent 32 years as owner, managing director and principal surveyor of the marine and cargo surveying company, Sparks & Company NV, in the port of Antwerp, one of the largest steel handling ports in the world. For the last 12 years he has been Director of Sparks & Company (Maritime Consultants) Limited, London. Publications and professional bodies with which he has been associated are:

- Author of Lloyd's of London Press publication *Steel Carriage by Sea*, now in its 4th edition.
- Author of Anchorage Press publication *Steel Pre-shipment Surveys*, together with the North of England P&I Association.
- Author of *Guidelines for Masters, Surveyors and Correspondents on the Carriage of Steel Products*, published in March 1998, together with Skuld P&I Association, Oslo.
- Attributes: steel section in *Lloyd's Survey Handbook*, 7th edition (1999); steel section in *Bulk Carrier Practice*, published in 1993 by the London Nautical Institute; steel section in *Thomas's Stowage*, published in 1999 by Brownson & Ferguson.
- Fellow of the London Nautical Institute (FNI).
- For 30 years, member of the UK P&I Club's "Advisory Committee on Cargo Claims" (now an ex-member).

STEEL
CARRIAGE BY SEA

By

A. SPARKS, F.N.I.
Master Mariner
Marine and Cargo Surveyor

FOURTH EDITION

LLP
LONDON HONG KONG
2003

Informa Professional
(a trading division of Informa (UK) Ltd)
Mortimer House
37–41 Mortimer Street
London W1T 3JH
professional.enquiries@informa.com

EAST ASIA
Informa Professional Publishing Ltd
Suite 1802, The Centrium
60 Wyndham Street
Central
Hong Kong
informa.asia@informa.com

First published 1988
Second edition 1995
Third edition 1999
Fourth edition 2003

British Library Cataloguing in Publication Data
A catalogue record for this
book is available from the
British Library

ISBN 1-84311-256-6

Text typeset in 10/11 Century Schoolbook by
MFK Mendip
Frome, Somerset
Printed in Great Britain by
MPG Books Ltd
Bodmin, Cornwall

Preface

I do not foresee any drastic changes in the carriage of steel by sea over the next decade and bulk carrier vessels will continue to be the main form of steel carriage by sea, and in this edition I believe that everything, from the steel mill to the receiver's stores, which would be useful to the lawyer, surveyor, ship owner, ship master, claims settlers, importers and exporters of steel etc, has been dealt with. Over the years the book has been very popular, and I recall one meeting I attended when there were seven copies on the table.

Some rearrangement in this edition has taken place with an Appendix being added containing tables and information that I have found necessary to be at hand, as a Marine and Cargo Surveyor and also in expert witness work. Diagrams and photographs appertaining to the formation and appearance of cargo sweat and internal sweating of steel coils has been dealt with as something new. Cargo plans and procedures in loading steel cargoes have also been added, as well as more in-depth opinions on the loading of incompatible cargoes, and the actions to be taken when loading of such cargoes are inevitable and unavoidable.

There is an explanation of "mill damages" and some expansion to many sections already in the book, e.g. in the section on reinforcing bars I have dealt extensively with the effects of sea water wetting and how this affects the use of the material in the cutting and bending procedures. I have also identified the organisation referred to as "CARES". Some new sections have been added, such as "Steel Prices", and the section on "Ventilation of Steel Cargoes" has been rewritten and brought up to date. Further developments on tank-top strength and the loading of steel coils are also included.

One of the basic intentions behind this book is to provide a means of ready reference to information connected with the carriage of steel by sea. It is to be emphasized that, in those parts where suggested methods of procedure are concerned, e.g. formulation of descriptive clauses for iron and steel products, the stowing, lashing and securing of cargo etc, these are methods that have been extensively used over many years, and are still used with continued satisfaction. They are presented as a guide to what has been, and can be, achieved, but need not be dogmatically adhered to without variation. On the contrary, conditions and situations differ from port to port, and from country to country, but the basic principles remain unchanged.

One of the main concerns in transporting cargo by sea is to avoid damage being sustained at any time, and especially when the cargo is in the care of the carrier. In order to achieve this, and prevent claims arising, careful handling, good stowage and care in custody, are of paramount importance. In such circumstances, if claims prevention takes first priority all else should automatically fall into place. For this reason, the underlining theme of this work is the prevention of claims.

London, February 2003 A. SPARKS F.N.I.

Contents

List of figures

List of colour plates

(between pages 146 and 147)

I. Steel—an overview

THE MANUFACTURE OF STEEL

IRON* is obtained from IRON ORE, which is principally a compound of iron and oxygen together with minor amounts of other elements, and such unwanted earthy matters as clay, stones and sand, etc. STEEL is an ALLOY consisting basically of iron and CARBON, and alloy steel is steel into which has been introduced other metals such as chrome or NICKEL, to mention but two, over and above a certain percentage.

It has been estimated that 5% of the earth's crust is composed of iron. Iron ores obtained from the earth in Europe are in most instances weak ores containing about 25% pure iron, so that the mining and extraction of the iron from such ore is very costly. High grade ores in abundant quantities are shipped into Europe by large bulk carrier ships from Canada, South American countries, Africa, India, Australia and Sweden.

The process of winning the iron from the ore takes place in a blast furnace and is first commenced by crushing and grading the ore down to a standard size. Lumps and dust below desirable dimensions are referred to as fines. The fines are not wasted, but are sent to a sintering plant where, together with quantities of coke-breeze and also LIMESTONE, they are roasted into a clinker form agglomerate called SINTER, which can form part of the CHARGE in the blast furnace. Equally important as iron ore to the burden of the blast furnace is the introduction of coke, which is produced in coking ovens on the premises of the steel plant. The type of coke used yields from a special type of metallurgical coal that is low in SULPHUR, and is chosen for this purpose and also for its strength and cohesive properties.

The blast furnace is a tall cylindrical construction the size of which varies considerably, and can be over 30 metres high, with a diameter of up to 14 metres at its widest point. Its construction consists of an outer shell of steel plates that has an internal refractory lining consisting of fire bricks. The smelting operation within the furnace is continuous and has a life span of up to, and from, three to five years, whereafter the furnace is closed down for the renewal of the lining before being recharged.

The charge, consisting of iron ore or pellets, coke, sinter and limestone, is introduced into the furnace in alternate layers through a double bell charging system at the top of the furnace. A blast of hot air, previously

* More details of subjects printed in SMALL CAPITALS can be found in II. ABC of Steel Carriage.

heated in hot-blast stoves, is introduced at the base of the furnace, so causing the coke to burn fiercely and sufficiently raise the temperature within the furnace to enable the iron to liquefy. The limestone acts as a flux in scavenging the charge of most of the undesirable extraneous materials which eventually float on top of the liquid ore collected in the base of the furnace, in the form of SLAG. The slag is run off through a SLAG NOTCH after which the iron pours through a TAP HOLE in the base of the furnace.

The molten iron is cast into INGOTS or into the form of PIG IRON for the convenience of handling, storage and transportation, or taken in a LADLE CAR direct to the steel-making furnace. The iron produced is not pure and contains up to 5% carbon, up to 3% silicon as well as small varying percentages of manganese, sulphur and phosphorus.

Having made iron, the next step would be to turn the iron into steel, and this is achieved in a special type of furnace. Taking Britain as an example, over the past three decades steel has been produced by means of the Open Hearth Furnace, BASIC OXYGEN FURNACE and the ELECTRIC ARC FURNACE. By the year 1980 no steel was being made by the open hearth method, and use of the basic oxygen furnace was well ahead of the electric arc furnace method of making steel (see under MINI MILLS).

The electric arc furnace is used for converting cold SCRAP metal into steel of a desired specification. The furnace consists of a circular container, the hinged lid-like roof of which can be swung clear in the horizontal plane. The scrap metal charge is dropped through the open top into the bottom of the furnace, whereafter the roof is closed and carbon electrodes are lowered down through the roof to be positioned just above the charge. An extremely powerful electric current is sent through the electrodes and an arc is struck with the charge, so causing the scrap to melt through the heat generated.

Through a door in the side of the furnace, additional elements are introduced. After a period of time samples are taken and analysed, and when the desired specification has been achieved the furnace is tilted to pour off the slag through the furnace door. Thereafter, the furnace is tilted in the opposite direction and the molten steel is removed through the tapping spout.

The basic oxygen furnace can be charged with hot metal or proportions of SCRAP, pellets and molten metal. A water-cooled oxygen lance is lowered through the top and is positioned just above and close to the surface of the molten metal. After this, it receives a blast of pure oxygen. The oxygen combines with the carbon and other impurities, most of which are carried off in the form of CARBON MONOXIDE GAS. The furnace is tilted nearly horizontal, samples are drawn off and analysed, and when the correct chemistry of the metal has been achieved the molten metal is removed through the tap hole. Thereafter, the converter vessel is upended and emptied by pouring the slag into a receptacle provided for this.

The molten steel, after having been poured from the electric arc furnace or the basic oxygen furnace, is cast into ingots which can be stored until required for the purpose of hot rolling. In other instances, the molten steel is deposited in a TEEMING LADLE to be continuously worked by a process of continuous casting into slabs or other construction shapes. It is in the steel-making furnace, also referred to as the converter vessel, that MANGANESE, NICKEL, ALUMINIUM, SILICON and other additives are introduced to produce CAPPED STEEL, KILLED STEEL, SEMI-KILLED STEEL and RIMMED STEEL, etc., and also to impart a special quality to the material necessary for special uses.

At this stage SLABS, BILLETS, BLOOMS or INGOTS will be produced and these can, in most instances, be considered as material with a metallurgically characteristic composition of mild steel. This slab, billet, bloom or ingot form lends itself admirably to storage ashore, on the premises of the steel mill or on board ship for export. To satisfy the needs of manufacturers, principally the vehicle construction and fabrication industries in various countries, and also to conform to the exigencies surrounding commercial manipulation in the unending search to comply with the rules of economics, the large quantities of mild steel slabs, billets and blooms are being continually transported in seaborne trades all over the world.

The ingot is rolled down into a slab which is rectangular in shape and may weigh somewhere between 10 and 30 tons. Thereafter, the slab is processed further by hot working to produce the following:

—Hot rolled steel plates, hot and cold rolled steel sheeting in coils, steel strip for welded pipes and tubes.

—Blooms and billets, from which is made (by the process of hot rolling) profiles, such as beams, angles, flats, squares, hexagonals, octagonals, round bars, rails, reinforcing bars; hot drawn wire, and wire rods, also seamless pipes and tubes, which are manufactured by forcing a mandrel through the end and length of a round billet. These products of the rolling mill are in most instances processed further to suit special commercial requirements (see Fig. 1).

THE STRUCTURE OF STEEL

Steel, being basically an alloy of iron and carbon, is a solid substance which is classified as a metal. The structure of all metals is crystalline, meaning that the actual composition of the material is composed of very many small irregular-shaped crystals of varying dimensions. It is usual when describing crystalline metallic structures to use the term grain in place of crystal. The perimeter surface areas of the grains are therefore called grain boundaries.

Iron being the metallic element in steel, the carbon constituent—relatively extremely small in quantity—will be disregarded for the moment whilst the structure of iron is considered. The grains in the iron are composed of thousands of unit cells the axes of which in each specific cell are all orientated in the same direction, but the direction of the axes varies from one cell to another. The atoms in the unit cell occur in a three-dimensional pattern which is called a SPACE LATTICE, which may be body-centred or face-centred.

—Body centred cube (bcc) alpha iron

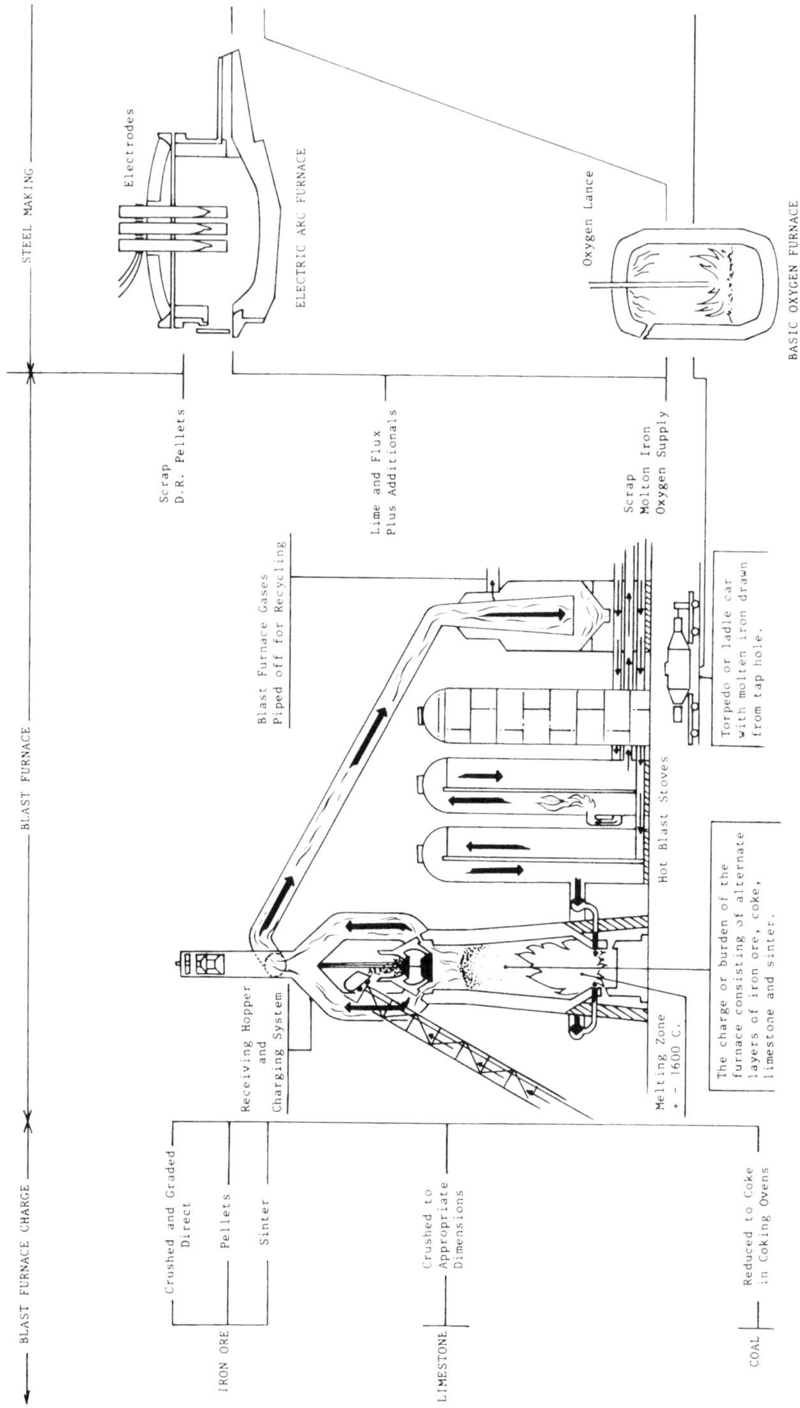

BLAST FURNACE CHARGE
BLAST FURNACE
STEEL MAKING
IRON ORE
Crushed and Graded Direct
Pellets
Sinter
LIMESTONE
Crushed to Appropriate Dimensions
COAL
Reduced to Coke in Coking Ovens
Receiving Hopper and Charging System
Melting Zone + - 1600 C.
The charge or burden of the furnace consisting of alternate layers of iron ore, coke, limestone and sinter.
Hot Blast Stoves
Blast Furnace Gases Piped off for Recycling
Torpedo or ladle car with molten iron drawn from tap hole.
Scrap D.R. Pellets
Lime and Flux Plus Additionals
Scrap Molton Iron Oxygen Supply
Electrodes
ELECTRIC ARC FURNACE
Oxygen Lance
BASIC OXYGEN FURNACE

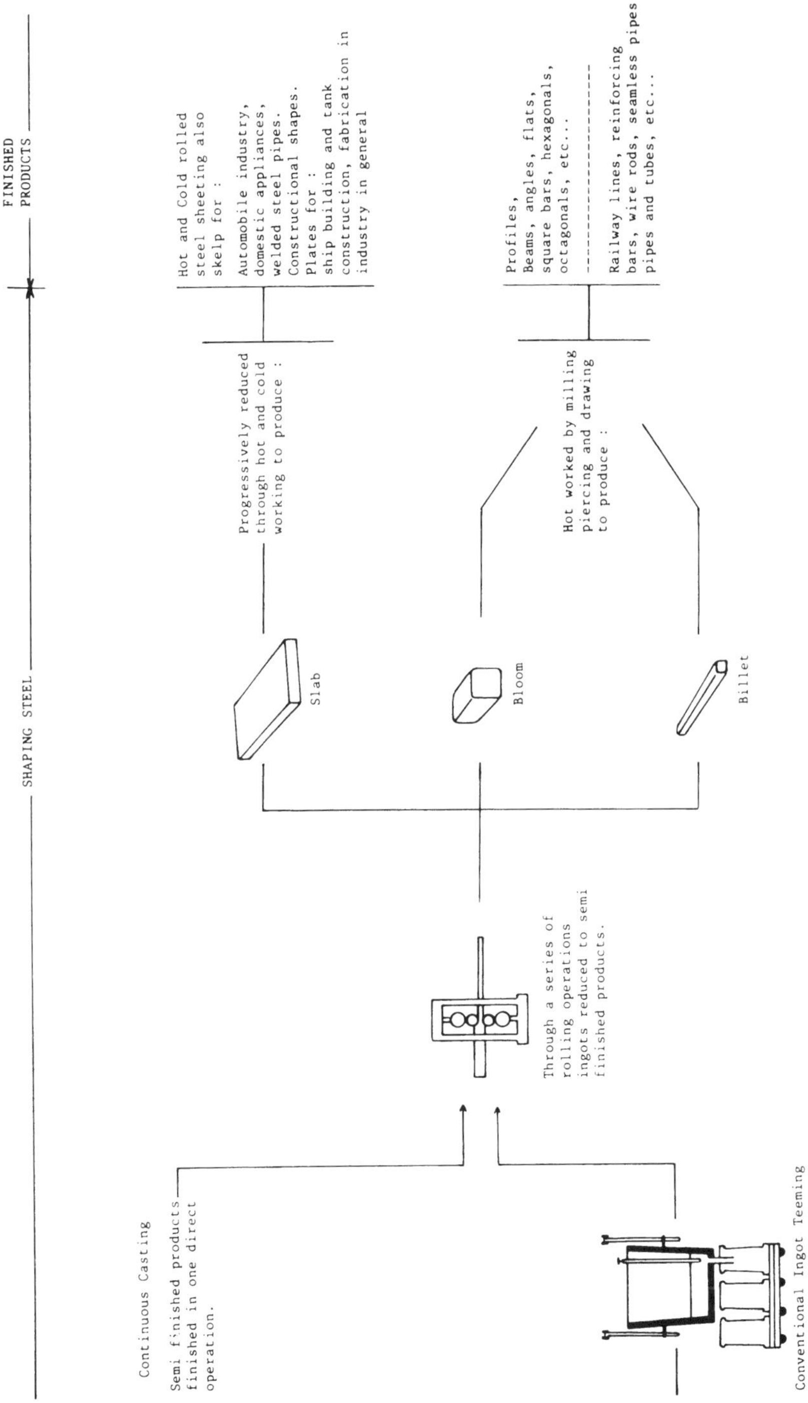

Fig. 1. Steel making—from iron ore to finished product.

—face centred cube (fcc) gamma iron

Iron is ALLOTROPIC, so that with certain changes of temperature or pressure the fcc will change to bcc and vice versa. The introduction of carbon to the iron to form steel does not interfere with the intrinsic properties of the behaviour of iron when an alloy or iron and carbon (steel) is subjected to heat, as will be realised from a study of the following equilibrium diagram. The diagram is a useful basis (Fig. 2) for the study of the behaviour of plain carbon steel when subject to various temperatures.

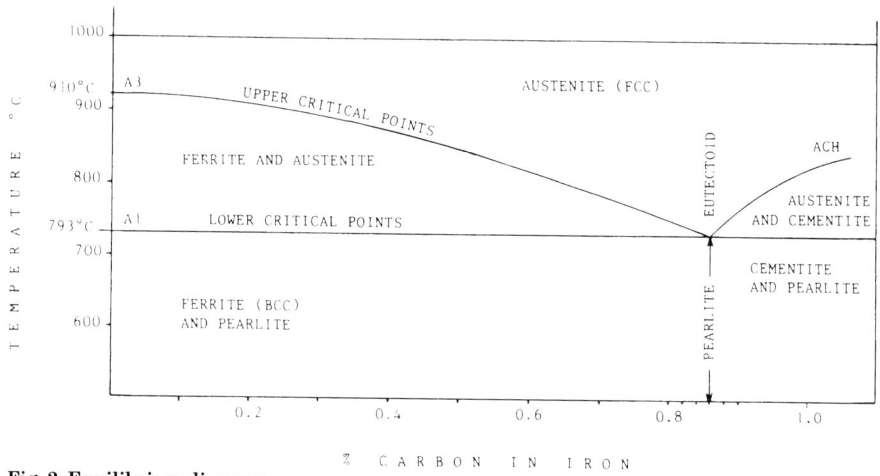

Fig. 2. Equilibrium diagram.

When the temperature of the steel is below the lower critical limit the unit cell is in the body centred form. As the temperature increases above the lower critical limit, either side of EUTECTOID, the atomic structure of the metal begins to change from the body centred cube. The diagram shows how comparatively rapidly this change takes place with varying carbon contents.

Once the temperature of the steel has passed the upper critical limit the change to the face centred cube is complete, and the metal is homogeneous throughout with all carbon being held in solid solution. The steel in this condition is known as austenite. During the increase from room temperature to the condition referred to as austenite, various physical changes take place related to the inherent constructional properties of the metal, which will reassert their presence when the metal is permitted to cool slowly. As the temperature of the steel falls and the upper critical limit is approached, the change from the fcc to bcc commences.

It is evident from the equilibrium diagram (see Fig. 2) that with carbon contents of less than 0.83% (eutectoid) the structure of the steel begins to change and develops into a combination of austenite and pure iron (ferrite). To the right of the eutectoid position, between the upper and lower critical limits the composition of the metal is austenite combined with cementite, a chemical compound of iron and carbon (Fe 3C). Definite and permanent structure is established when the temperature has decreased to, and passes below, 723°C. To the left of eutectoid on the diagram the composition is ferrite plus pearlite. Pearlite is composed of ferrite and cementite. In the final section to the right a combination of pearlite and cementite makes up the structure. When the change takes place at 723°C with a carbon content of 0.83% the structure becomes purely pearlite. These developments in the structure of the metal at various temperatures impart special qualities to the steel. They are recognisable in the structure of the grains upon examination.

The reason for these changes over different temperature ranges, and varying carbon contents in the steel, can be found in the fact that between the upper and lower critical limits the carbon is precipitated out of solid solution when cooling takes place. The significance of all this is that, by regulating the rate at which steel cools, after the application of heat, structures in the steel can be produced which are more suitable than others for specific requirements of industry (see also HOT WORKING, COLD ROLLED STEEL SHEETING and HEAT TREATMENT OF STEEL).

The ultimate in temperature manipulation is what is known as QUENCHING. With this method the temperature of the steel is lowered so rapidly from above the critical limit through the various stages as shown in the equilibrium diagram, that the change of the space lattice structure is not completed and most of the carbon remains in solid solution. This produces a type of steel called martensite which is extremely hard and brittle.

THE CARRIAGE OF STEEL PRODUCTS BY SEA

In introducing this subject it is considered appropriate to dwell, somewhat concisely, upon a few facts in connection with the subject of steel and man's dependence on it. In the event of our having to replace STEEL with some other materials the only real apparent choice appears to be timber. The tools which we need to shape timber to our requirements would be made of IRON or steel, and one does not have to ponder very long upon this fact to reach the conclusion that life as we know it would be vastly different if steel products were no longer available.

A large proportion of all manufactured goods are completely or partly composed of steel. Those which are not mostly require tools or machines made of steel in order that they might come into existence at all. It is estimated that iron and steel products may form about 90% of all the metal used in the world, and mild steel probably amounts to 85% of all steel manufactured. Practically all steel cargoes transported in seaborne trades are composed of products manufactured from MILD STEEL. Therefore, it can be considered that all products henceforth referred to are made from mild steel.

It is obvious that steel and products manufactured therefrom will continue to play a very important part in our lives, reaching into the foreseeable future and no doubt beyond. Therefore, it is reasonable to assume that the same existing pattern of international trade in steel products will continue to exist for a very long time. Commercial interests, in their never-ending search to find better quality products and achieve greater profit margins, are constantly negotiating international deals, which will ultimately result in merchant ships, partly or wholly laden with steel cargoes, criss-crossing the oceans and seas of the world with the aim of satisfying a vast network of market conditions in a large number of countries. This international trade in steel is of paramount importance in the stabilisation of steel prices and plays its part in, and has a favourable influence upon, the economies of many nations.

Steel cargoes

As mild steel products are so high on the list where ships' steel cargoes are concerned, the small amounts of special steels carried can be disregarded, with the exception of STAINLESS STEEL, galvanised steel products, and tin plate, of which, nevertheless, the basic product is mild steel. In general, the types of steel products loaded into ships are:

HOT ROLLED STEEL SHEETING: coils and plates in bundles. Mainly unwrapped and unprotected against moisture and the development of rust. Hot rolled steel sheeting which is pickled and oiled is always wrapped.

COLD ROLLED STEEL SHEETING: coils and packages of plates. These goods are always wrapped since any imperfections are inadmissible. The plating is usually oiled to prevent the development of rust. There are instances where unoiled material is shipped, but this is risky.

STRUCTURAL STEEL: such as mild steel beams, angles, channels, rounds, bars, etc., all unwrapped and unprotected against moisture and the development of rust.

GALVANISED STEEL SHEETING: coils and packages of plates which in view of the delicate nature of the material are always wrapped. The surface of the plating may be oiled or chromated. Second and third choice material is usually shipped unwrapped.

ELECTRICAL STEEL SHEETING: shipped in coils and packages. This plating is of an extremely delicate nature and is always well wrapped with total protection in mind.

COLD ROLLED STAINLESS STEEL SHEETING: in coils and also in packages, always wrapped. This is delicate material with a mirror-like surface.

HOT ROLLED STAINLESS STEEL SHEETING: shipped in coils and bundles, but unpacked and unprotected.

TIN PLATE: in coils and in boxes. Being fragile, expensive and highly susceptible to damage, the goods are always well packed.

RAILWAY IRON: the rails which form rail tracks. Unwrapped and unprotected against moisture and the development of rust.

GALVANISED PIPES: often chromated—shipped in unprotected bundles.

PIPES AND TUBES: coated for protection and uncoated. Usually unwrapped, in many instances ends are fitted with plastic cap or edge protectors, shipped in single pieces and in bundles.

SHEET PILINGS: coated for protection and uncoated. Never wrapped.

SLABS, BILLETS, BLOOMS: hot rolled semi-finished material, unwrapped and unprotected against moisture and the development of rust.

REINFORCING BARS: hot rolled semi-finished material, unwrapped and unprotected against moisture and the development of rust.

MERCHANT IRON: short length material of small cross-sectional dimensions. These goods are usually shipped in bundles, but are in most instances unpacked and unprotected. In general, hot rolled material consisting mainly of rounds, half rounds, light beams and channels, T bars, hexagonals, octagonals, square bars, etc.

WIRE RODS: hot rolled, usually shipped in coils of around 1,500/2,000 kgs. They are shipped in an unwrapped condition and completely unprotected from the elements, for which reason they are on average always affected by rust.

The main bulk of cargoes carried consists of steel sheeting in coils, packages and bundles, also structural steel and merchant iron.

Although some of the lighter weight steel products, such as short lengths of merchant iron, lighter weight coils, packages and bundles, etc., may be carried in general purpose sea containers and also lash barges, the majority of such seaborne tonnage forms all of, or is carried as part of, a ship's cargo, when the balance of the deadweight on board such ships will most probably be completed with general cargo or some other type of cargo. The greater proportion of steel cargoes are shipped in BULK CARRIERS, which vessels, owing to unobstructed spaces in the holds and large deck openings in the form of hatchways, lend themselves admirably in this respect to the transport of steel products. Full steel cargoes of 30,000 to 45,000 tons or more are not uncommon, and in such circumstances the only cargo on board is usually steel products. Nevertheless, tweendeck ships are sometimes used.

Main transport problems

The main problems surrounding the transport of steel cargoes can be itemised as follows:

(1) Determining the pre-shipment condition of the cargo with regard to sufficiency of packing, existing physical damage, the rust condition of cargo and in relation to all of these apparent defects, when warranted, compiling appropriate accurate DESCRIPTIVE CLAUSES to be inserted in the relevant MATE'S RECEIPTS AND BILLS OF LADING.

(2) Effecting careful and proper STOWAGE, DUNNAGING, LASHING AND SECURING OF CARGO, also dealing with damage which might arise during the actual loading and stowage operation.

(3) Taking the necessary steps to care for the cargo while it is on board the ship, e.g. VENTILATION OF STEEL CARGOES.
(4) Carefully discharging the goods at the final port of destination to avoid damage so as to deliver the cargo in the same apparent order and condition as it was when taken on board the ship.

The bill of lading

When goods are carried by sea it is expected that the master of the ship will deliver the cargo at the port of discharge and that such goods will be received in like order and condition as they were found to be when they were taken into the custody of the vessel at the port of loading. The satisfaction of such terms is vital to the interests of all sharing in the venture, as any deviation from the above-mentioned conditions will in most instances involve damage to the cargo. In such circumstances, the apparent condition of the goods when delivered will not concur with the description of the cargo as indicated in the relevant bill of lading, and this can result in claims being entered against the carrier. In the normal course of events, when claims are involved, all participants suffer some degree of loss which may be directly financial, but future goodwill and general credibility might well be affected.

The master of a ship is, by virtue of the Carriage of Goods by Sea Act, obliged to describe, by entering qualifying remarks in the bill of lading, the apparent order and condition of the cargo at time of loading on board the vessel, if the appearance of the cargo at that moment does not concur with the description of the goods appearing on the relevant bill of lading. The description of the goods as appearing in the bill of lading is supplied by the shipper and could in the meantime become erroneous due to the goods having sustained damage or having developed a rust condition which is not accounted for in the description provided. The shipper's description of the goods, as supplied to the carrier, may also have originally been erroneous by not accurately describing the goods offered for shipment.

Rust is a natural phenomenon of steel, and expert appraisal, involving a sound knowledge of the CORROSION AND OXIDATION OF STEEL SURFACES, is necessary in determining whether the presence of the rust has resulted in the development of a condition whereby the prime, sound condition of the material is affected. In every instance where rust is apparent an accurate descriptive clause should be entered in the bill of lading. Such clauses should not only be confined to the rust condition of the material but also to any apparent physical defects or contamination by extraneous matters or material. When the apparent condition of the goods at time of loading is such as to warrant the insertion of additional remarks in the relevant documents, very often shippers offer the carrier or master of the ship a letter of indemnity. Acceptance of such a letter is illegal, and it is not beyond the bounds of possibility that the carrier, or the master of the ship, in extreme circumstances, might be subject to serious repercussions for being party to a fraud. Further to this, the acceptance of such a letter could prejudice the shipowners' P & I Association cover. Shippers require clean bills of lading in order that LETTERS OF CREDIT may be negotiated without difficulty. It is therefore obvious that for their requirements to be met, shippers should ensure that sound and undamaged cargo be presented for shipment. Damage sustained by steel cargoes is often caused through handling damage.

Handling damage

Damage caused through rough handling, or improper handling, can add up to some importance. The only counter-measure which can be taken with regard to this type of damage is to handle the cargo with more care. Some examples of how these damages are caused may be:

(a) Allowing a steel coil to land too heavily when being lowered may cause it to become oval in shape. It should be delivered perfectly circular in shape; if not, and distortion is too excessive, the unit may have to be considered as having no value greater than that of scrap.
(b) Damage to the packing of steel coils, and also other packed steel, which results in plate edge damage and exposure of the contents to the possible development of rust.
(c) Stevedores in their unending efforts to be more productive and increase their tonnage records often overload slings, so causing flanges of beams and other items of cargo to be bent, chafed, scored, etc.
(d) Incorrect stowage, whereby crushing damage results in coils becoming distorted (oval in shape) and bends the webs of beams, also causes lengths of steel to become warped or bent. The remedy for this is to ensure that goods are properly stowed by using the correct type of dunnage selectively placed within the stow.
(e) Various pre-shipment accidents do occur on the quayside when collisions between cargo being manipulated and the peripheries of stacks of cargo awaiting shipment cause damage.
(f) Movement and chafing in the stow on board the ship when working in a seaway. This can only be remedied by correct stowage so preventing slackness from developing. Efficient securing of cargo will help to avoid such damage.

The above more or less covers the incidence of a class of damage involving tearage, buckling, bending, chafing and crushing of cargo. The following is considered to be of greater importance and concerns damage to steel products caused by rust.

Rust damage

Alleged damage to steel cargoes can involve dry corrosion such as the effect of tarnishing of copper surfaces, which causes them to turn green in colour, aluminium surfaces that become discoloured through oxidation, and also galvanised surfaces that develop dullness in appearance shortly after manufacture. With regard to the last mentioned, there have been many claims entered against the delivery of galvanised steel having a dull surface appearance. Receivers have agreed that the goods are undamaged but contend that cargo so dull in appearance, when placed in stores alongside locally manufactured bright material—which has not been subjected to the rigours of a sea voyage—is unattractive to potential purchasers of the material, who prefer the brighter galvanising and will only purchase the dull galvanised steel at a discount.

All steel products that are wrapped must be considered delicate material and regarded as finished products. Any damage whatsoever, or howsoever caused, will render the affected parts, or even the entire article of cargo, unusable for its originally intended purpose. Therefore, such goods are usually well protected by what is intended to be packing impervious to the effects of humidity. Nevertheless, if water is permitted to run across the wrappers it may penetrate the overlaps of the packing and eventually soak through the inner wrapping to contact the product within. When packing is applied to steel products the intention is that they be delivered to the end user free of rust, undamaged, and in like order and condition as they were when they left the production line.

Fresh water rust (loading during rain)

Ships' masters are often persuaded to load wrapped cargo during periods of rain in order to avoid the vessel being idle. In many instances, letters indemnifying the master against damage developing as a result of loading during rain are often offered to the master and accepted. As previously stated, participants in such practices, which are fraudulent, could in some manner or form find themselves in an unenviable situation.

Steel products that are not wrapped are often deposited uncovered on the open quay while awaiting shipment, and this applies especially to hot rolled steel products. The same goods when discharged from the ship are quite often stored in the open on the ultimate receiver's premises. They are also often transported uncovered on rail and road wagons and also in open river and canal barges. Where hot rolled steel products are concerned, outside, unprotected storage is acceptable practice in the industry, and as a result such goods are usually affected by visible surface rust. As such rust emanating from fresh water contact or the effects of humidity within reasonable degrees of development should not impair the merchantable condition of the material, it is not uncommon to load such cargo during rainy weather. Notwithstanding the above, whatever cargo is loaded, whether it be steel, bags of coffee or bulk material, the ideal situation, which should always be the objective, is to load cargo in a dry condition and to deliver it in like condition. The reaction of the receivers in receiving hot rolled steel products in a wet and rusty condition will vary according to local mentality, market conditions and for what use the steel is intended. If it is imported for immediate re-rolling there may not be any objections raised, but if it is to be stored for a period before being re-rolled its potential eventual deterioration may be a factor to be considered. Even if the goods are not for re-rolling and it might be agreed that at the time of delivery, in spite of the wetness and rust, the surface of the plating is undamaged, receivers contend that their customers are used to receiving such material in a blue condition (Prime Condition). Japan has been a case in question where large amounts of steel are produced for use in the home market and are generally stored under cover after manufacture; therefore, the goods remain in prime condition. Nonetheless, even in Japan and other countries, where the received condition of hot rolled steel must be near to prime condition if claims are to be avoided, there are instances on record where wet and rusty plating has been received without complaint. One can only conclude that market conditions dictate what the situation will be.

In view of the above, the carrier must take considerable care when loading hot rolled steel sheeting in coils or in bundles. If the goods are in prime, or near prime, condition by appearance, or provided with covered storage prior to shipment, they should not be loaded during rain. If the goods are on the open quay, exposed to the elements, loading during *light rain* should be undertaken only provided a preshipment agreement exists whereby the shipping documents will be claused for preshipment rust and wetness. Further to this, goods stored outside, even when loaded during dry weather and in apparent dry condition, should be closely scrutinised at time of loading, as usually free moisture trapped between the turns of plating squirts out when each coil is lifted owing to flexing of the windings. In such circumstances the shipping documents should be claused "Wet before shipment". Even under these conditions, the loading of wet cargo must be given careful consideration with regard to other cargo which might share the same stowage space. The wet cargo will raise the relative humidity of the atmosphere in the cargo compartment, which may initiate and promote the development of damage to other adjacent cargo.

N.B. In connection with uncovered storage and the effects of acid rain, see alphabetical section: CORROSION AND OXIDATION OF METAL SURFACES and DESCRIPTIVE CLAUSES.

Hot rolled steel products, when leaving the production line, are covered with a brittle basic oxide coating often referred to as MILL SCALE, which gives the steel a characteristic blue appearance. The mill scale is only an attachment to the product, not part of it, and will eventually become detached. Therefore, in the interests of the preservation of the goods, and before a protective coating can be applied, the mill scale must be removed. The mill scale is initially protective to the steel surface but partly detached. Areas of fragmented scale can be instrumental in the development of pitting and rapid deterioration. The old-fashioned method of removing the mill scale was to store the goods in the open, permitting the surfaces of the steel to weather until they were so rusty that the mill scale eventually dropped off, or became detached to such an extent that it could be removed by wire brushing. Today, when time is of the essence, the mill scale is disposed of by passing the steel through an acid bath—a process referred to as pickling—or otherwise the goods are shot blasted. These two processes efficiently eliminate all mill scale and any rust or extraneous matters.

After this, the treated surface of the steel is smooth and fine in appearance. This work having been completed, protective coatings can be applied with confidence. As explained later, under the heading of CORROSION AND OXIDATION OF METAL SURFACES, wire brushing rusty surfaces in preparation for the application of protective coatings is no longer recommended. This is because removal of the rust by brushing is no guarantee that the root cause of the development of the rust has been completely eliminated. In view of the facts here presented, allegations that rust damages hot rolled steel products might be treated with scepticism; otherwise, why would the goods be stored outside prior to shipment and allowed to weather for sometimes appreciable periods before shipment and after delivery at final port of discharge? Hot rolled steel products *can* be loaded into vessels during periods of light rain. They should not be loaded during *heavy* rain because an open hatchway on board a ship presents a large catchment area for rain. If too much rain water collects on the tank top in the bottom of the hold, the cargo could be standing in water either temporarily or permanently. Undesirable impurities, picked up by the rain water from the tank-top plating or from the bilge wells, could eventually cause the steel to become damaged during the course of the voyage. Previous remarks regarding loading of hot rolled material during rain cannot be too strongly emphasised, especially if incompatible cargoes will be involved. However, right across Europe, including all of the old Eastern bloc countries, and in many other parts of the world, hot rolled steel is invariably given outside storage. Even after the sea voyage and after delivery, claims are entered for rust and wetness to material which is free from pitting or scaling as it was, at time of shipment, virtually undamaged.

Cold rolling of steel is an extension of the process of hot rolling. A very great amount of cold rolled steel sheeting is consumed by the automobile industry, although such material is extensively employed in the manufacture of many goods, such as refrigerator panelling, all types of household goods, kitchen and office furniture, galvanised material and tin plate to mention but a few. Cold rolled steel sheeting is invariably well packed for protection against physical damage. No moisture should be allowed to come into contact with the packing, and such cargo must be considered as being of an extremely delicate nature when viewed in the context of its entirety, or as forming part of a ship's cargo. Generally speaking, hot rolled steel products will not be adversely affected, over a reasonable period of time, by wetness caused through contact with fresh water. Cold rolled steel and all wrapped steel products should never be permitted to come into contact with moisture. Most steel products, similar to those which usually form part of a ship's cargo, will rapidly deteriorate and their good condition become impaired, if they are wetted with seawater, or fresh water which has become contaminated by an aggressive pollutant.

Salt water rust

It has been established that the process of rusting and corrosion of steel involves electrolysis. Consider a steel plate, the surface of which is covered with a fine film of rust. Actually the surface of the plate is affected all over by a network of infinitesimally small electric cells. Each cell has a part which acts as an anode, and a part acting as a cathode. The degree of electron activity within each cell will depend upon the conductivity of the electrolyte present. The least active type of cell would be made to operate through an electrolyte composed of fresh water. This is one reason why ships are preferably taken to areas of fresh water when being laid up for an appreciable period of time. Salt water is a much more active electrolyte, and is well known to be hostile to steel surfaces as it causes rapid development of rust and corrosion. A solution of sulphuric acid would be even more destructive. Any rust formation on cold rolled steel surfaces, or on the surfaces of any packaged steel, is damaging and impairs the prime condition of the material, but this is not necessarily so where hot rolled steel surfaces are concerned. On the other hand, hot rolled steel surfaces attacked by rust caused through contact with seawater or any moisture charged with a medium producing a strong electrolyte, will corrode and damage the surface of the steel. For example, if seawater comes into contact with coils of hot rolled steel sheeting during a sea voyage, upon discharge from the vessel the steel should immediately be put through a pickling process. This would remove all the rust and those elements which caused the rust and ensure that with appropriate protection no further rust would develop. On completion of this reconditioning process the goods must be considered as being in sound merchantable condition, although it cannot be denied that it is reconditioned material.

If salt water-contaminated steel, as referred to above, is put into store and not given immediate attention, the rust condition can become so firmly established that one pass through the pickling bath will not restore the plating to its original condition. In such circumstances, the surface of the plating will be affected by etch marks, thus indicating that the root cause of the original development of the rust has not been removed and the potential for further deterioration exists. Such material may be treated with a protective coating which might in due course develop a localised bubbly appearance in the areas of the contaminated parts of the plating, and a rust condition would eventually become apparent. In many instances salt water contamination seriously involves the carrier's liability, and very heavy insurance claims can eventuate from this. It is in the circumstances imperative that the vessel's STEEL HATCH CLOSING APPLIANCES are kept in a condition which is beyond criticism.

Preparation of cargo holds

The need to prepare a ship properly for the reception of a steel cargo cannot be too strongly emphasised. Cargo holds are customarily washed out with salt water, seeing that so much of this chlorine-laden liquid is always available. The final washing should always be carried out with fresh water in order to remove salt crystals left on the metal work from the seawater washings. The fresh water should be taken from the on-board evaporators, as water obtained from city mains is usually heavily contaminated with chlorides. If this is not done, ship sweat contaminated with salt crystals may contact the steel cargo during the voyage so causing serious damage. In a bulk carrier special attention should be given to the underside of the steel hatch panels, which are often neglected.

SULPHUR is the enemy of steel products. Too much sulphur makes for brittleness in the finished product. Sulphur-bearing materials in contact with steel can, under certain conditions, cause devastating corrosion to develop. Inspection of the cargo holds of a ship which has spent any appreciable period of time carrying sulphur in bulk shows that a deterioration of the vessel's structure has advanced at a phenomenal rate. This serves to indicate how important it is to clean out the cargo holds meticulously before loading a steel cargo.

Many of the bulk cargoes carried by seagoing vessels contain sulphur, for example coal, iron ore, phosphates. Most of the compound fertilisers shipped in bags and in bulk have a chemical composition which is damaging to steel surfaces. It is necessary that the removal of all rests of previous cargo be considered of prime importance to the loading of a cargo of steel products.

Care of the cargo

Owing to the high density per unit volume of steel cargoes, localised concentrations of weight can be formidable. Consideration has to be given to good stowage, which is of paramount importance if chafing, warping, bending and crushing are to be avoided. The correct placing of dunnage, used to bind the cargo into a solid block, even out pressures in the stow, and to facilitate reslinging of the cargo, should be kept in mind during loading operations. Although securing by means of wire and timber may be necessary, it cannot be too strongly emphasised that correct stowage is the first requisite, failing which all else is to little or no avail.

The Carriage of Goods by Sea Act, in addition to provisions relating to seaworthiness, states that the carrier shall properly and carefully load, handle, stow, carry, keep, care for, and discharge the goods carried (see under the heading MATE'S RECEIPTS AND BILLS OF LADING). From the time the goods are shipped they are in the care of the master of the vessel until discharged at final destination, and he will be expected to exercise due diligence in every respect necessary for the preservation of the cargo. The ship should be in a proper state to receive the cargo, which must be loaded with care. The cargo must be properly stowed, lashed and in general well secured for the intended voyage. Further to this, the master, in certain circumstances, may later be obliged to prove that all measures and precautions humanly possible towards the preservation of the cargo were considered and satisfied before the voyage commenced.

Ventilation of the cargo

As steel cargoes are so susceptible to attack from rust, for which the presence of moisture is necessary, careful consideration has to be given to the temperature of the steel at time of loading and also to ever-changing climatic conditions that can be experienced during the voyage, which will dictate how the cargo is to be ventilated. If the cargo of steel is loaded in a port where atmospheric temperatures prevailing, for seasonal averages, are considerably lower than those temperatures which will be experienced as the voyage progresses, no ventilation should be given to the cargo.

This is because the temperature of a mass of steel in a ship's hold adjusts only very slowly to increasing outside atmospheric and sea temperatures. While these outside temperatures continue to increase, the temperature of the steel will probably be lower than the dewpoint temperature of the outside air, and if the cargo is ventilated, cargo sweat will inevitably develop. Where wrapped steel products are concerned, the avoidance of cargo sweat is of great importance. Sweat water trickling or flowing across the surfaces of the wrappers can penetrate the overlaps of the packing and gain access to the product within. When to open the ventilation should be decided by observance of the outside meteorological conditions prevailing and compared with the actual temperature of the steel cargo.

When a cargo of steel is loaded in a warm climate or is being ventilated on board the ship in relatively warmer temperatures than are expected to be experienced as the voyage progresses, cargo sweat should not develop. The incidence of ship sweat will arise if atmospheric temperatures fall, and in order to combat this, full ventilation should be given to the cargo compartments in order to keep the temperature of the atmosphere within the cargo space similar to that of the outside air. There are on record many sea voyages where steel products have sustained heavy damage through incorrect ventilation procedures.

N.B. For full details see alphabetical section: VENTILATION OF STEEL CARGOES.

Incompatible cargoes

Owing to their high density, steel cargoes have always been favoured as bottom weight in vessels loading miscellaneous cargoes of which steel forms a part. In fact, previously, there were certain trades where the freight for the steel paid for the expenses of the voyage. Freight earned on the general cargo and other cargo overstowing the steel was pure profit. Steel overstowed with other cargo raises problems which are in certain instances insurmountable. The problem arises from accompanying cargoes that are by their very nature incompatible with steel, such as bagged chemicals, hygroscopic cargoes (which would include timber, tobacco, jute) and many other cargoes which are capable of conditioning the atmosphere within a ship's hold.

In many trades the amount of steel and other cargo obtainable, or offered to the carrier, is insufficient to complete a vessel's deadweight if the preservation of the steel is to be of prime consideration. Neither the steel nor other cargo available will result in the voyage being a viable proposition if a choice has to be made between two types of cargo because they are incompatible when stowed together. The problem presented, or which may be presented, is that at a moment when the steel cargo should receive no ventilation, the hygroscopic cargo stowed in the same compartment may very well require ventilating in the interests of its own conservation.

The steel trade from the Far East to Europe is a good example of steel products sustaining damage through being carried in combination with other cargoes, or as a result of opening hatches at ports on the way to load other cargoes. Quite often, part cargoes of steel have been loaded in Japan or Korea for northern Europe during the northern winter. The vessel might then proceed to Taiwan to load plywood or perhaps to the Philippines, Singapore and/or Malaysia to load timber. As a result of this unfavourable itinerary—from the point of view of the preservation of the steel—between closing hatches in Japan and re-opening at the next loading port, in a few days there can be a temperature difference of as much as 20°C. This relatively warmer air, coming into contact with the steel cargo, will cause heavy cargo sweat to develop. The transportation of forest products in combination with steel is traditionally regarded as bad stowage. There are certain trades when at specific times of the year parcels of timber and steel can be stowed together and transported with success even though elements of risk are always present. Any such projected voyage must be given careful consideration in anticipating how great the hazards are which might be involved. Hot rolled, semi-finished, unwrapped products are less of a risk than wrapped material. In fact, the risk should not be taken with wrapped material.

A sensitive cargo

Where sea transport is concerned, items comprising parcels of steel products are usually heavy and often large in size. During handling, such items of cargo are easily damaged. The very nature of the commodity makes it sensitive to the development of rust through contact with moisture. Steel cargoes can, in general, be considered as highly susceptible to damage claims by third parties. Stowage of steel is more complicated and requires more attention than most other dry commodities. Shifting of such cargoes at sea can result in heavy damage to the ship and cargo, and in extreme cases the loss of the vessel and crew has been involved. The condition of the cargo as presented for shipment often gives rise to discussions which revolve around the clausing of bills of lading. Qualifying remarks covering the rust condition of the material may be demanded by the carrier, whereas the shipper requires clean documents in order that letters of credit may be negotiated without difficulty.

The steel trade, in general, is highly competitive and, as such, only the most astute operators can survive even at the best of times. Profit margins are usually far from being phenomenal, or even comfortable, and the misfortune of one participant in a venture will probably be capitalised upon by another. The commercial climate so created pervades all aspects of the business, from which the carrier, to a certain extent, does not escape.

LOADING, DISCHARGING AND CARGO PLANS

Loading

In many instances complete, or near complete, cargoes of steel products are loaded into bulk carriers that are withdrawn from the bulk cargo trades. In such circumstances the crews of these vessels have little and often no experience of the loading, stowage and carriage of break bulk cargoes. Therefore, charterers find it expedient to appoint a super cargo, experienced in the loading of steel cargoes, to make up a stowage plan, confer with the stevedores, and supervise the loading of the cargo, in conformity with the governing charter party, and generally assist the master in the techniques of loading such cargoes.

As far as the carriage of the cargo is concerned, according to International Maritime Law (see Article 111, subsection 2, of the Hague and Hague Visby Rules, in accordance with the signatories to the rules) "the carrier shall properly and carefully load, handle, stow, carry, keep, care for, and discharge the goods carried". This must be kept well in mind, as in the event of cargo claims arising for delivery of damaged cargo entered at the point of discharge, it is evident from the wording of that part of the Rules quoted above that the vessel is, at the least, provisionally in the front line. Such claims are first entered against the carrying vessel that may ultimately be found partly or totally responsible for the delivery of damaged cargo. It is to be emphasised that the ship's master is not considered to be an expert on merchandise; however, he is expected to familiarise himself with the type of cargo to be loaded and this is where his P & I correspondent and P & I surveyor can be helpful.

It is normal procedure for the Master to receive a message giving advance details of the cargo to be loaded. This information should, and usually is, provided prior to the vessel's arrival at the first loading port and in sufficient time for the Master to reply with any queries or requests for additional information. It is essential that the description of the cargo provides:

—The type of steel to be loaded
—If steel coils are involved the average weight per individual coil and aggregate weight per lot and height to be loaded (i.e., number of tiers).
—Long steel, dimension, amount and weight.
—For all steel, in general, dimensions amount and weight.

The preparation of the cargo holds is important and the relevant charterparty will specify that the compartments to receive the cargo are to be "clean, dry and in a suitable condition to receive the intended cargo". If the holds have to be washed out seawater will undoubtedly be used, however, it is imperative that the final wash should be made with fresh water, ideally by water condensed by the vessel's evaporators. Salt crystals remaining from salt-water washings have been known to contaminate ship sweat that may contact the cargo. Fresh water from city mains is usually contaminated with chlorides.

The super cargo will usually arrive at the loading port before arrival of the vessel, having all necessary information with him, supplied by the owners via the charterer's questionnaire, with regard to the vessel. He will study the type of steel cargo to be loaded, confer with the stevedores and draw up a preliminary pre-stowage plan showing where the various lots of cargo will be stowed, also indicating the particulars of each lot (see Fig. 3). Upon arrival of the vessel the charterer's representative, the super cargo, or an appointed surveyor, will inspect the cargo holds, which will be accepted or refused. If refused the vessel will remain off-hire until the charterers are satisfied with their condition for receiving the intended cargo. The notice of readiness to load will eventually be accepted and loading will commence. Taking all into consideration, after a reasonable period of time after the arrival of the vessel, the super cargo accompanied by the stevedore's representative will present the provisional preshipment loading plan to the master for his approval. The master should ensure that the surveyor appointed to perform a pre-shipment survey of the cargo, on behalf of the carrier/owner, attends this meeting.

The vessel's master will always be responsible for the stability, trim and safety of his vessel and crew. Therefore it will be incumbent upon him to ensure that at all times the vessel is never allowed to trim by the head or to develop a list. A too heavier trim by the stern is also undesirable as certain types of cargo stowage, in block stowage form, stowed at the forward ends of the holds, may collapse, e.g. wire rod cargo. Further to this, a more satisfactory orientation of the stowage is achieved when the vessel is nearer to even keel, than for example with an exaggerated trim by the stern.

With regard to the actual stowage of the cargo, attention is directed the appropriate sections in this book. However, various types of steel products can be loaded into individual cargo compartments and it is in respect of this that the master should pay particular attention. There are two types of cargo to be considered as follows:

—Semi finished products which are unwrapped, unprotected and are usually stored outside exposed to the elements prior to shipment. Such goods are either wet or have been exposed to wetness at some time. In many instances such cargo apart from a rust clause is described as being "wet before shipment" (WBS).

—Wrapped cargo which should never be allowed to come into contact with moisture or be exposed to appreciably fluctuating ambient air temperatures. This type of cargo must always remain under cover and while awaiting shipment remain in covered lighters, covered rail wagons and closed sided warehouses. The types of cargo involved have already been defined, see **Steel cargoes** on page 8.

Usually, and frequently, serious problems do arise, if dry wrapped cargo is stowed in one end of a cargo hold, e.g. (CRSS) wrapped "cold rolled steel sheeting" in coils and other cargo (HRSS) unwrapped "hot rolled steel sheeting" described on the relevant mate's receipt and bill of lading as "wet before shipment", is loaded into the same compartment. In view of the fact that the moisture in the WBS cargo will raise the relative humidity of the ambient air in the cargo hold, and may cause condensation damage to the CRSS cargo, such cargoes must be considered incompatible and should not therefore be co-mingled, by stowing them together in the same cargo hold.

For the master of the vessel, such a stowage situation results in him being faced with a dilemma. On the one hand, to permit such a stowage can result in heavy indefensible claims for the ship owner. On the other hand, wet and dry cargo will in all probability have to be mixed inside the vessel's holds if the voyage is to be a viable proposition for the chartereres. In most instances the WBS cargo will form a major proportion of the cargo to be shipped, and for either wet or dry cargo to be shut out would be financially disastrous for the charterers. The master may approve the provisional cargo stowage plan, as presented, for "disposition of the weight of the cargo only". He would be within his rights to reject the incompatible cargo stowage. However, he could suggest a change of stowage, by stowing the wrapped cargo in different cargo holds to that of the WBS cargo. Usually this will not be a possibility, especially if more than one discharge port is involved. In the case of a five-hold bulk carrier, on average only three of the five holds can be worked at the same time. To work more than three holds can be self-defeating to some extent, depending upon the layout of the quayside sheds, types of cargo being handled at the same time, and also the disposition of the goods ashore. The tonnages to be loaded and discharged must be spaced out in the holds so that all holds being worked are completed as simultaneously as possible. Further to this, maintaining the vessel in suitable trim has to be kept in mind at all times.

The master will have to involve his local P & I Club correspondent. He should also contact the owner before making a final decision in relation to the problem, as local interests may influence P & I Club correspondent's opinions. In all probability there will be no change in the situation, and basically what this will mean is that the owners have been in contact with the charterers and have reached some form of agreement, which will prevent cargo having to be shut out. Such an agreement will not in anyway relieve the master/owners from their responsibility re delivery of damaged cargo, as discussed earlier. In such circumstances, if the master has to sign any cargo plans they should be signed under protest and refer to his letter of protest to the charterers re the incompatibility of the stowage of the cargo. A "letter of indemnity" should not be accepted in exchange for a clean bill of lading, but unfortunately often is so accepted: the P & I Associations do not agree with this practice, as it is considered illegal. The first pre-shipment stowage plan will usually not be the last. There will be changes to the originally proposed layout of the cargo and about three or four plans may be involved, each new plan cancelling the former. See Fig. 3 for an example of a useful type of plan for a steel cargo.

It is customary and recommended that a cargo surveyor is appointed, either direct for owners account, or by owners P & I Association, to perform a PRE-SHIPMENT SURVEY of the cargo. It is important that the master checks with the surveyor that he has carried out the necessary SILVER NITRATE

TESTS for the presence of chlorides. In connection with this, during winter periods, when sub zero (freezing) temperatures prevail, the stevedores invariably treat the outside storage working areas with sodium chloride; the ground adjacent to the stacks of cargo becomes impregnated with this chemical. After thawing out, mud and water is splashed up onto the cargo so that random silver nitrate checks on cargo in the lower stacks, especially on base coils near to the quay surface, should be carried out. This is important, as any salt reaction on cargo found at the discharge end will in all probability be the reason for a claim being entered against the carrier. Preferably before loading commences, or as soon after as is found possible, the steel hatch closing appliances should be surveyed by a surveyor competent to carry out such a survey for watertight integrity and condition. He should in due course issue a HATCH CONDITION SURVEY REPORT. Seawater damage to cargo caused by leaking hatches, which are found to be defective, places the vessel's owners in a claims situation, which is in most instances indefensible.

A situation often arises whereby the charterers obtain the owner's agreement to go 50/50 in share of the pre-shipment surveyor's fees. In such circumstances the surveyor is representing both charterers and owners. Normally, charterer's and owner's interests are parallel but conflicts of interest often do arise, e.g. disagreements over stowage of the cargo. Such an arrangement can be very problematic for the surveyor, as disputes can also develop over the clauses covering the condition of the cargo. One area where friction will undoubtedly arise will be over suspension of loading for periods of rain. The master will consult the surveyor re suspension of loading, when perhaps rain is threatening, or is already falling, i.e. light rain or drizzle. The decision as to whether loading should stop rests with the master. If the surveyor advises the master to stop loading and charterer's representatives disagree with the surveyor, who is working for both sides, the situation can become very delicate.

It is of importance that during the course of loading notations are made in the deck log book accurately reporting the times of commencement of loading, any stoppages and for what reason, and all times connected with the working of cargo. These notations and registered times should coincide with those mentioned in the statement of facts, which will be presented to the master for signing at completion of loading. As regards stoppages for rain, vessels' masters, of chartered vessels, are reluctant to stop loading cargo until rain is actually falling, when it takes up to half-an-hour to cover all of the hatches. The reason is, that if the master stops work when rain is only threatening and it eventually does not rain, charterers will claim that the stoppage was unnecessary and for that period the vessel is off-hire.
Care of the cargo while on board the vessel is the responsibility of the master. His aim should be to deliver the cargo in the same order and condition as it was when taken on board the vessel, and as described on the relevant bills of lading. All cargo should be kept dry. The ventilation of the cargo should be taken very seriously, as failure to correctly ventilate the

cargo and to keep the proper ventilation records, can result in moisture damage claims being difficult to defend. For information on ventilation procedures, see VENTILATION OF STEEL CARGOES.

Discharging

At the port of discharge it is advisable to have appointed a competent marine and cargo surveyor to report that, upon arrival, the hatches were properly battened down and that all wedges and securing arrangements were in place and properly applied. He should witness the opening of hatches and at that moment report that all cargo to sight was dry and in apparent good order and condition, if that be the situation. The wrapped cargo is the most important, as it is easily damaged through contact with moisture caused by sweating through fluctuating temperatures. If a coil cargo is loaded at a port where atmospheric temperatures are around freezing point (sub-zero) and the vessel is at sea with this cargo for about fifteen days on voyage to an area where atmospheric temperatures are around 22°C, or more, there is every possibility that cargo temperatures may be well below the temperatures prevailing at the discharge port. Consequently, when hatches are opened and the cargo is seen to be wet from cargo sweat, the surveyor should check the surface temperature of the cargo. The master should put the charters/receivers on notice with a letter disclaiming any responsibility for any eventual cargo claims in respect of sweating/moisture damage of the material.

The surveyor should check the condition and cargo worthiness of any barges and or railway wagons etc where direct over side transshipment takes place. The surveyor must follow the discharge of the ship and not fail to be on board when the final layers of cargo are being removed from the vessel's holds; this is when bottom damage in the vessel shows up. Finally, the stevedores will, at completion of discharge, present the master with their out turn report, listing various damages, which they claim were seen to exist during the course of discharge. If the master is asked to sign such a document he should sign it "*for acknowledgement of receipt only*".

Cargo Plans

An example of a type of cargo plan, facilitating the disposition and placing of cargo prior to loading, is shown in Fig. 3. Much information about ship and cargo is required to be at hand, and avoidance of mistakes and blunders will be more easily prevented if as much technical information is available on the plan itself. Initially, as a guide, the decision as to how much of the intended cargo shall be loaded into each cargo hold may be decided by using the bale measurement of the vessel per cargo hold and calculating out, what percentage it represents of the vessel's total bale measurement. On the plan,for example, it will be seen that the cbm of No. 4 hold is 6,308 m^3 in the top right hand corner of the hold on the plan. This amounts to 18.7% of the total bale measurement for this hold and in such circumstances if 27,600 tons of cargo is available for loading then 18.7% of

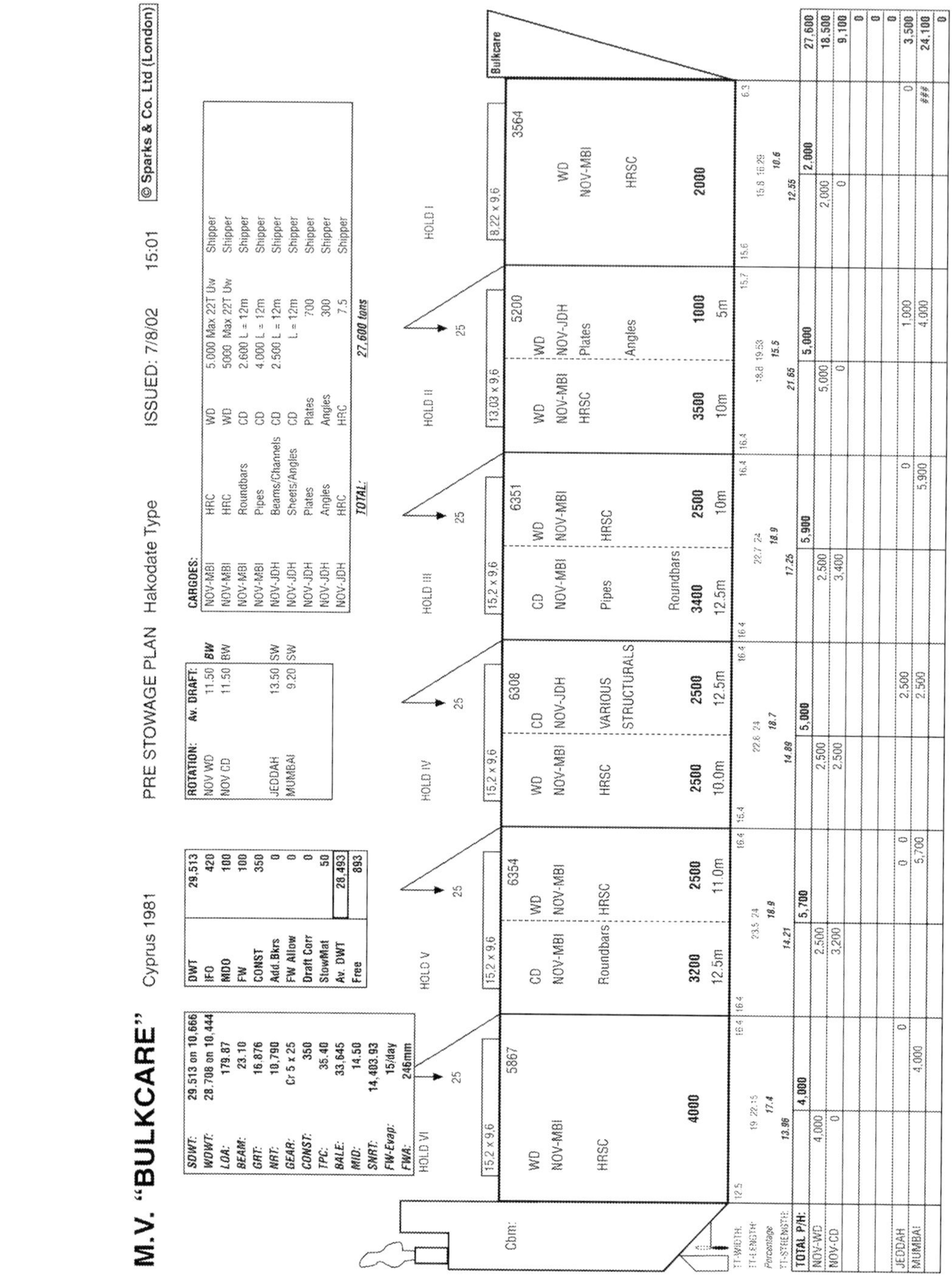

Fig. 3. A type of cargo plan, showing the disposition and placing of cargo prior to loading.

27,600 m/tons = 5,161 m/tons of cargo can be loaded into this cargo compartment. Similar calculations will be made for the other holds and all of the tonnages obtained will in due course have to be adjusted for trim and various changes in cargo bookings etc. At the top of the plan the name "Hakodate Type" is the name given to a standard type of bulk carrier built in Japan. The following refer to the various abbreviations in the boxes at the top of the plan:

SDWT	Summer deadweight
WDW	Winter deadweight
LOA	Length over all
BEAM	Within the confines of the ship's side frames
GRT	Gross registered tons
NRT	Net registered tons
GEAR	Deck cranes 25 tons safe working load
CONST	Stores and spare gear. Mud in ballast tanks, etc. Variable but determinable by use of the deadweight scale with the vessel in a light condition.
TPC	Tons per centimetre immersion
BALE	Vessel's total bale measurement
SNRT	Summer net registered tons
FW-Evap	Fresh water evaporators. The vessel can condense 15 m/tons of water per day.
FWA	246 mm. The difference in draft between salt water and fresh water.
DWT	Deadweight tons
IFO	Intermediate Fuel Oil
MDO	Medium Diesel Oil
FW	Fresh Water
Add Bkrs	Additional Bunkers
Draft Corr.	Draft Correction
Stow Mat	Stowage Material, Dunnage and securing materials
Av. DWT	Average deadweight
Free	Tonnage light of average deadweight.

The vessel loaded at Novorossiysk in the Russian Black Sea area for Jeddah (Saudi Arabia) and Mombai in India. At Novorossiysk the vessel loaded at WD (Western Dock) and CD (Central Dock). The figures in the hatchways represent the dimensions of the hatchways. Taking No. 4 hold again it will be seen that the heavy figures of twice 2,500 are the tons loaded, in total 5,000 m/tons. The smaller figures beneath indicate that the cargo in the after end extends 10 metres forward and the cargo in the forward end extends 12.5 metres aft, total 22.5 metres, which from the figures beneath is the length of the tank top plating shown as 22.6 24: 22.6 is the free length of the tank top, some obstructions are in way of the balance 1.6 m. The end transverse measurements of the tank tops are seen to be in the case of No. 6, 12.5 m at the after end and 16.4 m at the forward end. Nos. 3 to 5 are box shaped at 16.4 m. On the left hand side of the plan beneath the propeller all else is more or less self-explanatory.

II. ABC of steel carriage

ABSORPTION

A gas (such as moisture vapour) taken up by chemical or molecular action. In such circumstances the gas permeates the whole body of the solid. Such a situation will occur when there is a higher vapour pressure of the ambient air surrounding hygroscopic cargo than the vapour pressure of the actual cargo.

ADSORPTION

The condensation of a gas (such as moisture vapour) on the surface of a solid. A good example would be moist air passing through a silica gel bank which will adsorb moisture so drying the moist air.

AGE OF RUST

Where maritime steel claims for moisture damage and rust are concerned, it is often helpful if the exact age of the rust can be determined by expert appraisal. Photographs of the rusty steel surfaces are produced and in some instances spurious opinions are advanced. In one case, from photographs, a metallurgist who was commissioned to give such an opinion reported that the vessel could not be held responsible for the damage since the rust, as shown in the photographs, developed prior to shipment. He did not seem to realise that it was the court that would decide who was liable for the damage; furthermore, it was known that the goods came directly from the production line to the ship and also that the bills of lading were signed "clean on board".

From original photographs (copies can be misleading) the best assessment possible would be to state that light-coloured rust indicates that rusting commenced recently, but precisely how long it has been established is impossible to determine. As time goes by the colour of the rust deepens and becomes dark brown, from which it can be stated that the rust condition had been established for some time or some considerable time. The speed at which rust develops depends upon relative humidity, purity of the atmosphere, temperature, chemistry of the material and storage conditions; all of these vary from one location to another.

A better judgement can be made if some, if not all, of the history of the goods is known. This, combined with survey reports and climatological reports from the port of shipment, may help to formulate a better opinion. However, the precise time of commencement of rusting is impossible to determine; an approximation is possible if the necessary information is available.

In many instances, where the age of the rust is concerned, not all metallurgists' opinions are reliable. Admittedly all metallurgists have studied the theory of corrosion but might not have had the day-to-day experience of viewing and evaluating the condition of the various rusty surfaces; there are many different branches in the metallurgical profession. There are metallurgists who specialise in corrosion and whose opinions would be more reliable.

ALLOTROPY

A term used to describe the property possessed by certain chemical elements (carbon, sulphur, iron, etc.) existing in two or more distinct forms which are identical chemically but have different physical properties. Under certain conditions, one form may change to another (with changes of temperature or pressure).

ALLOY STEEL

May be defined as plain carbon steel to which metallic elements have been added in large enough percentages to alter the characteristics of the steel. Carbon steel does contain minor amounts of other elements besides the basic iron and carbon, such as silicon and manganese. However, unless these elements are increased in percentage beyond the usual small percentage found in carbon steel, the steel is not considered an alloy or special steel. Alloy steels are divided into two types. Low Alloy Steel, under 10% of added elements; High Alloy Steel, over 10% (usually 15/30%) of added elements.

ALUMINIUM

The most important source of this metal is through the processing of bauxite ore, of which there is a great abundance in the earth's surface. The characteristics of the metal are its lightness in weight when compared with other metals (S.G. 2.7), resistance to corrosion, high electrical and heat conductivity, and its chemical resistance. With regard to steel making it is used as a deoxidising agent, hence the term al-killed or aluminium killed steel. Aluminium coatings are applied to the surfaces of steel through a patented process of alloying the surfaces of carbon or alloy steel with aluminium by diffusion. This process is referred to as "aluminising", which produces a coated steel offering good resistance to heat and corrosion. The ever-increasing demand for coated steels has resulted in greater tonnages of "aluminised" steel being transported by sea.

Pure aluminium resists corrosion because a hard, tenacious surface coat of aluminium oxide (Al_2O_2) forms on the surface of the hot metal as it comes off the production line; as such it has a relatively low strength. However, when the metal is alloyed with other elements, usually chromium, silicon, copper, manganese, magnesium, the protective film is not formed. The process of alloying imparts to the material certain desirable characteristics such as improved workability, strength, etc. All the alloys offer good resistance to atmospheric corrosion and also to many of the corrosive substances/materials usually encountered, but alloys with copper are less resistant to those containing silicon. "Alclad" is a name given to alloyed material supplied in the form of sheet, tubing and wire which has a surface coat of pure aluminium in order to combine the strength of the alloy with the corrosion resistance of the pure aluminium coating.

Aluminium in contact with the atmosphere and well ventilated will not deteriorate but depending upon the purity of the air the surface appearance can become very dull and unattractive in appearance. Experience suggests that a decreased oxygen supply together with a high relative humidity will result in the development of corrosion as is shown in Fig. 23, which was an alloy aluminium coil stored on a wooden pallet. The pallet sustained wetness and the high relative humidity of the air penetrated between the turns of plating to cause the white rust damage shown. Similar shipments of alloyed aluminium sheet have been lacquered for protection. Fresh water wetting to billets, slabs, etc., is usually of no consequence, but on board ship rust-stained sweat water and contact with rests of previous cargo can result in claims in the form of cleaning costs. Contact with seawater can cause localised pitting damage in material where the main alloying constituent is copper. Aluminium and alloys of aluminium are shipped in coil form, usually on pallets, or in packages.

ASSESSMENT OF LOSS CAUSED BY DAMAGE

Guidance on the assessment of loss can only be discussed here in the very broadest of terms. Although various types of defects, such as bending damage and rust damage, can be categorised to a certain extent, degrees of damage and circumstances differ widely. Therefore, the following should only be considered as a guideline to steps which may be taken towards the rectification of damages which do occur during the seaborne carriage of steel products. With regard to the quantum of allowances which are granted as compensation and also costs which might have to be taken into consideration, geographically these will vary considerably.

Cold rolled steel coils

Damage sustained by this type of material is usually caused by either:

(a) Free moisture contacting the packing, e.g. rainwater, ship sweat, penetrates the overlaps of the packing and contacts the material within.

(b) Cargo sweat forming on the packing trickles down and is caught up in the protection ring overlaps, whereby this free moisture penetrates to the material within.

(c) If the coils are subjected to appreciably fluctuating atmospheric temperatures they will sweat internally between the turns of plating. However, if the sweating is not prolific the protective coating of oil applied to the surface of the sheet prevents damage developing in the short term, in the longer term emulsification of the oil with the water will lead to the development of rust.

(d) Free moisture penetrating the packing, apart from causing localised rusting of the plating, will result in an increase in the relative humidity of the air within the unit. If the situation is such that the dew point temperature of this air exceeds the temperature of the steel, which is usually the case, condensation will take place upon the plating in the form of beads of sweat water.

(e) Galvanised sheet steel wrappers are extensively used for the packing of coil cargoes and may be new, or comprised of mill waster sheets. Galvanised surfaces become rapidly affected by oxidation and are usually affected by oxidation films, white rust and dried-up moisture streaks, all of which will in most instances in no way be representative of the condition of the contents of the coils. The steel comprising the turns of plating may sweat internally, with the moisture gravitating down to the base of the coil where it is exuded, so causing the edges of the turns of plating, at the base of the coil, to rust in a fan-like pattern from the inner edge of the centre core to the outer circumference turn of plating. The pattern of rusting is confined to the bend of the plating, so that the distance covered by rust is less horizontally where the plating angle is acute (near the inner circumference turns) than at the outer circumference (see Fig. 4). If the moisture is free moisture from an external source and drips down from the center core, there will be a vertical streak of rust and it will not form a fan-like pattern. Care should be taken to distinguish between the two causes and effects.

Presuming only a small proportion of the turns of plating are affected, e.g. up to one third of the weight of the coil or less, the rust affected sheeting could be removed and the remaining sound plating used in the normal manner. That part of the sheeting which is rusty would have along its length large patches of rust, whereas the plating situated between these rust patches might be in good condition.

Depending upon circumstances, it would be usual practice to determine the average quantity of damaged plating involved, and apply to this a depreciation in conformity to the outlets for second choice material. On the other hand, if contact with the moisture which caused the damage is not of recent origin, the plating may be so severely rusted that it may have to be considered as having no more commercial value than that of scrap. Further to this, depending upon the efficiency of the oil film and possible condensation forming between the turns of plating in that part of the coil which would otherwise be considered sound material, i.e. in the unrusted part of the coil, the resulting damage might be much more severe.

Seeing that cold rolled steel sheeting in coils is ordered to a required specification, reconditioning by reprocessing the material is, apart from not being a viable proposition, an impossibility as far as concerns returning the goods to a sound merchantable condition and at the same time maintaining the specifications of the purchaser.

Packages

In those instances where moisture has penetrated the packing and infiltrated the individual steel sheets, the surfaces will be affected by rust, so that the sheets having these rust patches upon their surfaces cannot be used for their originally intended purpose. The individual sheets so damaged would have to be sorted out and re-sold into the industry as either second choice material or, where the damage is too extensive, possibly reduced to the value of scrap. For reasons already stated above, the goods cannot be reconditioned in such a manner that they can be returned to a condition compatible with the original specification.

Handling damage

Where steel coils are concerned, the most common type of handling damage is buckling of the inner and/or outer plate edges, also tearage of side packing thus resulting in deep score marks or bending of adjacent plate edges. The usual method of assessing this type of damage is to count the number of turns of plating affected and by means of a simple calculation involving measurement of the diameter of the coil and the density of mild steel (which is 7.85), the quantity of plating can be determined. Depending upon the degree and extent of the scoring or buckling of the plate edges, an allowance may be granted to cover the costs of cutting out or removing the damaged turns of plating, also the loss of value involved through it being unsuitable for its originally intended purpose on account of the defects. Where packages of steel plates are concerned the same procedure would be required.

Hot rolled steel coils

Where moisture of an aggressive nature, e.g. salt water, has penetrated between the turns of plating, reconditioning is always a possibility. If the progression of the rust can be arrested in time, through acid bath pickling, the goods can be returned to prime condition. If the type of plating concerned is destined for cold rolling then a pickling process will automatically be involved with no loss on account of the rust. On the other hand, if the steel plating is to be cut into plates and directly made up into bundles, then the costs of the acid bath pickling operation might have to be taken into account before the cutting and bundling is carried out, as there would undoubtedly be delay before the goods were marketed. That is to say that too much time would elapse before the sheet surfaces were cleaned and a protective coating applied.

Where goods are to be cold rolled, and if the aggressive pollutant has been in contact with plating over a too long period of time, more than one pass through the acid pickling bath may be necessary. Depending on whether the plating is pitted, or if after pickling etch marks are showing upon the surface of the material, then the situation would have to be reassessed in the light of whether reconditioning would be so extensive as to put the goods off specification, or as to whether such action would eventually be a viable proposition. Likewise with goods which are to be cut into plates for bundling and probably for the fabrication industry, heavy damage might prove to be irreparable when the original specification of the material is compared with what a reconditioning operation would produce. What can be done in such circumstances depends largely upon how soon after contact with an aggressive pollutant rectifying action can be taken.

Handling damage

Where steel coils are concerned, bending damage to the plate edges, providing that it is not too severe and has not stretched the material beyond its point of elasticity, can be and usually is removed by a roll (planing unit) built into the decoiling machine. This roll is there for the purpose of flattening the plating and it will flatten bends on the edges. Where deep score marks and severe bending is concerned, it must be considered whether the edges of the plating are mill edges or slit edges, and what degree of tolerance is involved. For example, plating breadth of 1 m may be ordered with a tolerance on each side of 5 mm. Therefore, if the score marks do not extend more than 5 mm into the width of the material then there should be no loss involved.

Galvanised steel coils and packages

When galvanised surfaces have become affected by voluminous deposits of white rust, re-galvanising is not beyond the bounds of possibility. Whether this can be accomplished will depend, to a great extent, upon the quantity involved, severity of the damage and where the re-galvanising operation is to take place. Usually costs will preclude such a course of action. However, where small quantities of material are involved, such an operation will not in any circumstances be a viable proposition, and the affected parts of the plating, whether it be from coils or from packages, would be considered reduced in value to that of scrap. Where a large quantity, of damaged material is concerned, this can also be regraded to the value of second choice material and the value of this will be determined by market conditions prevailing at the time.

Patterns of damage

Galvanised sheet steel wrappers are extensively used for the packing of coil cargoes and may be new, or comprised of mill waster sheets. Galvanised surfaces become rapidly affected by oxidation and are usually affected by oxidation films, white rust and dried up moisture streaks, all of which will in most instances in no way be representative of the condition of the contents of the coils. The steel comprising the turns of plating may sweat internally, with the moisture gravitating down to the base of the coil where it is exuded, so causing the edges of the turns of plating, at the base of the coil, to rust in a fan like pattern from the inner edge of the center core to the outer circumference turn of plating. The pattern of rusting is confined to the bend of the plating, so that the distance covered by rust is less horizontally where the plating angle is acute (near the inner circumference turns) than at the outer circumference, see Plate 4. If the moisture is free moisture from an external source and drips down from the center core, there will be a vertical streak of rust and it should not usually form a fan-like pattern. Care should be taken to distinguish between the two causes and effects.

Fig. 6 is of galvanised steel sheeting in coil. It is considered that a condensate formed internally during cooling, after the goods left the production line, with resulting formation of white storage stain developing over some considerable period of time. The entire coil was so affected. In addition to the prong like projections, the beginning of further finger-like projections can be seen alongside the center strip of white rust. Fig. 4 shows the straight drip down of moisture and rust from the centre core of hot rolled coils, and Fig. 5 shows internal sweating in a hot rolled coil. The effects of internal sweating / condensation to hot rolled surfaces is totally different to that of galvanised surfaces, as two different metals are involved. Further to this, where hot rolled surfaces are concerned, in the short term, when so affected, the mill scale offers protection and prevents rust developing.

Fig. 7 shows a diagram of a wrapped steel coil wetted through contact with "cargo sweat". For some reason the temperature of the coil has been lower than the ambient air possibly on a voyage from a cooler to a relatively warmer area. The surface of the packing begins to sweat, moisture forms, as a condensate, and runs down the surface of the packing to become caught up in the edge protectors (protection angle). The free moisture penetrates through the kraft paper packing to contact the actual material within. The moisture introduced causes the air within the interstices of the coil to become supersaturated and sweat globals form throughout the coil on the surfaces of the plating—sometimes totally, sometimes partially—and with emulsification of the passivating oil. The lower part of Fig. 7 indicates what the appearance of the strip would be when the coil is decoiled, with characteristic patches of rust at intervals where the free sweat water has penetrated the base of the coil.

Structural material

Bent plates may be flattened by re-rolling. Bent flanges of beams, channels, etc., can be heated and straightened. Nevertheless, the possibilities of rectifying the defects will depend on the severity of the damage with regard to the bends or dents exceeding the elasticity of the material.

Where rust conditions require urgent attention shot blasting will in most instances rectify the situation. The charges for shot blasting are calculated on the basis of tonnage, type and dimensions of material involved, and some companies issue an annual booklet of shot blasting tariffs. In any case, in most modern industrial countries there are existing rates for this work which are easily obtainable.

BASIC OXYGEN FURNACE

This is the main method of producing steel in bulk and the modern furnace can take about 40 minutes to convert iron and scrap into steel. The principal material used is hot metal; however, 70% molten iron and 30% scrap metal are often used together in this system.

The furnace is tilted and charged with scrap metal and then with molten iron which is introduced direct from the blast furnace. The vessel is then returned to the upright position, after which a water cooled oxygen lance is lowered into the furnace and oxygen is blown into the metal at great speed. The oxygen combines with carbon, sulphur, phosphorus and other elements, so eliminating these unwanted impurities from the molten charge. During the oxygen blow period lime

Fig. 4. Hot rolled steel coils. Note drip-down from the centre is not fan-like.

Fig. 5. Hot rolled steel coil. Note internal sweating.

Fig. 6. Decoiled hot rolled steel coil.

is added as a flux to help carry off oxidised impurities in the form of slag, which floats on the surface of the charge. The next step is to refine the metal by adding various elements in determined quantities in order that the desired composition of the steel is reached. Thereafter, the furnace is tipped into the horizontal position and the molten steel tapped off and run into a ladle. After all the steel has been removed, the furnace is tipped into an upside-down position and the slag is run off into a slag ladle.

The advantage of the basic oxygen furnace production of steel over previous methods, such as the open hearth furnace, is that the pure oxygen used prevents nitrogen remaining in the molten steel. Further to this, the method affords a high control over the quality.

BEAMS

"I" beams and wide flange "H" beams are mainly used as bearing piles or in structural work. Unwrapped, they are shipped unprotected in either single pieces or according to size in bundles secured with wire. The surfaces of the flanges are somewhat rough and often have a lined appearance, imparted to them by the process of milling. This may sometimes be mistaken for a defect.

Usually, the goods are offered for shipment in lengths of 20 feet (6.10 m), 40 feet (12.19 m), and 60 feet (18.29 m), but in-between lengths are not uncommon. Claims can arise due to severe pitting caused by lengthy exposure to the elements (see Corrosion and oxidation of steel surfaces) and such pitting can, in certain

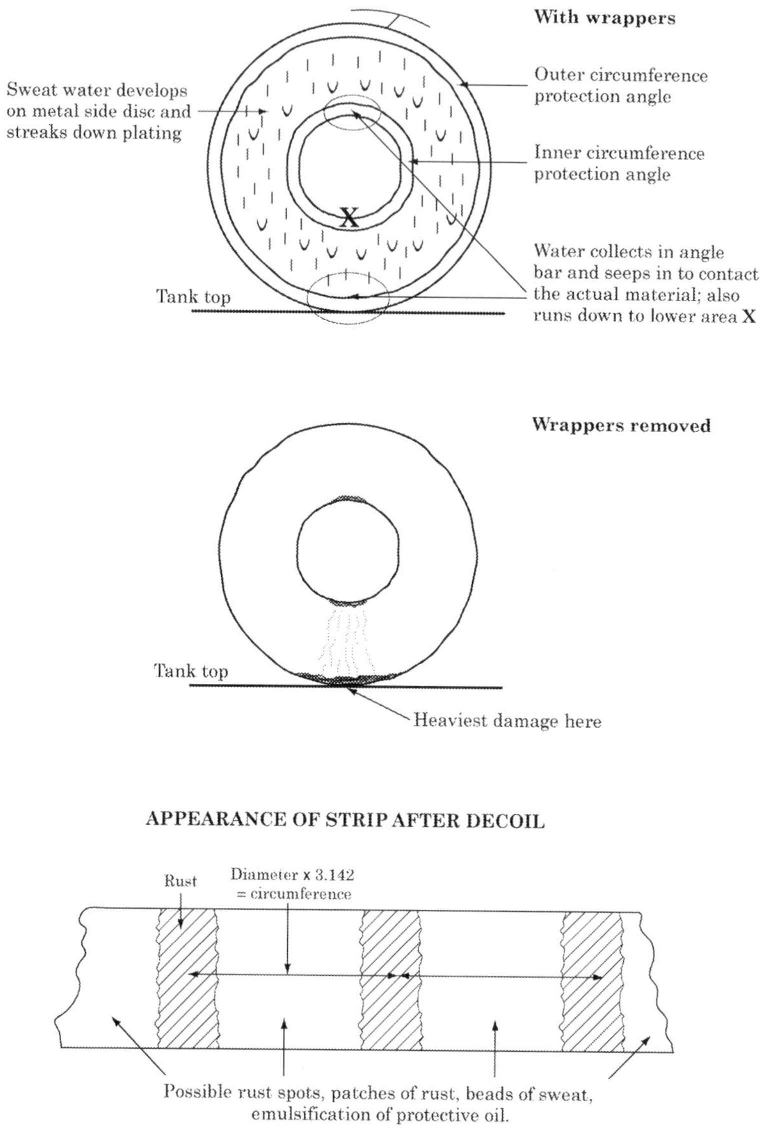

Fig. 7. Cargo sweat in wrapped steel coils.

circumstances, impair the continuity of the strength of the girder. Beams are usually handled by the use of heavy chain slings or wires which cause bending of the beam flanges when slings are overloaded (see Fig. 8(b)).

It is particularly important that, when being stowed, the flanges of the individual beams overlap alternately, and not progressively in one direction (see Figs. 8(a), (c)

Fig. 8(a). Flanges incorrectly overlapping for stowage purposes on-board ship.

Fig. 8(b). Bent flanges caused through overloading slings. Note: beams wired into bundles.

Fig. 8(c). Goods stacked ashore awaiting shipment. Note: correct and incorrect overlapping of flanges.

Fig. 8. "H" beams.

Fig. 9. "I" beams.

and 4). If the flanges are not properly aligned in the stow, pressure exerted from the overstowing cargo can cause the flanges and webs of the understowing beams to become distorted. When beams, wired or secured together in bundles, are seen to be incorrectly arranged, as explained above, mates' receipts and bills of lading should be appropriately claused to cover the situation. One such endorsement, for the bill of lading, might be: "Beam flanges in bundles incorrectly overlapping for stowage purposes".

Where loose beams are concerned, it will be incumbent upon the master of the ship to ensure that the flanges of the beams are correctly overlapping in the stow on board the vessel.

If the pre-shipment inspection of beams indicates that the flanges are dented or bent, and if beams are seen to be warped in their length, the appropriate remarks should in due course be entered in the relevant bill of lading. The type of remark used in such circumstances might be: "Three beams, flanges locally bent in four positions" or "One beam, warped and bent along entire length".

Where "I" beams are concerned, owing to their rather exaggerated disproportions in relation to width of flange as opposed to depth of web, circumstances may demand that they be stowed with the webs horizontal instead of vertical.

Claims for fresh water rust conditions are rare where all semi-finished products are concerned, and especially these large beams.

BILLETS

After the steel is made to a predetermined specification it is poured from the converter into an ingot mould. These ingots if they are not to remain in the ingot form are taken to a soaking pit where they are held at a high temperature until they are transferred to the rolling mill where they are rolled into a semi-finished billet which measures about 150 mm/150 mm. These items of semi-finished steel are manufactured in batches of varying lengths. The finished billet will eventually be subjected to further hot rolling into various profile forms such as beams, channels, angles, flats, squares, etc.

Billets transported by sea are unwrapped and unprotected against rust. They are usually stored in the open and exposed to the elements and in general will be found partly or wholly covered with a thin film of rust brought about through weathering.

Fresh-water rust which is uniform and of recent development is of no consequence. Each billet is heavy, and in a ship's hold, providing the top tier of billets is chocked off in the gaps between individual pieces where necessary, should be adequately secured against movement during the voyage.

Owing to the weight involved, such goods should be stowed in the bottom of the ship if possible, and the dunnage beneath the lower tiers should be placed directly above the transverse floors to prevent spot overloading of the tank-top plating.

BLACK ANNEALING

An annealing process which imparts a black colour to the surface of the steel. Used in tin plating and other materials, e.g. black annealed wire.

BLAST FURNACE

The blast furnace (see Fig. 10) is used to extract iron from the iron ore and could be referred to as an indirect method of reduction to near pure iron. Sizes of the entire structure vary considerably. However, they are of an immense size, reaching a height in many instances of more than 70 m. The actual furnace consists of a cylindrical shell constructed from steel plates, the inside walls of which are lined with heat resistant fire bricks (refractory bricks). The actual furnace can be as high as 35 m with a diameter of about 12 m at the base.

At the top of the furnace there is a double bell charging system through which the burden, consisting of iron ore, coke and limestone, is built up and consistently replenished. Adjacent to the aforementioned charging system are large exhaust pipes through which the hot gases rising to the top of the furnace are directed away into a dust cleaner, gas cleaning plant (spray chamber) and gas holder to be

Fig. 10. Blast furnace at Sidmar works, Ghent in Belgium.

returned to heat the blast air (in the hot blast stoves) to be eventually readmitted to the furnace, after which, these spent gases are exhausted to the atmosphere.

A blast furnace can operate continuously for up to three to five years and produce thousands of tons of iron per day. The uninterrupted iron producing period of the furnace is referred to as a campaign, which must, in due course, come to an end in order to renew the refractory linings which deteriorate as time goes by.

Simply explained, alternate layers of iron ore, coke and limestone are, as required, charged into the furnace through the double bell charging system. The hot blast of air introduced at the base of the furnace causes an extremely high temperature to be maintained within the furnace, as a result of which the coke burns and the carbon monoxide gases given off combine with the oxygen in the ore (iron oxide) to pass vertically upwards to the top of the furnace, and leave behind near molten iron. While this situation has been developing the limestone has scavenged the charge of extraneous materials which accumulate in the form of slag.

The molten iron settles at the base of the furnace with a layer of molten slag on top. Through a tap hole the molten iron is run off, to be cast into pig iron or transported in the molten state direct to the converter for refining into steel. The slag is poured off and retained for other uses (see under Slag).

The furnace takes its name from the blast of hot air and gases forced up through the bottom of the furnace through the charge. Extremely high temperatures are involved and at the base of the furnace temperatures rise to more than 1,700°C. The temperature reduces gradually throughout the furnace to about 300°C at the top.

It is of interest to note that a medium-sized blast furnace can produce between 5,000 and 8,000 tons of iron in each 24-hour period, requiring about 7,500 tons of iron ore, 3,000 tons of coke, 750 tons of limestone and 2,000 tons of air.

BLOOMS

Semi-finished products, square in cross section, rounded on edges and directly produced from the rolling of ingots. A single bloom may be 23 × 23 × 6,000 cm with a weight of about two and a half tons—dimensions greater than this may be encountered from time to time. It is from such goods that structural steel is manufactured, e.g. wide flange beams, “I” beams, channels, and also rails and pipes, etc. There are also round blooms, but whether round or square they are shipped in single pieces (see Fig. 11).

BULK CARRIER

In seaborne trades where large amounts of steel are to be transported, the general purpose bulk carrier is most widely used. The characteristics of this type of ship are that the engines and accommodation are at the afterend of the ship. The cargo carrying compartments are encompassed by that part of the hull forward of the machinery and accommodation. In general, these ships used for carrying steel are equipped with five cargo holds and therefore have five deck openings (hatchways). Usually they are fitted with the necessary cargo gear to load and discharge cargoes

Fig. 11. Blooms.

but gearless bulk carriers are not uncommon. There are no tween decks fitted in the cargo holds of a bulk carrier.

As the name implies, this type of vessel was originally designed with the intention of carrying bulk cargoes, such as coal, grain, iron ore, concentrates, etc. Nevertheless, owing to the great length of cargo hold available in which to stow large, heavy, long lengths of steel, and also to large deck openings which facilitate rapidity and ease in loading and discharging the ship, constructionally bulk carriers lend themselves admirably to the carriage of steel.

The type of vessel employed in the steel trades is usually between 20,000 and 30,000 tons deadweight, and the following are some particulars of such a ship.

Gross tonnage	19,831
Net tonnage	13,973
Deadweight	34,186 m/tons
Overall length	185.5 m
Summer draft	11.51 m
Extreme breadth	26.7 m

In the vessel from which these figures were taken the five cargo holds had an uninterrupted and unobstructed length of between 25 and 33 m. The exception was No. 3 Hold, which could be used as a ballast hold, which had an overall length of 15 m.

The cellular double bottom tanks are of extremely strong construction and it is not unusual for tank tops to be capable of officially withstanding something like 27 metric tons per square metre. This means that if the official figure, i.e. tons per square metre, is exceeded, such spot overloading can result in localised distortions of the tank top plating. When spot overloading is imminent such a situation is avoided by using suitable dunnage to spread the concentration of weight (see under Tank-top strength).

CAMBER

This is a condition that affects strip steel sheeting. During the rolling process one side of the sheet may be more pressed than the other and worn rolls may be responsible for this. The fact is that every steel coil is affected by camber to some degree. If a coil of strip steel were unwound on a flat surface it would form an arc, as one side will be slightly longer than the other. In most instances the effect is not great and does not result in any difficulties being experienced, however, exaggerations of the defect can result in two types of defect called the sweep and the snake. The snake is the type of camber, which results in unrolled coil, which oscillates from side to side. It can be found in mill coil but is more common in more slit coils as it is generally caused in the slitting process. During the process of passing through a rolling mill the strip may be squeezed in one area more than the other and this causes the strip to become elongated in some areas more than others. This can cause a variety of shape defects including “centre buckle”, “edge wave” and “camber”. These are related problems as they are examples of side-to-side differential. The coils may be rolled unevenly because the rolls are worn, roll deflection is not controlled, the temperature across the width of the rolls is inconsistent, or because the upper and lower rolls are not parallel.

CAPPED STEEL

This steel is much the same as rimmed steel except that the action of rimming is not allowed to fully develop. The material is used for cold rolling (see under Rimmed steel).

CARBON

This is a non-metallic element found in all organic compounds and which when blended with iron forms plain steel. The carbon content is the deciding factor as to whether the steel will be denominated as low, medium or high carbon steel, all of which are basically an alloy of iron and carbon. The higher the carbon content of the steel the harder and stronger it becomes at the expense of ductility. Not only is the importance of carbon in steel making confined to its role as indicated above. Carbon has an affinity for oxygen; it is the basis of coke which is used as the fuel in the blast furnace. The carbon in the coke scavenges the oxygen from the iron ore in order to produce the iron. The resulting carbon monoxide gas is piped off through the top of the blast furnace.

CARBON MONOXIDE GAS

Hot carbon monoxide gas passing upwards through the blast furnace drives off moisture in materials charged in at the top of the furnace. It creates a chemical change so reducing oxides of iron, helps to calcine the flux (limestone), melts the slag and iron, reduces oxides of manganese, silicon and iron, and removes sulphur from the molten iron.

CARBON STEEL

Steel that owes its properties chiefly to the carbon content of the metal rather than the presence of other alloying elements, which are seldom present in appreciable amounts. Carbon steel is an alloy metal consisting of iron and carbon. The carbon content is usually greater than 0.2%. The manganese content is usually 0.3% to 0.8% but not greater than 1% or the silicon content greater than 0.05 to 0.03%.

CAST IRON

This is an iron and carbon alloy containing between about 2 to 4% carbon and, similarly to steel, it also contains small amounts of manganese, sulphur and phosphorus, but is higher in silicon content. Its high carbon content makes the metal very brittle so that it cannot be rolled or forged. Therefore, it is cast in moulds by pouring the molten metal. Cast iron is hard and durable, and some of the uses to which it is suited are the manufacture of pipes, iron baths, machine parts and stoves, etc. This material rusts less easily than forged iron and steel.

CHANNELS

As the name implies, these goods are lengths of steel, the profile of which is in a laterally extended "U" shape, this being clear from Fig. 12(a), (b), (c). The type of channel here referred to is used in structural work. With the exception of the remarks made about overlapping of flanges for stowage purposes, all that has been said in the sections dealing with "beams" also applies to channels. The goods are wired together in bundles and are usually shipped in lengths of 20 and 40 feet.

CHARGE

The blast furnace charge is the collective name given to the iron ore, coke, sinter, limestone, DR pellets, etc. The charge is fed into the furnace through a double bell and hopper arrangement at the top of the furnace.

Fig. 12(a). Method of nesting channels.

Fig. 12(b). Bundles of channels bent in handling.

Fig. 12(c). Steel channels awaiting shipment. Note: contamination by unidentified powder.

Fig. 12. Channels.

CHEMICAL COMPOSITION OF STEEL

This refers to the constituent make-up of the steel which will provide certain qualities for whatever purpose it is intended. The higher the carbon (C) content the harder the steel; conversely the lower the carbon content the softer and more ductile the steel becomes. Sulphur (S) is something that cannot be eliminated completely and promotes brittleness in steel. Silicon (Si) is an important alloying element in both ferrous and non-ferrous metals. It is also extensively used as a deoxidizer in steel making. Manganese (Mn) is always present in commercial steels (see MANGANESE, page 111). PHOSPHOROUS (P) (see page 136). Molybdenum (Mo) is used to control the properties of the steel during smelting and is used as an alloying element. Chromium (Cr) is associated with high tensile strength and resistance to atmospheric corrosion and the action of acids. Nickel (Ni) is a chief alloying element in steel making increasing strength, hardness and toughness.

The above are the main elements used in steel making which are referred to in specifications in the ordering of steel. The percentages necessary in the manufacture of steel for whatever purpose, e.g. mild steel plates, railway iron, pipes, etc, are specified in the standards produced by various countries, see under STANDARDS AND QUALITY CONTROL, page 178.

The chemistry of some materials might be as follows.

	C	**Si**	**Mn**	**P**	**S**
Steel plate sheet and strip	0.20	0.1	0.50	0.050	0.050
Railway iron—rails	0.65	0.10	0.80	0.040	0.040
Flats, rounds and squares	0.22	0.50	1.60	0.050	0.050
Steel pipes and tubes	0.22	0.35	0.80	0.045	0.045
Structural steel—plates, bars	0.26	0.15	0.85	0.040	0.050

The above table gives an idea as to the quantities present in the chemical composition of the various steel uses. The standard tables provide tremendous variations, all of which contribute in producing steel with specified tensile strength, yield stress, hardness, weldability, plastic deformation properties etc. Chromium, Molybdenum and Nickel are added to produce stainless steel.

After the specification has been drawn up and agreed, during or after manufacture the purchaser may appoint a surveyor/controller to check the material, after which a quality control certificate will be issued. The quality of the steel is confined mainly to the chemistry of the material for which obviously the vessel can never be held responsible. Contracted damage sustained after manufacture has nothing to do with quality—it concerns condition.

CHROMATE COATING

The metal is immersed in a solution of potassium dichromate, or some other suitable chemical, for the purpose of inhibiting or reducing corrosion (see under passivation). It is not uncommon for galvanised surfaces to be chromated and this would apply especially to material which is to be transported by sea.

CHROMIUM

Obtained from chrome ore. Chromium is a metallic element which when added in small quantities to carbon steel increases hardness, tenacity, resistance to wear and corrosion. A familiar use of the metal is in chromium plating of domestic hardwear fittings and in the bodywork parts of motor cars. Its main use in industry is in the manufacture of various alloys, as it imparts the properties of hardness to steel and is resistant to corrosion. It is widely used in high speed steels and the manufacture of stainless steel.

COATED STEELS

A form of protective surface,or decorative finish. Protective surfaces, usually against corrosion or abrasion, may be applied by electrodepositing, chemical deposition dipping, welding, spraying or thermol evaporation in vacuo or ore in inert gas atmospheres. The most well known coated steel in the field of marine cargoes is galvanized material, which is mainly applied hot dipped or by electrolytic methods. There is also paint coated steel in coil form which with both types here referred to can be applied to one or both side of the plating. The galvanised steel sheet is extensively used in the manufacture of car bodies.

COBBLE PLATES

These goods are usually a collection of over-rollings or end cuttings from various orders or a collection of plates from other sources; they do not usually follow any specifically defined specification. Surveyors tend to regard the goods as they might scrap steel, but it is a mistake to do this. The goods are shipped in many varying states of physical correctness or incorrectness. One such shipment consisted of bundles of plates comprising four and sometimes five plates per bundle. The outer two plates were each six metres in length; the inner plates were either two or four metres long. These plates were hot rolled and were offered for shipment in prime condition; they were for re-rolling. On the other hand, large single plates in a rusty and waved condition were referred to on the shipping documents as cobble plates.

COKE

The type of coke used in the production of iron is referred to as metallurgical coke. Such material is chosen owing to its low sulphur content and coke may be looked upon as the skeleton of coal. Coke is produced from coal in special coking ovens installed at the steel works. In the blast furnace, coke has a triple role insofar as it is used as a fuel to raise the temperature within the furnace. It physically supports the burden and its inherent porosity permits the gases to penetrate upwards to the top of the furnace. Its third function, which is of extreme importance, is that the carbon monoxide gases produced by burning off the coke combine with the oxygen in the ore so reducing iron oxide to iron.

COLD ROLLED STEEL SHEETING

Manufacturing process

This material is manufactured by re-rolling and cold reduction of hot rolled steel sheeting. The first step taken in the process is when a hot rolled steel coil is unwound and passed through rolls which flex the plating in such a manner as to break and fragment the mill scale. The plating is then pickled in an acid bath for the purpose of removing the mill scale completely, as well as rust and other extraneous matter which might be present. Thereafter, the material is washed to remove any traces of acid, so leaving the surface of the sheet clean, fairly smooth and dull grey in colour.

In order to prepare the steel for cold working by improving its grain structure, the plating is annealed by heating to a high temperature in a special type of furnace. It then passes on to the cold reduction mill, which consists of a number of sets of rolls, five strands in tandem (see Fig. 13) where the plating is under the pressure of the rolls and under tension between strands, progressively reduced in thickness to the required gauge.

In order to rectify the adverse effects of cold working, the coils are again annealed in sealed furnaces, into which a special oxygen-free atmosphere is fed, to avoid the formation of scale upon the surface of the plating. Finally, the sheet is given a skin pass through a single strand of rolls, which is referred to as temper rolling. In this operation the sheet is further reduced, but only very slightly, and then re-coiled. The object of this is to produce the required mechanical properties and surface finish. In general, the purpose of cold rolling is to impart to hot rolled material a finer finish in the form of a fine smooth surface, accuracy of thickness and

Fig. 13. Installation of five strands of rolls in tandem for producing cold rolled steel sheeting. Observe spare interchangeable rolls in the foreground. Sidmar works, Ghent, Belgium.

dimension and also improved tensile strength. As a rough guide, in general, plate thicknesses of 0.5 mm or even less are involved. Coil weights are between 5 and 15 tons per unit with a width of plating from about 900 to 1,250 mm. Coil diameters of from 1,000 to 1,400 mm are about average. Apart from being wound into coils the continuous strip, or band of steel, is also cut into separate sheets which are made up into oblong packages.

Packing

The finished coil, with a fine clear surface, is oiled for protection against the development of rust, secured with a steel strapping band applied around its circumference, completely wrapped in moisture-resistant paper packing, after which an outer metal cover is applied for further protection. Flat metal bands, usually not less than four, are passed transversely through the core piece of each unit and also not less than three additional bands around the circumference to secure the packing in place and also to prevent telescoping. The banding is also applied to ensure that the coil remains tightly wound, see section under "Packing".

The effects of contact with moisture

Cold rolled steel sheeting can in a certain sense be considered a finished product. Any rust on the plating is not acceptable. In consideration of these facts, during loading or transport and discharge of the material, it must be kept completely dry at all times. Damage, resulting in heavy claims, can arise through contact with rain during loading and/or discharge from the ocean carrier. On voyage, ship and especially cargo sweat have been responsible for heavy losses. Likewise, penetration of seawater into cargo compartments can be particularly devastating. Free moisture flowing across the outer metal envelope can pass beneath the overlaps of the packing and eventually penetrate the contents. As explained in similar circumstances with hot rolled coils, but with more serious results, through capillary action, moisture, which has penetrated and contacted the goods, is drawn between the turns of plating (see Plate 1).

So-called protection or strengthening rings, which consist of an angle iron placed around the outer and inner circumference edges of the packing, have proved to be an innovation which promotes rather than reduces the incidence of damage. Strengthening of the edge packing does provide greater protection against handling damage, but the benefits to be derived from this are far outweighed when the packing comes into contact with free moisture. The moisture trickles or runs across the surface of the packing to be caught up and accumulated in the edge protections This water seeps through, penetrating the inner packing and eventually contacts the plating of the coil (see Plates 2 and 3 and *Patterns of damage*, p. 32).

Rust

In the event of the plating developing a rust condition, whereby part of the coil so affected will be rejected and considered unusable for its originally intended purpose, loss in value will result. Depending upon the degree and overall extent of the rust a decision will have to be made as to whether the plating qualifies as second choice material or whether it should be degraded to the value of scrap. One thing is certain, such material affected by rust cannot, in most instances, be reconditioned.

Pre-shipment condition and storage

Owing to the delicate nature of cold rolled steel sheeting, and the risks involved in permitting it to come into contact with moisture, covered storage at all times is of great importance. Varying extremes of temperatures should be avoided in order to guard against condensation taking place upon the wrappers.

The question is often raised as to whether rust on the wrappers is indicative of the contents being similarly affected. The possibility exists that it might be, if the rust is in the form of drip marks across the surface of the packing, suggesting that at some time the coil has been in contact with free moisture.

A well-established dark coloured rust might indicate lengthy pre-shipment storage (old stock), which has resulted in the protective coating of oil draining to the bottom of the unit and escaping through the packing. Given time, the very thin film left behind on the surface of the plating coagulates and protection is lost in some areas, so resulting in the good quality and appearance of the surface of the sheeting becoming impaired. A thin film of light coloured rust all over the packing, or in patches, need not bear any relationship to the condition of the contents.

In many instances waste sheets are used to wrap the coils, and these may be affected by patches of rust before they are applied. Some mills paint the wrappers of the coils, so imparting to the goods a favourable appearance. Further to this, painted packing may cover existing rust or prevent rust development. It avoids pre-shipment remarks, appertaining to a rust condition, being inserted in the bills of lading at time of loading into the ocean carrier. The fact is that uncoated steel sheet packing or metal envelopes, although possibly bright and free from rust when applied to the coil, will rather quickly develop a fine light coloured film of surface rust through contact with the atmosphere. This film of rust will gradually thicken and darken in colour as time goes by, so imparting to the coil packing a completely rusty appearance.

Galvanised steel sheeting is now extensively used for the metal export outer packing of steel products, especially steel coils. A galvanised surface exposed to the atmosphere in industrial areas rapidly develops blemishes, white oxidation marks, dullness and often a dirty appearance. Such a condition can be misleading, as far as the condition of the contents is concerned, as most experienced steel surveyors would agree.

Industrial uses

This type of steel sheeting is used extensively in the automobile industry for the manufacture of car bodies. It is used for household appliances such as refrigerators, bread bins, electrical home appliances, radiators, metal furniture, drums, etc. It is also the basic material from which tin plate and galvanised sheets are made. Also, there is a growing demand for pre-coated cold rolled steel plating, which has a diverse range of uses. To suit some customers' requirements, coils are finished on slitting lines which cut the wide steel band down into narrow widths (see also COLD ROLLED STEEL SHEETING).

Handling damage

This is a type of damage which originates from what might be termed "rough handling". When a unit is permitted to strike some object which pierces the packing or rips it open to expose the edges of plating, and as a result of which the plate edges become dented or scored to some degree, rust rapidly develops on the exposed plating. Depending upon the extent of this damage, claims can be forthcoming. During lifting, if the proper lifting gear is not used, the inner core turns of plating become locally buckled and/or dented. If these defects are pronounced, damage to the plate edges can be involved.

A type of handling damage often encountered with cold rolled steel coils is often referred to as ovalisation, which can be caused by lowering a coil at such a speed that it lands heavily. This can result in the unit being forced out of the round so that it becomes oval in shape—pressure in stow can also produce similar results (see Fig. 14). In such a condition the coil will not fit into the decoiling machines if the ovalisation is too pronounced. In some instances, a jack is placed in the centre core and the unit is forced back more-or-less to its original roundness. This manoeuvre need not always meet with success; consequently, the receiver may abandon the coil to cargo insurers who will arrange for it to be disposed of by sale which eventuates in loss.

Packages

To suit customers' requirements cold rolled coils are often cut into short sections, in general measuring roughly 1,000 × 2,000 mm. The stack of sheets is wrapped in

Fig. 14. Ovalised coil.

moisture-resistant paper and finally fitted with an outer metal envelope. Each package is then secured with flat metal strapping bands placed both longitudinally and transversely around each unit. Beneath each package there is usually a framework of strong wooden bearers, which impart to the package rigidity and facilitate manipulation by fork-lift trucks. The goods are particularly prone to damage in handling by the activities of fork-lift trucks, the forks of which pierce the sides of the packing and damage the adjacent plate edges. Packages of cold rolled steel sheeting should at all times be kept dry (see Fig. 15). (For details of the stowage of this material on board ship, see the section on Stowage of steel sheeting in coils and packages.)

CONSTRUCTIONAL SHAPES AND HOLLOW SECTIONS

All of the profiles shown in Fig. 25 under the heading of MERCHANT IRON, with the exception of flats, angles, beams and channels, "T" and "Z" bars, are hot or cold drawn and in some instances hot rolled to form hollow sections. Such goods have a great diversity of uses and, for instance, heavier sections might be used in buildings, buried in the floors, to carry electric cables, etc. Such material can have high value and is decidedly of a delicate nature (see Fig. 16).

CONTINUOUS CASTING

See under TEEMING LADLE on p. 257.

Fig. 15. Packages of cold rolled steel sheeting.

Fig. 16. Box sections.

CORROSION AND OXIDATION OF STEEL SURFACES

Rust is the corrosion product formed when steel reacts with oxygen and water ($4Fe + 2H_2O + 3O_2 = 2Fe_2O_3.H_2O$). Although rust is primarily hydrated ferric oxide, it contains other compounds. Analysis of a range of rusts has indicated that air-formed rust can contain, and generally does contain, 5% of compounds other than $2Fe_2O_3.H_2O$. These derive in part from the steel, which contains elements other than iron, e.g. copper, silicon, manganese, etc., and in part from atmospheric contaminants and pollutants, mainly sulphates such as ammonium salts, which are generally present in the rust. There are many stages in the progress of rusting, which, considering the sea transport of steel, fall into two categories:

(a) A rust condition which in certain circumstances is unavoidable, i.e., normally to be expected, and the development of which has not reached such an advanced stage as to impair the original sound value of the goods.
(b) A rust development which has damaged the material so causing it to have depreciated in value. In such circumstances, and in all probability, the true condition of the goods would not be reflected by the typed description—as supplied by the shipper—in the relevant bill of lading, neither would the associated invoice show the true value of cargo so affected.

In consideration of the effects referred to in (a) and (b) above, it is to be remembered that, in the context of the transport of steel products by sea, these remarks concern hot rolled steel goods that are unpacked and unprotected against the development of rust. The situation with cold rolled goods, or any wrapped steel products, is viewed quite differently, because the wrappers are applied to protect the product and to ensure that there is no rust development upon the surface of the material. In the event of oxidation taking place, for one reason or another, and irrespective of the severity of such rust, the goods are in most instances considered to be damaged. Therefore, conditions discussed in (a) and (b) above, for hot rolled material, will not apply where wrapped material is concerned.

How and why rust develops

Only some of the noble metals, such as gold and platinum, are found pure in nature. Most metals are found as compounds in the form of ore. Iron, being the basic material used in the manufacture of steel, is found in iron ore. As an example, a particular type of iron ore known as "hematite" (Fe_2O_2) may yield only 25% pure iron. From the moment the iron is parted from the oxygen in the blast furnace it is thereafter always in an unstable, unnatural state, and will be continuously striving to convert to the combined condition.

Iron, and alloys of iron, in the process of returning, either wholly or in part, to its natural state, passes through a process of electrochemical degradation due to a reaction with the environment. This destruction of the metal is accelerated by the presence of acids and gases.

The type of rust phenomena associated with ships' cargoes is in most instances brought about through the effects of wet corrosion. In such circumstances, the rusting of steel surfaces is associated with electrochemical cells for each of which, in order that they might be active, the following components are necessary: an anode, a cathode, a conductor connecting the anode to the cathode, and an electrolyte. With regard to the last mentioned item, this may be represented by fresh water, rain water, humidity in the atmosphere or a more active electrolyte such as acid rain evolving from pollutants in the atmosphere, or salt water. As salt water is so often responsible for the occurrence of rust damage in steel cargoes carried by ships, the following should be considered in relation to wet corrosion or, as already stated, the electrochemical degradation of steel surfaces.

Although fresh water can act as an electrolyte, it is a poor conductor of electricity. Salt water offers less resistance to the flow of electrons, whereas sulphuric acid has even greater qualities in this direction. The following illustration represents the working of a simple galvanic cell, the electrodes being represented by two metal plates (see Fig. 17).

The compound sodium chloride (NaCl), which is common salt, is as the name implies, a combining of chlorine with sodium. These two elements readily combine and in so doing there is the development of an electron imbalance between the atoms involved with the creation of ions being both positively and negatively charged. When dissolved in water the ions are dispersed to move freely through the solution and become conductors of electricity. When the electrodes are of different metals this increases the potential difference within the cell, and the further apart in the galvanic scale the two metals are placed, the greater again will be the strength of the current which develops.

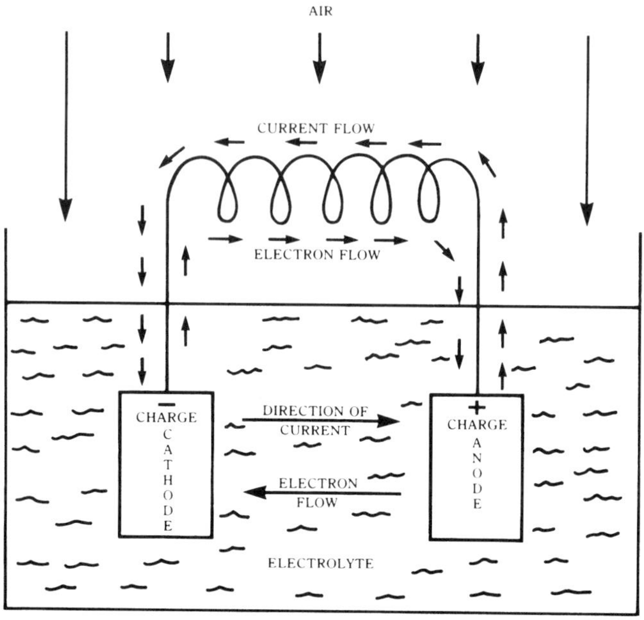

Air supplies oxygen which dissolves in an ionised solution (the electrolyte). One electrode loses electrons to the solution and so serves as a positively charged anode. The other electrode accumulates an excess of electrons and becomes the negatively charged cathode. Electrons flow within the cell, to create an electric current, as it attempts to equalise. Similar action takes place within a corrosion cell, the anode slowly dissolves.

Fig. 17. Action of a simple battery.

Rust development through contact with the atmosphere

When rust develops on steel the surface of the metal is covered with a network of minute galvanic cells similar to that described above. It is, of course, obvious that two different metals are not involved. The presence of particles upon the surface of the steel, together with even the minutest irregularities in the surface, including composition and character of the metal itself, will be sufficient, in the presence of moisture which is more often than not in the form of vapour in the atmosphere. This vapour will act as the electrolyte, and the steel linking the two areas representing the cathode and the anode, through the crystalline lattice structure of which electrons can freely pass, will complete the circuit, so forming an electric cell. The red dust film which can be seen upon the surface of the steel is the residue resulting from the gradual destruction of the anode.

Iron and steel surfaces when left undisturbed and exposed to the environment will develop a fine uniform film of rust. The surfaces of steel plates are a good example of this and the ever-changing polarity of adjacent particles ensures that no localised penetration develops. The greater the quantity of moisture available the more rapid a rust condition becomes established, and the greater the electrical conductivity is, the more severe the rust condition which develops. If sufficient humidity is available, in the early stages rust development is rapid but slows down later as the residue thickens and limits the oxygen available. Further to this, as the process advances, certain conditions which develop within the tiny electric cells further restrict their activity.

In areas where the relative humidity of the atmosphere is 60% or less there is a negligible degree of rusting of steel surfaces. At very low percentages of relative humidity no rusting will take place at all through contact with the atmosphere. It has been determined that when the relative humidity is over 65–70% but below 80% there are effects which cause oxidation of steel surfaces and promote a slow development of rust. Relative humidities in excess of 80% cause more rapid and severe rusting. It is to be emphasised that the contributory effects of humidity to the development of rust will be considerably retarded or enhanced in accordance with the atmospheric temperatures prevailing and the degree of air pollution to which the goods are exposed. Steel stored in the open in humid tropical climates will rust much more rapidly than the same type of material given similar outside storage conditions in a climate where sub-zero temperatures prevail. In desert areas such as Arizona in the U.S.A., the African deserts, etc., there is so little moisture present in the air that steel will not rust.

The effects of air pollution

Air pollution might be defined as a stage in the pollution process of our surrounding atmosphere at which pollution concentrations become objectionable. Such conditions reach critical levels in industrial areas where hot rolled steel is provided with outside storage. In coastal regions all over the world there is an accumulation of industry in the areas of sea ports where steel is often stored in the open awaiting shipment. Emissions from factory and incinerator chimneys, also the activities associated with large nearby metropolises cause considerable quantities of dust particles to settle upon exposed steel surfaces. Near to the sea the relative humidity of the air is nearly always above 80% so that the presence of dust particles combined with a fairly saturated moisture film in contact with the steel surfaces creates a potential for the electrochemical reaction necessary for the promotion of rust.

Air pollution, in industrial areas, is caused mainly through the burning of fossil fuels such as coal, petroleum products, natural gas and other combustible materials. The disposal into the atmosphere of this form of waste, which has a high proportion of sulphur dioxide and nitrogen dioxide, causes it to react chemically with the moisture in the air to form sulphuric and nitric acids. In the event of precipitation taking place, the rain water is highly acidic and is referred to as "acid rain". The term used for the measurement of the acidity in the atmosphere is "pH", which is defined as the negative logarithm of the hydrogen ion concentration in kg per cubic metre. The scale used is between the numerical figures 0 to 14. A pH value of about five would be considered as indicating high acidity, and the lower the number the higher the acidity present in the atmosphere.

Mild steel surfaces exposed to acid rain rust more rapidly and much more severely than do surfaces exposed in non-polluted areas. Seeing that the very act of producing steel causes the atmosphere to be polluted, and that steelworks usually form part of the heavily industrialised areas, and also taking into consideration that a lot of hot rolled material is stored in the open exposed to the elements, it is often queried as to how important acid rain is with regard to the rusting of mild steel. The fact is that exposure of steel surfaces for an indefinite period of time to any type of moisture-charged atmosphere, where temperatures are above freezing point, will result in damage through rust. Therefore, the length of outside storage must be limited in accordance with the condition of the atmosphere with regard to the degree of pollution and relative humidity of the air. Hot rolled steel sheet can, on average, in temperate climates, spend up to six months exposed to the elements without sustaining fresh water rust damage. Structural steels and merchant iron must not be exposed for more than two/three months or less. Users of hot rolled wire rod have been known to carry a six month inventory all stored in the open and exposed to the elements. A large importer in central Europe carries a stock of hot rolled coils, on average between 450,000/500,000 m/tons, all stored in the open for up to nine months.

Differential aeration corrosion

In some instances the surface of the steel is not smooth, as often arises with hot rolled steel sections, but is affected by what is termed in the industry "roughness", which is surface irregularities in the form of a line-like pattern produced by the milling operation when the goods are manufactured. As the effect of this, or other surface unevenness, such as when mill scale is not intact but locally fragmented, another corrosive situation will come into existence which is referred to as "differential aeration corrosion" and causes localised pitting of the surface of the material.

The effects of differential aeration become most pronounced when there are cavities in a metal surface within which oxygen concentration is relatively lower. Such areas become anodic to surrounding metal parts to which there is a higher diffusion of atmospheric oxygen. In such circumstances, greater electromotive force is generated within the cell than would be expected under normal conditions with smooth surfaces, and the potential for greater localised corrosive penetration of the metal exists, which is referred to as pitting. Scratches, score marks and similar damage to smooth surfaces can produce the same effects as can extraneous materials such as an accumulation of dust spots, sawdust and rests of previous cargoes which collect spotwise upon the surface of the material (see Fig. 18).

Passivation

Passivators can be defined as inhibitors of corrosion. One example of natural passivation is when the oxide film formed becomes sufficiently developed as to stifle, to a considerable degree, further activity of the electrochemical process associated with corrosion and rusting of the surface of steel and other metals. Artificial passivation might be achieved by coating the surface of the metal with a chromate passivator or by painting the surface of the material. Before applying such corrosion inhibitors all traces of previous corrosion cells must be removed, also abnormal and sporadic unevenness of the surfaces, as well as chlorides and

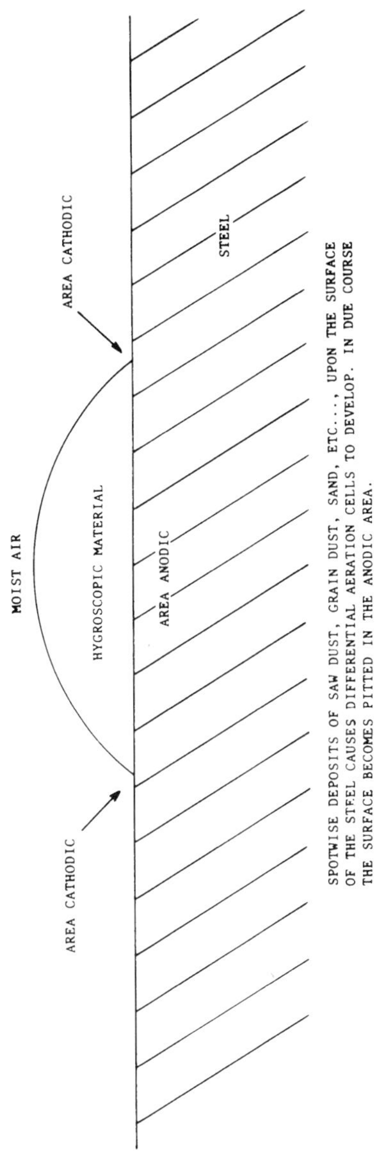

Fig. 18(a). Breaks in oxide scale on mild steel.

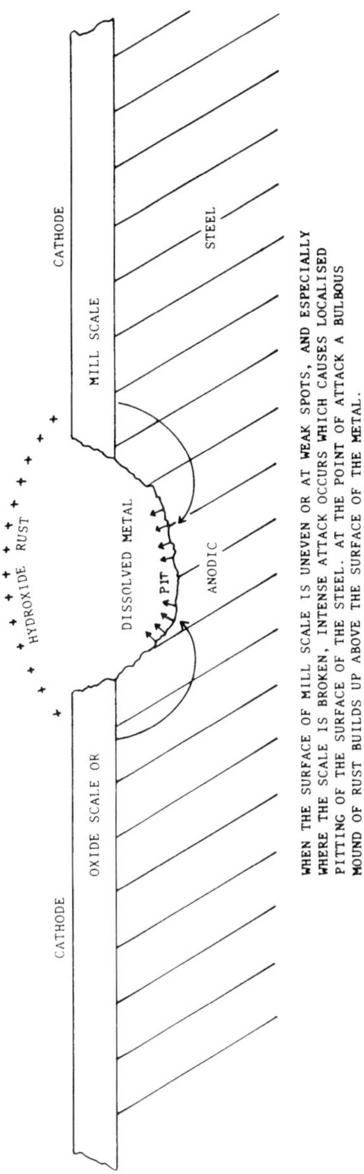

Fig. 18(b). Pitting caused by differential aeration cell.

other extraneous matters adhering to the surface of the material or materials to be treated. If the surface to be treated is not suitably prepared, given time, a rust condition will again become established. Given the above facts it can be appreciated how problematic rust development can become upon cold rolled steel surfaces.

CORTEN STEEL

This is a mild steel with a very low carbon content but the presence of nickel, chrome and copper imparts to the metal an improved resistance to corrosion. It has been extensively used in the manufacture of sea containers, but where repairs are concerned, welding of this material is more time consuming than is the case with ordinary mild steel.

DESCRIPTIVE CLAUSES—PRE-SHIPMENT CONDITION OF CARGO

The effect of oxidation of steel surfaces is generally described by using the single word "rust" or "rusty". Where goods show visible signs of rust at time of shipment it would, therefore, seem both appropriate and logical to describe the apparent condition of such cargo as being "rusty" and clause the relevant bill of lading accordingly.

Depending upon what type of receipt document is used customarily, e.g. shipping order, mate's receipt or exception list, it will in all probability by qualified with the word "rusty" or "partly rusty" whenever a rust clause is considered applicable. Such a descriptive clause is very much within the interests of the carrier, as in most instances the goods will be similarly described when they are received at the port of destination. Even if the rust condition when shipped deteriorates further during transit, and is eventually such as to render the goods unusable for their originally intended purpose, the carrier will, in all probability, and depending upon circumstances, endeavour to fall back on the argument that the goods were discharged in the same condition as when loaded. Failing complete exoneration, the carrier will undoubtedly develop some credible argument, revolving around the pre-shipment rust condition, which will produce some measure of satisfaction.

From the point of view of the receiver, the above-mentioned situation, for obvious reasons, is far from satisfactory. This fact was demonstrated some years ago in the port of Antwerp when shippers collectively declared that, in view of the difficulties they experienced in negotiating letters of credit when bills of lading were claused with the single word rusty, more accurate descriptive clauses must be used. They also stressed the point that any casual or flippant approach to the compilation of descriptive clausing was something that they could not accept.

There are many differing rust conditions, none of which at any one given moment would be completely identical between any two parcels of cargo, or even selected specimens of steel. Nevertheless, the differences need not be so great as to prevent the setting of parameters which can be used in the form of descriptive clauses for such material. One of the recognised standards and authorities on this subject is the Swedish Standard SIS 05 59 00 which defines four grades of rust as follows.

Grade A Mill scale intact, unbroken, and covering the entire surface of the steel.
Grade B Mill scale partly broken and/or fragmented in places, exposed steel surfaces showing rust development.
Grade C All mill scale disappeared from the surface of the steel, which has a rusty appearance.
Grade D Complete surface of the steel rusting with significant corrosion development and pitting.

(See under the heading Mill scale for definition.)

Grades A, B and C are supposed to indicate that any apparent rust has not advanced to such a degree as to require the entailment of additional costs in surface preparation before the goods can be used in the normal manner. Rust grade D would signify that the surface of the steel has become damaged through the development of rust and additional costs for surface preparation would be involved.

With regard to the time taken for the development of the above-mentioned grades, it has been stated by one authority on the subject that in the United Kingdom the following passages of time might apply and should be considered as applying to the grades as above mentioned.

Grade A Complete coverage by mill scale would not last more than a few days.
Grade B The situation here would last several weeks before further deterioration became apparent.
Grade C Condition as indicated would last from three to nine months.
Grade D Condition as indicated would last more than six months before becoming more seriously affected.

Geographically, the above might apply to some areas, but in itself the statement is unreliable because the degree and rate of development of rust is dependent upon (a) the composition of the steel itself; (b) temperature; (c) humidity; (d) surface condition; (e) pH of the atmosphere to which the goods are exposed and (f) frequency of rainfall—conditions will differ from area to area.

To mention yet another document dealing with degrees or grades of rust, there is "The American Rust Standard Guide". This comprises a series of photographs clearly showing various stages in the development of rust, each photograph being coded. On numerous occasions it has been suggested that this document's codes could be used in the steel trades between Europe and the United States of America for the clausing of shipping documents. These propositions are discussed later under the heading of Pre-shipment surveys.

In arriving at the type of clause which might be used, it is worth discussing a list of clauses which appears in a booklet produced by the International Chamber of Commerce. Under the heading of "Iron and Steel" a total of 37 clauses have been listed that might be used as descriptive clauses. As an example, three of these propositions are discussed as follows:

Clause "Arrived in open trucks, wet before shipment"

That part of the clause "wet before shipment" refers to the apparent conditions of the cargo, and would therefore be an acceptable clause. "Arrived on open trucks" has nothing to do with the condition of the cargo and this part of the clause should normally be inadmissible.

Clause "Shipment during rain"

This is a clause which has nothing to do with the apparent condition of the cargo—it is completely unacceptable. Such a clause entered in a bill of lading could, and probably would, be extremely damaging to the interests of the shipowner (see under Mate's receipts and bills of lading).

Clause "Snowy bars"

We know what is meant by the expression, but such a clause is hardly a clear and accurate description of the situation. A more preferable description would be "All cargo covered with snow". It is considered that the system of choosing descriptive clauses, for iron and steel products, as used in the port of Antwerp is the most straightforward and workable arrangement yet devised. According to the London P & I Association's circular, dated 28 February 1964, which is still valid, the list of descriptive clauses referred to could be used by the carrier in any port in the world.

A copy of the above-mentioned circular can be found at the end of this entry, wherein it is stated that no descriptive clause should be used which might indicate the degree of rust where pre-shipment surveys are concerned. In other words, the clause used should not lead one to the conclusion that on account of the rust condition the steel is damaged, so qualifying for a lesser value. From what has been stated earlier the psychology behind the statement is quite clear.

The requirements of the circular are more difficult to put into practice. When a joint pre-shipment survey is held, the words damage and depreciation inevitably enter into the discussion when differences of opinion arise. Therefore, although the clause used might not indicate any degree of rust or damage, the thought behind its compilation will undoubtedly be based upon a degree of rusting. Even when a unilateral survey is held the surveyor will have to use some method of deciding whether or not he will use the descriptive clause "Rusty" or "Rust Stained", etc. This means that although the descriptive clause will not refer to the degree of rust it must, by virtue of the fact that it is inescapable, possess a hidden meaning which appertains to the actual good or bad condition of the cargo at the time of shipment. In the event of litigation or arbitration being eventually involved, the surveyor might be asked to appear as a witness, when he would have to explain how he had reached a decision as to what clause should be applied.

In such circumstances, as described above, the master of the vessel might also be obliged to testify as indicated. Although he would quite rightly state that he is not an expert in commercial products, and maintain that the rust condition of the steel appeared as he had stated in the bill of lading, he could be manoeuvred into a position whereby he is pressed into explaining how and why he decided on a certain clause from alternative clauses which were available to him. It might finally be contended that it was encumbent upon him to familiarise himself with the cargo which he received into his custody and, in view of the situation prevailing at the time, seek the necessary advice so enabling him to have decided more accurately and intelligently a suitable choice of clause.

Conclusions reached from the above leave no doubt that even though the clause chosen to describe the apparent rust condition of the steel does not indicate any

ESTABLISHED 1869

UNITED KINGDOM MUTUAL STEAM SHIP ASSURANCE ASSOCIATION, LIMITED.

PROTECTING AND INDEMNITY CLUB

THOS. R. MILLER & SON,
Managers
D. R. MILLER C. I. MILLER
E. T. JONES J. E. L. WRIGHT
N. F. LEDWITH J. D. MILLER
S. E. FOWLER J. C. J. SHEARER
M. D. MILLER M. B. SUMMERSKILL
D. V. CHEESMAN
Cables and Telegrams:
MUTUALITY, LONDON, E.C.3.
TELEX 23544.
Telephone AVENUE 4571 (9 lines).

14-20 ST. MARY AXE,

LONDON, E.C.3.

28th February, 1964

LETTERS OF INDEMNITY

With reference to our circular dated 22nd October, 1963, relating to Letters of Indemnity, experience has since shown that where signs of rust on steel shipments are apparent at the time of shipment, Mates' receipts and Bills of Lading need not necessarily in all cases be claused with the single word "rusty" as stated in sub-paragraph (c) of that circular. Some qualification to the word "rusty" may be justifiable in certain circumstances.

In appropriate cases, therefore, it is permissible for any of the following clauses to be used when describing steel shipments which show signs of rust or a similar condition on shipment:—

Partly rust stained.
Rust stained.
Rust spots apparent.
Some rust spots apparent.
Rust spots apparent on top sheets.
Some rust spots apparent on top sheets.
Top sheets rusty.
Some top sheets rusty.
Rusty edges.
Some rusty edges.
Rusty ends.
Some rusty ends.
Rust spotted.
Rust and oil spotted.
Wet before shipment.
Wet steel tubes.
Wet bars.
Rust on metal envelopes.
Covered with snow.
Pitted.
Rusty.
Rust with pitting.
Goods in rusty condition.
Edges bent and rusty.
Partly rusty.

When packed sheet iron is shipped the following two clauses may be used:—

Covers rusty/wet
Packing rusty/wet.

It must be emphasised that the clause which is used must accurately describe the apparent condition of the steel shipment and must also come within the clauses as set out above.

Apart from sub-paragraph (c) of our circular of 22nd October, 1963, which is accordingly hereby modified by this circular, the remainder of that circular, in particular sub-paragraphs (a) and (b), remains unchanged.

In no case should any reference be made to the degree of rust such as "atmospherically" or "superficially".

Yours faithfully.

The Britannia Steam Ship Insurance Association Limited,
Tindall Riley & Co., Managers.
The London Steam-Ship Owners' Mutual Insurance Association Limited,
A. Bilbrough & Co., Ltd., Managers.
The Steamship Mutual Underwriting Association, Ltd.,
Alfred Stocken & Co., Ltd., Managers.
The Standard Steamship Owners' Protection & Indemnity Association Limited,
Charles Taylor & Co., Managers.
Sunderland Steamship Protecting & Indemnity Association,
John Rutherford & Son, Secretaries.
Newcastle Protection & Indemnity Association,
Martin Fryer, Manager.
The North of England Protecting & Indemnity Association Ltd.,
W. Ferguson, General Manager.
United Kingdom Mutual Steam Ship Assurance Association Limited,
Thos. R. Miller & Son, Managers.
The West of England Steam Ship Owners' Protection & Indemnity Association, Limited,
R. S. Fort, Manager.

P & I Association's 1964 letter concerning letters of indemnity.

degree of rust, with many clauses to choose from, some credible method must be involved in reaching a decision as to what expression is applicable. Since 1964 the list of clauses referred to in the London Group P & I Association's circular has been used to describe the rust condition of steel surfaces where both import and export cargoes are concerned.

The official list of clauses is divided into two categories which are referred to in Antwerp as "restrictive" and "non-restrictive". Taking the view of cargo interests, a restrictive clause would suggest that the rust on the steel had developed to such an extent that there was deterioration in the surface of the material. Because of this, a restrictive clause would suggest that the declared sound value of the goods had become impaired, and they would no longer qualify as prime material. On the other hand, a non-restrictive clause would signify that prior to shipment there existed a rust condition—which normally develops on unwrapped and therefore unprotected hot rolled steel products—the extent of which had not impaired the surface condition of the material or its declared sound value.

Some prominent shippers have strongly condemned the list of recommended P & I Rust Clauses (see pages 65–67), often used for describing the pre-shipment condition of steel cargoes, as being antiquated, meaningless and completely out of date. They have suggested that the list be scrapped in favour of using a coded system of photographs such as those attached to the Swedish Standards on rust, The American Rust Standards Guide or the ISO 8501–1 : 1988. The reason these pictorial rust condition publications came into existence was for the purpose of promoting and negotiating sales of hot rolled steel products between buyer and seller. Whereas the interests of shippers and carrier are in some respects parallel, their interests in the pre-shipment clausing of bills of lading are diametrically opposed. There is no evidence to suggest that there will be a change in this situation within the foreseeable future.

There are a number of objections against using photographs for showing the apparent order and condition of steel cargo prior to shipment. There are day-to-day changes in the appearance of the rust condition of hot rolled steel cargoes while awaiting shipment, since during such periods, in most instances, the goods are stored in the open, exposed to the elements. This means that they can be rained upon at regular intervals, exposed to acid rain and dust from other cargo on the same berth and also subjected to fluctuating high range temperatures. As the rust condition develops it will commence as a light-coloured film of rust either partly or wholly covering the visibly exposed parts of the material. The colour of the rust changes and deepens at a rate dependent upon the aggressive nature of the environment, eventually reaching the final stage of rust staining before qualifying for the clause "rusty" or "partly rusty" as the case may be. All the various stages of rust development are covered by the rust stained clauses. A proposed photograph would only show the condition of one very small area of one piece of steel from a shipment of perhaps many thousands of tons of cargo. It is an impractical, unrepresentative, inaccurate and misleading method of showing the apparent condition of parcels of weathered steel. This photographic method of showing the condition of steel directly placed in covered storage conditions as soon as it leaves the production line may have something to be said in its favour. However, such material, after determination of its in-store condition, may be transported over great distances on open rail wagons before being stored for lengthy periods on an open quay awaiting shipment. Export cargoes from the old Eastern Bloc countries are a good example of where goods from central Europe are exported using through bills of lading. Admittedly, the pre-shipment survey is not carried out in the factory but in the port.

There was an incident where a ship master, following the advice of his surveyor, signed bills of lading which were claused P.R.S. This was meant to indicate that the goods represented by the bill of lading were, prior to shipment, in a "Partly Rust Stained" condition. The master of the ship questioned this unusual and uncustomary method of applying superimposed clauses to bills of lading; he was assured by his surveyor that there was nothing to fear, as all would be fully explained in the surveyor's report. Unfortunately, after discharge of the cargo, a fresh water rust-related claim was entered against the carrier, which claim eventually evolved into a court action. Notwithstanding the fact that everyone involved in the case knew what P.R.S. was meant to signify, the judge, in all his wisdom, decided that the bill of lading carried no plain language superimposed clause and was therefore a clean document.

The use of photographs, to show the pre-shipment condition of hot rolled steel products, referred to on the bill of lading by some form of coded figures and/or numbers may lead to serious implications not favourable to the interests of the carrier. Such a system would be totally inappropriate to describe the condition of cargo which was wrapped in steel envelopes, e.g. cold rolled steel products, galvanised material, etc. or any rust condition the configuration of which required to be described. Since the publication of the second edition of this book surveyors in a number of steel exporting countries have put forward suggestions for changing or modifying the P & I Association's 1964 circular letter relating to pre-shipment clauses (see page 63). It is a fact that the original list is now outdated and through use it has been found that many clauses are either superfluous to needs or are partially or totally duplicated. Some of the clauses have withstood the test of time and have been found irreplaceable with anything more descriptive, practical or useful. The opinions of a number of experienced surveyors around the world have been considered. The list of P & I suggested clauses has been reviewed and a number of P & I Associations have maintained their confidence in the clauses by issuing advice to their members to use the revised list as shown below. Other P & I Associations have not commented to any great degree and the old list for them remains unchanged. The following is the revised list of clauses.

Rust Stained

When 75% or more of the surface area of the item of cargo is covered with a fine, light-coloured film of rust, which, when removed by wire brushing, scraping or wiping, reveals a smooth, bright metallic surface.

Partly Rust Stained

When under 75% of the surface of the item of cargo is covered with a fine, light-coloured film of rust as described for "Rust Stained" above.

"Rusty"

When 75% or more of the surface area of the item of cargo is covered with a dark-brown-coloured rust formation, which, when removed by scraping or wire brushing, reveals an uneven, dull surface still rusty in appearance to some degree.

"Partly Rusty"

When less than 75% of the surface of the item of cargo is covered with a dark-brown-coloured rust formation as described for "Rusty" above.

"Rust Spots Apparent"

Localised, very slight penetration of rust in the form of rusty spots and especially in areas where the mill scale coverage is broken or fragmented. The spots are not prominently bulbous in appearance and when cleaned away leave a smooth steel surface.

"Rust Spotted"

Localised penetration of rust through the mill scale. Bulbous, revealing an uneven surface when removed by wire brushing. Parts of the surface without mill scale may be rust stained with rust spotting.

"Wet before Shipment"

Partly or totally wet surfaces apparent before shipment but may only become apparent when moved for shipment, e.g. lifts of structural steel, when lifted water runs out from inside. When lifting hot rolled coils, flexing of the windings often squeezes out water from between the turns of plating; in which case, although initially apparently dry, the goods are wet before shipment.

"Rust with Pitting"

Deep-brown-coloured rust which, often with rust scale formation, when removed by wire brushing or scraping, reveals pitting of the surface of the steel.

"Covered with Snow"

Visible surfaces partly or totally covered with snow and/or ice.

Other clauses not included in the P & I Association's circular that have proved to be useful are:

Areas of steel surfaces reacting to silver nitrate solution tests

Suspect areas should be tested. If they react the test is proof that chlorides or salt water contamination is involved.

Stained by an unidentifiable powder

These powders are often hygroscopic and contain chemicals which can be aggressive to steel surfaces. The colour of the powder may be mentioned. *(Surveyors should endeavour to take and retain samples of the powder.)*

"Rust Streaked" "Evidence of contact with free moisture—drip down rust streaks"

These clauses will apply mainly to the packing of wrapped material; they are very important because drifting free moisture can penetrate the overlaps of packing and contact the contents thus causing damage direct by contact and it can also create internal sweating of the material.

N.B. Where wrapped goods are concerned many of the above-mentioned clauses can be used, e.g. "Wrappers Rust Stained".

Restrictive clauses

Rusty, Partly Rusty, Rust Spotted, Rust and Pitting, Rust Spots Apparent.

Non-restrictive clauses

Rust Stained, Partly Rust Stained, Wet Before Shipment, Covered with Snow.

N.B. Combinations of some clauses may be used.

Because the degree of rust development can vary so much, the decision as to the line of division between the application of a restrictive and non-restrictive clause is difficult to define. Such decisions can only be made with confidence by participants who are experienced in such activities on a day-to-day basis. Qualifications to the basic clauses in order to make them yet more descriptive and accurate are permissible and the clause "Rust Stained", for instance, may be extended to "Heavily Rust Stained" within the sense of the definition. It is an indication that the rust condition of the material has advanced more in the direction of qualifying as a restrictive clause. Of course, by the use of the word "Heavily" reference is being made to the degree of rust. This will be a more accurate description; however, it is clear from the wording of the P & I Association's circular (see page 58), that they do not agree that any reference is made to the degree of rust. Another example of extension to any of the chosen clauses could be in the case of coils, e.g. "Wrappers Rust Stained and affected by drip down rust streaking".

In order to demonstrate the basic principles involved in reaching a decision as to whether a restrictive or non-restrictive clause should be applied, two easily definable situations can be examined in Plates 4 and 5.

The round bars shown in Plate 4 were deposited on the open quay approximately 24 hours before the photograph was taken. They remained uncovered, humid conditions with a high temperature prevailed, and it can be seen that the surface of the metal is covered with a fine film of bright red rust. In removing the rust with a wire brush or some suitable instrument, a bright, smooth, metallic surface is revealed. Such goods would be described as "rust stained".

The situation shown in Plate 5 is completely opposite to that of Plate 4. When the rust is scraped away a bright metallic surface is not revealed. The mechanism of rust development has seriously penetrated the surface of the material and a rust condition has become established which may, or may not, be easily rectifiable by further surface preparation. Steel in such a condition would be described as "Rusty".

The two examples shown illustrate conditions situated in the outer areas of the rust spectrum when viewed within the concept of adopting a restrictive/non-restrictive method of compiling rust clauses. The dividing line between "Rust Stained" and "Rusty" lies somewhere between these two examples. Experience is decidedly necessary in order to accurately formulate descriptive rust clauses, and to defend the decisions made, should this be necessary.

With regard to applying descriptive clauses to wrapped material there is one point of considerable importance. If the aspect of the packing suggests rust streaking such as might develop from dried up moisture runs, the bill of lading might be claused: "Rust Streaks on Packing". Such an aspect as indicated may very well mean that previously free moisture runs on the covers have penetrated the overlaps of the packing and contacted the steel within.

Galvanised surfaces

Coils and packages of galvanised steel sheeting are usually packed under metal envelopes consisting of the same material. The pattern of rust visible on the wrappers should be carefully examined and may be described as follows:

—Wrappers partly (or completely) stained by white oxidation marks.
—Wrappers partly (or completely) affected by white rust.
—Wrappers partly (or completely) affected by white rust (white oxidation marks) in way of apparently dried up moisture runs on the packing.
—Partly (or completely) affected by white rust in the form of streaks on the surface of the packing.

Handling damage

Wrapped cargo such as cold rolled steel sheets in packages and coils, packages and coils of tin plate, galvanised sheet and steel strip, various types of wire and wire rod, to mention but a few, have to be considered as delicate material. Any denting, buckling, tearage and/or scoring of the packing may very well be an indication that the adjacent plating wire beneath is similarly affected and therefore damaged. Such damage may be small, but nevertheless qualifies for a depreciation being granted at final destination. On the other hand, such damages spread across a large parcel of cargo can eventually amount to a large claim. The sum total of such claims covering a number of voyages can reach sizeable proportions. Such damages should not be neglected at time of shipment—appropriate descriptive clauses should be entered in the relevant bill or bills of lading.

The type of damage referred to usually eventuates from slings used to lift the goods prior to shipment. Depending on the weight of each unit, coils are lifted either singly or two at a time. In such situations, either inner circumference edge packing is dented, or both inner and outer edges of the packing. Also, usually every item of the parcel will have similar damage to the packing. In the case of rectangular packages of steel sheet, the longitudinal wooden stiffeners beneath may be incorrectly placed away from the edges. As a result of this, when lifts of three or four packages are handled with wire slings, denting of the lower edges of the packing, of the lower lift, is a common occurrence. Imprints in the packing, of the wire strands forming the wire, are easily detectable, so leaving no doubt as to the type of lifting gear used. The same situation will apply with unwrapped hot rolled material except that the actual damage to the steel will be visible.

It is to be noted that coils and packaged material should be always handled with braided wire slings, and/or, in the case of coils, "C" hooks. If wire slings or chains have to be used, protection pieces must be fitted in the area of the lifts' edges. Where structural steel and merchant iron is concerned, wire rope slings and chains may be used, but overloading of the slings should be avoided as this causes bending damage.

When a decision has to be made with regard to a descriptive clause, it has to be remembered that accuracy is of major importance. It is proven from experience that shippers prefer quantities and not percentages in the clauses—use of the expressions "some", "several", "numerous", "a few" must be avoided in expressing quantities. The following descriptive clauses have been used to describe the handling damages referred to.

Wrapped material

Steel sheeting in coils

(a) Inner and outer circumference edges of packing locally dented where handling gear marked.

(b) Coil No. (unit number printed on packing) side packing locally torn open in two positions near outer circumference edge. Visible plate edges bent and rusty.

N.B. Where wrapped material is concerned, the rust condition of any visible plating, comprising the actual contents, need not be so accurately defined. In such instances, any rust constitutes a damage. This need not be the case where hot rolled, unpacked material is involved.

Steel plates in packages

(c) (Number of packages affected) lower long side edges of packing dented where handling gear marked.

(d) (Number of packages affected) side packing pierced and torn open locally in two positions. Visible plate edges bent/scored and affected by rust.

N.B. The damage referred to in (d) is often caused by the action of the forks of fork-lift trucks, a fact which is quite evident. Any rust damage should be as explained under (b) for coils.

Unwrapped material

Steel plating in coils (hot rolled)

—Inner and outer edges of plating locally dented and/or buckled where handling gear marked. Outer edge plating (number) windings affected. Inner edge plating (number) windings affected.
—Edges of windings locally affected by deep score marks.
—(Number) coils, inner turns and/or outer turns of plating telescoped up to (measurement of extent of telescoping).
—(Number) coils, projecting plate edges torn and bent.
—(Number) coil or coils loosely wound, strapping bands slack and converging.

N.B. In certain circumstances, slack windings can be contributory to upsetting the stability of the stow on board the ship.

Steel plates in bundles (hot rolled)

(Number of bundles affected) lower long side of plate edges locally bent and/or buckled where handling gear marked—(number) plates affected.

N.B. Scoring or gouging of plate edges is also of importance and should be referred to in the clause.

Single steel plates

These goods are usually handled with chain or wire bridle type slings, the ends of the legs of which are equipped with a special type of lifting hook. Incorrect placing of dunnage in stacks ashore, before shipment, can cause permanent deformation. Clauses which have been used are as follows:

—(Number) plates, edges locally bent upwards in (number) places.
—(Number) plates, permanently waved along entire length.
—(Number) plates, waved and/or distorted at one end.

Structural steel

—(Number) pieces, flanges locally bent upwards in (number) positions.
—(Number) beams bent, flanges and webs distorted.
—*Applicable to beams only*. Flanges of beams incorrectly overlapping for stowage purposes.
—*Applicable to beams only*. Beams incorrectly nested for stowage purposes.

N.B. The above two clauses need not be used if the stevedores separate the beams and stow them in the ship with flanges alternately overlapping in the correct manner (see under Beams). It is highly unlikely that the stevedores would agree to separate and readjust the nesting of beams since it is too time consuming.

Merchant iron (small scantling material in bundles)

(a) bundles loosely secured.
(b) Bundles insufficiently strapped/secured.
(c) Individual items comprising bundles projecting on ends—(number) pieces heavily bent.

N.B. The condition as expressed in (a) and (b) above will eventually lead to the condition as mentioned in (c).

—In (number) bundles (number) pieces bent along entire length.
—(Number) bundles (number) strapping bands broken or missing—goods partly loose.

Pipes/tubes

Mild steel line pipes

—(Number) pipes dented on one end (both ends) pipe ends out of round.
—(Number) pipes locally dented in (number) positions.
—(Number) pipes, bevelled ends scored, nicked and/or indented.

N.B. In the case of damages to end bevelling, shippers might insist on "depth of defects not exceeding 3 mm" in excess of which cutting off the pipe end and re-bevelling might be necessary. It could likewise be appropriate to add "depth of defects exceeding 3 mm" (see under the heading "Pipes").

Cement lined pipes

—Inside cement coating—hairline cracks apparent.
—(Number) pipes cement coating chipped and/or broken on ends.

Specially coated pipes

—(Number) pipes, protective coating chafed and/or scored.
—(Number) pipes, rust stained (rusty) where protective coating broken or missing.

N.B. Some of the special coatings applied to pipes consist of epoxy, polyethylene, bitumous materials, coal-tar and asphalt mastics. Cement coatings are usually applied to provide weight to sink the pipes in water. It is extremely difficult to transport specially coated pipes without causing some damage to the coating. The marine transport operation is the most difficult in this respect.

Small diameter pipes in bundles

See remarks suggested under the heading of Merchant iron re projecting ends, bending, strapping, securing, etc.

Wire rods

—Bundles not pressed tight, strapping bands slack.
—Bundles loosely secured, slack and leaning to one side (out of vertical).
—(Number) bundles, strapping bands converging, wire windings spraying open.
—(Number) bundles, end windings pulled about, bent, twisted and generally mangled.
—(Percentage per bundle) of visible windings score marked, nicked, grooved.

Wire (wrapped)

—Wrappers locally torn, visible contents scored, chafed and rusty.

If not provided for in the title of the goods, as typed on the bill of lading, where unpacked material is concerned, the clause "Unpacked and Unprotected" should be written on the bill of lading (in general this applies to all goods).

Miscellaneous additional clauses

—All lifts stained with grease and/or oil spotted.
—(Number) coils, per coil 1 or 2 strapping bands broken.
—(Number) packages, per package 1 or 2 strapping bands broken.

N.B. With regard to strapping bands, the minimum rather than the maximum number of straps are applied per unit necessary to its security. It is therefore considered that, if any of the bands are broken or missing, the securing has become impaired.

—(Number) coils, packages, lifts, stained with an unidentifiable white powder.

Chloride contamination

(Number) lifts, coils, packages, heavily rust stained, areas reacting to silver nitrate tests.

Deck cargo

Goods shipped on deck at shipper's risk, vessel not responsible for damage or loss howsoever caused.

Sheet pilings

Rust visible where protective coating (if coated) broken, scratched, or missing. Interlocking grooves bent/dented at (number) positions.

N.B. If uncoated for protection, a normal rust clause will be applicable.

DIRECT REDUCED IRON (DRI)

By definition direct reduction would apply to any process other than the conventional blast furnace for extracting iron from ores. In order to offset the large investment, maintenance and operating costs involved with a blast furnace, a process was devised to separate the iron from the oxide in the ore, by treating it with suitable hydrocarbon gases.

In the developing countries of the world, as well as in traditional steel-making countries, what are referred to as mini-mills have come into existence. Briefly, a mini-mill is an installation for steel making, whereby the conventional blast furnace is dispensed with. In other words, the process of steel making commences at the converter stage, one step ahead of the blast furnace.

Natural high grade iron ore is crushed, partly freed from unwanted constituents by screening (normally less than 3% gangue) and the resulting iron ore fines are compacted into pellets or briquettes. There are a number of similar methods used to separate the oxide from the iron. In general the procedure is to subject the pellets or briquettes to hydrocarbon reduction gases whereby the oxygen is removed in converting the reducing gases to carbon dioxide.

The reduction process does not produce temperatures sufficient to melt the iron and the resulting material is metallic and sponge-like. The concentration is evident by analysis, which would reveal about 92% metalisation. The remaining 8% would consist of carbon, phosphorus, sulphur, aluminia, silica, etc. In its form as produced, it can be considered as virgin iron to be combined with scrap in any desired proportion in the electric arc furnace so by-passing the initial conventional procedure for which a blast furnace is necessary.

Direct reduced pellets are honeycombed, and the surface area ratio to mass is very high. Once the pellet is formed oxidation will be continuous, accelerated by high humidity and excessively accelerated through contact with moisture, especially seawater. It is absolutely imperative that the goods do not get wet. Wetting of the cargo through rainwater during the course of loading results in the evolvement of hydrogen gas, which is highly explosive. There are cases on record where too much moisture and wetness has resulted in the development of fire.

Such a complete separation of the oxide from the iron means that the pellets have high potential energy and the reconversion process into iron oxide is exothermic. While the goods are on board ship, a method must be devised for monitoring the temperature, percentage of oxygen and hydrogen during the entire voyage from start to finish. Probes into the bulk should be numerous and at varying levels. For the carriage of such material, the type of vessel employed would normally be equipped with steel hatch closing panels, which should be in perfect condition. All joints should be sealed in such a manner as to render them and the cargo space as airtight as possible. This is achieved by covering the joints with masking tape but better results have been achieved by using a silicon sealant.

The goods are so unstable that where pellets, lumps and cold moulded briquettes are concerned arrangements have to be made to carry the cargo under an atmosphere of inert gas. This gas may be introduced above the bulk or into the base of the bulk pellets. In any case, arrangements are elaborate, many precautions have to be taken and expert advice is necessary before the carriage of such a cargo can be contemplated.

Hot moulded briquettes are less of a problem but nevertheless stringent precautions have to be taken. Joint P & I Club circulars have been issued with regard to these cargoes, and information contained therein has been compiled with the aid of a technical panel. Loading of a direct reduced iron cargo should not be undertaken before the P & I Club circulars have been studied nor without expert assistance.

The precautions to be taken when carrying this material can be found on pages 67 to 70 of the IMO publication *Code of Safe Practice for Solid Bulk Cargoes* (BC Code) 1994 edition. However, on page 68 of the code there is a provision mentioned with which many would not agree. On the page referred to, A1 stipulates that the goods should be carried under an inert atmosphere containing less than 5% oxygen and less than 1% hydrogen of the atmosphere to be maintained. However, from what is stated in A2.B, it is understood that, providing the DRI has been treated or manufactured with an oxidation and corrosion inhibiting process in such a manner as to prove to the competent authority that such arrangements will provide effective protection against dangerous reaction with salt water or air under shipping conditions, the provisions of A may be waived. In relation to this the following case history should be considered.

Case history

A vessel loaded a cargo of DRI pellets at a Russian Black Sea port in five cargo holds. The pellets were perfectly dry when loaded and the competent authority waived the provisions of the BC Code page 68 section A on account of the fact that the pellets were alleged to be thermally passivated. The hatches were closed and completely sealed with masking tape; the deck ventilators were also sealed. During the voyage seawater entered one of the cargo holds, not a large quantity, and more or less immediately localised heating in the cargo commenced. The temperature increased to 280°C but no action could be taken as no contingency plans had been formulated for such a not-so-surprising situation developing. When the ship arrived at the port of discharge in America the atmosphere above the cargo was flushed out to replace the hydrogen and oxygen with nitrogen. Thereafter, for a period of three days oxygen levels, temperatures and L.E.L. (Lowest Explosive Level) percentages were monitored before it was considered safe to commence discharge. In such circumstances it is imperative that discharge is commenced without delay when hatches are opened and continues uninterrupted until all heated cargo is out of the ship. In the case in question well over one million dollars was involved in the claim which was eventually brought against the carrier.

From experience it has been proved that any passivation process is only effective to some degree against oxidation of the material through contact with the atmosphere. Once free moisture, either fresh or salt water, is involved, passivation is usually ineffective. It is believed that the only proven method of carrying this cargo safely is by maintaining the cargo holds in an inert atmosphere. The most effective method of providing an inert atmosphere is by injecting the inert gas at the bottom of the stow in order to force out the air within the bulk. It is believed that irrespective of the length or nature of the voyage the cargo should be maintained in an inert atmosphere.

When heated cargo is discharged it should be spread out to cool and not heaped. It must also be stored away from any other DRI pellets which are unheated, i.e. normal cargo. Even cargo from the same cargo hold where heating occurred should be stored apart from other cargo when discharged; there is always the possibility that heated pellets are mingled with what appears to be sound cargo which later may start to heat. After the heated cargo has been spread out the use of water to cool it should be avoided if possible. If water must be used, it is dangerous to take it from the city mains as this water is invariably brackish to some degree: chlorides in city mains water can promote heating of DRI pellets. Where seawater is involved laboratory tests will prove inconclusive evidence of this. Distilled water washings from sound/normal DRI pellets can show values for the six main solids normally found in seawater but not in the same proportion, thereby upsetting the results of any analysis performed for the purpose of proving that seawater was involved.

N.B. It is to be noted that the provisions mentioned above refer to cold moulded DRI pellets. Hot moulded DRI in the form of briquettes is substantially less reactive with water and therefore less hazardous than the cold moulded type (see BC Code, *Code of Safe Practice for Solid Bulk Cargoes*, pages 69–70).

DUCTILITY

The plasticity of a material shown when under tension, which is measured by the amount that it can be permanently elongated without destructing, e.g. the elasticity of wire to be drawn through a die must be known in order to achieve a successful operation.

DUNNAGING, LASHING AND SECURING OF CARGO

Dunnage should preferably consist of strong hard-wood planks and square sections of timber, which are used for the purpose of preventing the cargo from coming into contact with the component parts of the vessel's structure. It is also used for the purpose of protecting the vessel's metal work and cargo, and at the same time dunnage does assist to a certain extent in spreading the weight of the cargo upon the vessel's tank tops or tweendecks.

Dunnage in the form of square sections, small baulks of timber, is used generally throughout the stowage for the purpose of keeping the stow level, binding the vertical tiers of cargo into a solid block so as to prevent shifting and also to facilitate re-slinging during the course of discharge. Strong hard-wood dunnage is essential, as dunnage wood compresses and too much compression combined with breakage of dunnage throughout a stow can result in the cargo sustaining damage through unequal pressures being applied to lengths of cargo which, as a result of this, have not settled at the same level.

It is essential that dry dunnage is used, as wet dunnage is capable of giving off sufficient moisture to promote, in certain circumstances, the incidence of sweat damage to the cargo. In most major ports in the world, dunnaging of the cargo is carried out by stevedore personnel during the course of loading. The lashing and securing of the cargo is either performed by stevedoring gangs who specialise in lashing and securing, or by professional independent companies. One point which does arise from time to time, is whether the lashing and securing and dunnaging of the cargo can be considered as the same operation. The two operations are entirely different. Normally dunnaging is carried out as part of the stowage operation of the cargo, whereas lashing and securing is a separate operation performed by personnel who are not directly involved with the actual loading of the cargo or placing of the cargo in stow. If the stow is not kept level and the quality of the dunnage is poor, it will end up broken and fragmented as shown in the photograph (see Fig. 19). Dunnage with a moisture content above 14% should not be used because it may affect the environment within the cargo space.

ELASTICITY

This is the measure of the ability of a material under load, resulting in change of form, to return to its original form when the load is removed. Steel plates are often locally deformed in handling, the elasticity of the material is exceeded and the plate remains locally bent.

ELECTRIC ARC FURNACE

The electric arc process of making steel relies on cold scrap metal and DRI. One of the reasons for its popularity is the fact that only approximately four hours is required to convert scrap metal/DRI into steel.

The electric arc furnace consists of a circular shaped vessel with a removable roof and through which project three carbon electrodes which can be raised or lowered. The process consists of withdrawing the electrodes and swinging open the roof, after which steel scrap/DRI is deposited inside the vessel. The roof is swung back into place and the electrodes are lowered into the furnace. A very strong electric

Fig. 19. Disintegrated dunnage caused mainly through incorrect stowage of steel beams.

current is passed through the scrap/DRI and an arc is struck with the electrodes, which causes the scrap/DRI to melt.

Thereafter, lime, dolomite and fluorspar are added in order to scavenge out the impurities such as silicon, manganese and phosphorus to form a slag on top of the molten metal. Samples are taken to check the composition of the material and ferro alloys are added to adjust the chemistry of the steel.

When the exact composition of the metal is reached the temperature is controlled in order to achieve the correct temperature for casting. After this the slag is poured off, the furnace is tilted and the steel is poured off into a teeming ladle, then carried away and poured into moulds to form ingots. Also the molten steel can go straight to the continuous casting plant so by-passing the ingot stage.

With this method of steel making, a high grade alloy steel can be produced because in the process there is a high degree of control and refinement possible, and also impurities are reduced to a minimum. One other advantage is that the furnace can be operated entirely on scrap/DRI charges and steel can be produced without the assistance of a blast furnace.

ELECTRICAL STEELS

Grain-oriented and non grain-oriented steels belong to a group of materials referred to as electrical steels. The metal has a special composition and the aim in its construction is to produce a product which combines the properties of high thermal and electrical conductivity, magnetic permeability and low core loss. Grain orientation involves manipulation of the atomic structure of the metal with a view to achieving the special properties mentioned above. Coatings are applied, such as silicon, in order to produce high interlamination resistance.

Such steels are used in the construction of generators, transformers and many other forms of electrical equipment. The flatness of the sheet is of paramount importance in order to ensure an even and direct flow of current. When insulation coatings are applied, any bending of the sheet which might affect, even to the slightest degree, the continuity of the covering layer is inadmissible.

The goods are manufactured in the form of flat steel strip plating and are shipped in wrapped bundles (packages) or in wrapped coil form. The coils are usually fixed to a wooden skid and are in most instances placed upon it with their centre cores vertical. The goods are packed in much the same way as for cold rolled steel sheeting. The bundles of steel sheets or coils are first wrapped in plasticised kraft paper, after which an outer steel sheet envelope is applied and secured by means of flat metal strapping bands placed around each unit in two directions. The packages are fitted with a wooden frame beneath the bearers to assist in maintaining the rigidity of the package and to facilitate handling.

Electrical steel sheeting is an extremely delicate and costly product, which should be handled with the greatest of care. Any damage whatsoever will probably result in claims being entered against the carrier. As explained above, any deformation of the plating caused by distortion of the roundness of the coil, pressure of securing wires or general handling damage can result in the coil or package so affected being refused by the receiver. As there is a limited market for such material, losses can be proportionately heavy (see Plate 6 and Fig. 20(a) and (b)).

ELECTROPLATING

The surface of the plating is first prepared by shot blasting or pickling. The article is immersed in an electrolyte whereby the item is made the cathode of the cell so formed. Metals are deposited upon the surfaces of the plating from their salts by means of an electric current. It is an electrochemical process with the same basic mechanism as that discussed for corrosion (see Fig. 17, e.g. electrogalvanising).

EMULSION

A milk-like mixture such as will occur with the suspension of oil and water together, each liquid being suspended in the other. Strip steel plating with an oiled surface (in coils or packages) often sustains wetness in transit when moisture penetrates, or condenses upon, the turns of plating. This water combines with the oil to form an emulsion which in time coagulates. In such instances the protective coating of the oil is destroyed.

ENCLOSED SPACES

In relation to entering cargo holds at sea which are loaded with steel, as explained for ventilation purposes, no problems with this are foreseen. From experience it is known that compartments partly filled with hygroscopic cargoes such as cereal products, can be lacking in oxygen (normally at a level of 20%) after having been closed down, unventilated, for a period of time. SOLAS (IMO) refers to this under the heading of "Oxygen analysis and gas detection equipment", which equipment should be on board all bulk carriers. Steel is an inert cargo, which does not mop up oxygen other than by a slight, slow, oxidation of the exposed surfaces of the lifts. It only occupies a relatively small part of the cargo space and securing timbers are usually minimal in amount. No such problems as referred to, with steel cargoes, have been experienced or believed to have ever been reported.

END USES

The steel products carried by sea more or less invariably have to undergo further processing before they end up as a finished product. Such end uses are many and varied and a complete list would not only be extensive but superfluous to the intended purpose of this book. The following covers the main uses of the material:

(a). Flattening on one side, caused by handling.

(b). Condition of coil after wrapper removed.

Fig. 20. Electrical steel.

Hot rolled steel

Steel strip in coils may be imported for the purpose of decoiling, cutting into ±2,000 × 1,000 mm plates which are bundled and sold off to industry. Bundles of the same material, already prepared as mentioned above, will be imported and similarly disposed of. Coils may also be slit into strip for the manufacture of piping, trim pieces, hoop iron, fence posts, all types of fabrication where narrow strip bands can be utilised.

The plating is also used in the fabrication of truck and rail car bodies, steel doors, water and oil tanks, trim sections for all sorts of industrial uses, base plates platforms etc. Alternatively, hot rolled strip sheeting, so intended, may be processed into cold rolled steel sheeting.

Cold rolled steel

Similarly to hot rolled material, cold rolled strip may be imported for the purpose of cutting into ±2,000 × 1,000 mm plates which are packaged and sold off to industry. In strip form it is used for the manufacture of pipes or items exceeding the standard package sizes. Tremendous amounts of this plating are used in the fabrication of road vehicle bodies, household appliances such as refrigerator casings, bread bins, electrical home appliances, radiators, metal furniture, steel drums etc. In the finished product a protective coating is always applied to the surface of the material.

Galvanised steel sheeting

Used in the manufacture of car bodies and trim pieces in the automobile industry. It is extensively employed in the manufacture of ducting and trunking for air-conditioning and ventilation installations, roofing, guttering, water tanks, pipes and parts used in the electrical industry.

Structural material

Large sections, e.g. beams, channels, angles, are usually employed in the construction of bridges and high rise buildings. Small scantling material, i.e. flats, angles, channels and other shapes mentioned under the heading of MERCHANT IRON (see page 123) are used for the manufacture of steel ladders such as fire escapes. They are also used in factories for making racks, the frames of steel doors, railings, window bars, steel furniture—the list is endless.

Electrical steels

See under the same heading on page 77.

Stainless steel

In seaborne trades this material is usually shipped in the form of hot rolled coils. However, small individual shipments of cold rolled plate in packages are not unknown. The cold rolled plate (see pages 176–177) is used in the manufacture of household appliances (e.g. pots and pans), kitchen sinks, cladding in buildings such as visible panelling, tanks containing liquids which are of an aggressive nature when in contact with steel surfaces.

Sheet pilings

See under the same heading, page 166.

Wire rods

This material is cold reduced or hot rolled and then galvanised or, depending upon use, plastic coated. However, there are uses where it is used in its original form such as for road mesh. Its diverse uses are as follows:

Fencing, netting, wire panelling, piano wire, springs and clips, parts for engines, spark plugs (high carbon steel), woven covering for cables, staples, nails, pins, screws, bristles for brushes, cord for reinforcing rubber products (e.g. tyres, wire rope and cables), steel wool, road mesh, welding rods, the wires in pre-stressed concrete, chicken wire, etc.

Steel plates

Large single heavy plates are usually imported for the purpose of tank building or ship building.

Reinforcing bars

See under the same heading, page 149.

Pipes and tubes

See relevant sections on pages 136–146.

Slabs, billets and blooms

Theses are semi-finished products which are destined for re-rolling into all the various types of material as mentioned above.

EUTECTOID

A mixture of two or more constituents that forms on cooling from a solid solution and transforms on heating at a constant minimum temperature. Eutectoid steel contains approximately 0.83% carbon.

FINISHING TRAIN

After the semi-finished product, e.g. slab, billet or bloom, has been sufficiently broken down in size through the roughing train, it passes through the scale breaker and then on to the finishing train. The finishing train is similar in appearance to the roughing train except that it usually consists of five to seven rolls in tandem. What finally emerges from the sets of rolls of the finishing train is a product which has been reduced to the final shape and thickness required—the finished product.

FLATS

Flat bars are rectangular in cross section and shipped in various lengths, although 6 m is a popular length for these goods. They are usually shipped in unwrapped bundles which are secured with a number of wires or flat metal strapping bands. This is hot rolled material and therefore is usually stored in the open prior to shipment and in consequence of which some rust development is apparent.

The goods are used extensively in the structural and fabrication industry. Small scantling material can be used in the construction of railings, stairs, etc., and also in various types of inside work, and experience has shown that contrary to general practice, small dimensional material should be protected from the elements as the development of too much rust can provoke the displeasure of receivers.

Bundles which are not even-ended (individual bars projecting) are susceptible to bending damage, and loosely secured bundles can present problems in stowage and in handling. See bundled flats in Fig. 21.

FLOW LINES. See LUDER'S LINES

FULL HARDENED STEEL

In the rolling of strip steel plate, hot rolling improves the quality of the material by reducing the size of the grain structure, whereas, cold rolling has the opposite effect. As a result of this, cold rolled steel is in a "full hard condition" when it leaves the production line. It may remain in this condition if it is to be used for, e.g., roof plating, or other uses where ductility is not an essential aspect of the material. For other uses, the material must be more ductile and this is achieved by evening up the stresses in the material etc., by annealing. A process of quenching accomplishes other degrees of hardness.

GALVANISED STEEL

Coils—packages—structurals

The demand for high quality rust preventive steel sheeting and coated strip is increasing annually. Coatings based upon zinc are popular, because they provide the additional benefit of a sacrificial element in areas where the steel might be exposed. Galvanising is easily the most common metallic coating and it has been reported that the annual growth in demand for galvanised sheet is well above crude steel growth rates.

The galvanising of steel sheet can be defined as the application of a zinc coating

Fig. 21. Bundled flats.

over the surface of the steel to protect it against exposure to the atmosphere and consequent rusting. As discontinuity or ruptures develop in the coating, and should holes occur in the zinc, electrolytic cells form in these areas when the surface of the sheet is moist, and the iron is cathodically protected by sacrificial corrosion of the zinc. The main methods of applying the zinc coating to the steel sheeting are as follows.

Hot dip method of galvanising

A length of steel C.R.S.S. strip sheeting, in coil form, is subjected to a continuous process, whereby the surface of the plating is coated with molten zinc. The various steps necessary to achieve this are briefly as follows:

(a) The surface of the plating is cleaned by passing it through a weak alkaline solution or by vapour degreasing.
(b) The plating is subjected to a pickling process for the purpose of removing scale and surface rust by passing the sheeting through a hydrochloric acid bath.
(c) The prepared sheet is passed through a zinc ammonium chloride flux which serves to evaporate the surface water on the sheet and removes residual oxides.
(d) The fluxed surface of the plating enters a zinc bath, whereby the zinc coating is applied to the surface of the steel. The thickness of the zinc coating is controlled in accordance with the service intended.

The hot dip method of galvanising is characterised by what is referred to as "spangles". These are to the naked eye a clearly visible crystalline pattern in the surface of the zinc coating. This is the most widely used method of galvanising steel sheeting, which is being increasingly used in the automobile industry and also for a host of other applications, where a more efficient and lasting protection for the steel is required, owing to the plating being subjected to relatively more hostile conditions, e.g. roofing, water tanks, gutterings, etc. Zinc coated steel is also used as part of the manufacture of many products, such as structural steel works, tubes and pipes, wire, nails, dry batteries and by the building industry in general.

Electrolytic galvanising

With this method of galvanising, the action of depositing the zinc on the surface of the steel is by electrolysis. The advantages over the hot dip method are:

(1) Greater economy in the use of the zinc.
(2) The thickness of the zinc coating can be controlled with great accuracy.
(3) A coating of fine zinc can be applied.
(4) As opposed to the hot dip method a more uniform coating is possible.
(5) Greater adherence properties.
(6) A good base surface for applying paint.

The plating is dull grey and unattractive in appearance as opposed to the bright perspective of the zinc surface produced by the hot dip method.

For stowage on board ship see STOWAGE OF STEEL SHEETING IN COILS AND PACKAGES.

Deterioration after galvanising

On leaving the production line the applied zinc surface is bright—in the case of the hot dip. However, as the plating cools, the surface of the zinc is attacked by the atmosphere and becomes covered with a very thin layer of zinc oxide and zinc hydroxide which is transformed by carbon dioxide in the atmosphere into a protective layer of basic zinc carbonate. The forming of this protective coating, as it develops further, tends to cause the zinc surface to take on a dull appearance. Given time, the surface also attracts dust particles from the atmosphere so that the material looks very dull and dirty.

The efficiency and lasting properties which will provide long-term protection to the steel beneath will depend upon the environment. Zinc coatings will deteriorate faster in marine atmospheres and those areas where air pollution is prevalent than under climatic conditions where the air is purer and drier. Under certain conditions where the atmosphere is not affected by a high degree of pollution, and the plating surfaces are situated in air streams where they are subjected to periodical fresh water wetting and drying, it has been known for the plating to survive for 30 years or more.

White rust

An aggressive liquid, such as seawater or chemical dilution, when moisture is involved, e.g. moist air, causes a thick voluminous white dust to form on the surface of the plating and this effect is generally referred to as "white rust". If such a condition continues, the galvanised coating will in due course be consumed and eventually the atmosphere will reach the surface of the steel beneath the zinc and red rust will appear.

The development of white rust on galvanised surfaces seems to favour an atmosphere where the air is highly moisture laden and the ambient air is still. Such conditions can prevail in an unventilated ship's hold, and is responsible for the heavy white rust deposits and damage whereby structural material has to be re-galvanised after delivery.

Storage stain

It has been explained how in freely exposed atmospheric conditions a protective layer of basic zinc carbonate forms on galvanised surfaces. If, owing to close stacking of plate surfaces, such as would be the case with sheets in packages or strip sheeting in coil, there is moistness or the presence of free moisture, the protective layer of basic zinc carbonate may not form owing to restricted access of air and subsequently an insufficient supply of carbon dioxide necessary to the formation of the protective layer. Instead, the initial corrosive action forming zinc oxide and zinc hydroxide will be continuous and voluminous and the damage it produces is known as "storage stain".

Type of packing

Galvanised sheeting in coil form, and also plates stacked one on top of the other to form a bundle, are wrapped in kraft paper, after which a metal envelope of galvanised waste sheeting is applied. This is strapped up as a single unit and secured with flat metal strapping bands.

The galvanised metal wrappers are sometimes new and bright in appearance, but often affected by drip marks through having been in contact with liquids before being used to cover the coils or packages. Wrappers are often affected by white rust, and there may be evidence of sporadic areas of oxidation having developed at some time.

During the transport from the factory, by lighter or lorry, and in storage awaiting shipment, even bright new wrappers can deteriorate considerably in appearance. Further deterioration can and will take place during the ocean voyage. In view of this, merchants taking reception of such goods are highly suspicious of the condition of the contents of the coils or packages, owing to the condition of the wrappers. Claims are often entered and surveys are held. In some instances there is ample evidence to support the fact that moisture has penetrated the wrappers and gained access to the actual goods, so initiating the development of damage to the material. In some instances, the suspicion aroused by the condition of the wrappers has proven to be false as the contents are found to be in good order and condition.

Galvanised pipes (small diameter). See **Small diameter pipes** on p. 142

Ambiguity surrounding cause of damage

In many cases the aspect and nature of the damage gives rise to doubts as to what the basic cause of the damage might be. Upon removal of the wrappers, the edges of the plating are bright and the entire unit shows no stains or blemishes of any description. When the coil is unrolled, or the stack of sheets turned over in the case of packages, the galvanising is seen to be heavily oxidised with large voluminous areas of white rust. Quite often, all surfaces throughout the coil or packages are affected. In other instances, defects on plate edges are relatively minute, and bear no relation to the enormous centralised damage found within the unit being examined. A special feature of this type of defect is that, in general, the damage is mainly confined to the surface centre of the plating and may not extend completely to the edges.

Conditions referred to above, have more than likely originated from the fact that when, after manufacture, the sheet is wound into coil, or the sheets have been stacked in the case of packages, air is entrapped between the plating. When this take place, if the material possesses a high temperature, the air entrapped between the plating is capable of holding more moisture than it can when the goods cool. The moisture held by the entrapped air, upon cooling, precipitates moisture onto the surface of the plating. This leads to the development of a corrosive condition known as "storage stain", which is not visible externally (see Fig. 6).

The damage as described above usually develops in the mill or factory and probably before the packing is applied. Only a very small amount of moisture is involved, so dampness is enough. The damage is not through an external cause, which would usually be associated with an increase in the relative humidity of the ambient air surrounding the coil. It might be said that the cause of the damage had inherent vice characteristics; however, such damage could be prevented by a controlled climate in the place of storage.

Hot dip galvanised and electro-galvanised strip steel sheeting in coil

Unoiled

Depending upon the climatic conditions to be experienced in the area where the material is stored prior to shipment or after delivery, e.g. large ambient air temperature fluctuations, it may be necessary to provide climate controlled storage conditions in order to avoid damage. This certainly must be the case during most months of the year in the northern hemisphere. Many claims against the carrier fail based upon storage conditions after discharge of material from the ship.

In those instances where plating surfaces are unoiled, appreciable fluctuations of temperature will cause the goods to sweat, with resulting damage to the zinc coating of the wrappers and the material, which will result in the development of condensation and free moisture within the unit, white oxidation marks and/or the development of white rust as explained earlier; except that in this instance cooling of the material is not the cause. The cause is associated with fluctuating atmospheric temperature, increased relative humidity and dew point temperatures. The minuscule gap between each individual turn of plating is filled with an extremely small volume of air, the inherent values of which should be maintained in a state which is compatible with the ambient air. If atmospheric temperatures fluctuate appreciably with a subsequent increase in the relative humidity of the ambient air, there will be a corresponding increase in the vapour pressure. In such circumstances, as the packing is not a hermetic seal, a pressure will be exerted outwards/inwards so causing interaction towards equalisation between the ambient air—effects of vapour pressure—and the air between the windings of the coil/coils. If the dew point temperature of the air in the interwinding space exceeds the temperature of the steel/zinc surface a point of supersaturation of the air may be reached and moisture will be precipitated onto the surface of the sheet/zinc coating and this will continue to precipitate moisture as long as these unstable conditions prevail. As moisture is precipitated it will be replaced by moisture flowing in from the ambient air; conditions which will continue until either the temperature of the plating rises above the dew point temperature of the ambient air, or the ambient air dew point temperature, owing to a change in ambient relative humidity, falls below the temperature of the steel. For reasons stated, although the mechanics of the development of "storage stain" are involved, the resulting damage is quite different.

As an example of proof of increased relative humidity infiltrating steel cargo it is a well-established fact that free water infiltrating a ship's hold will raise considerably the relative humidity within. As an example, a shipment of 2,000 tons of chromated, electro zinc plated steel and chromated hot dip galvanised steel coils was carried by a coaster which was fitted with one long cargo hold. During the course of a short voyage air temperatures were constant. During the voyage one joint in the steel hatches, at the after end of the hatch, leaked and a dribble of seawater entered the compartment causing damage to a number of coils in the after extremities of the cargo hold. This ingress of only a very small quantity of seawater resulted in an increase in the relative humidity of the air in the cargo hold, as a result of which all other cargo untouched by any free moisture was affected by prolific sweating internally. The sweat water, i.e. fresh water, condensed in globules on the surface of the plating and as it formed it continuously gravitated to the area of the bottom of the coils, causing white and red rust to develop. These effects are much more pronounced on board a ship at sea than in a warehouse ashore, as vibration and the working of the ship in a seaway cause the turns of plating to tighten and loosen, bend, flex and respire; although only to a small degree such action is undoubtedly very effective in drawing air between the coil interwinding spaces.

It is to be noted that chromating is, when properly applied, an effective defence against "storage stain" but is completely ineffective once free moisture becomes involved, such as when the plating begins to sweat.

Oiled plating

Applying a coating of oil to the surface of the zinc is a good defence against "storage stain". A lot of the oil will slowly gravitate towards the lower extremities of the coil; however, the surfaces will maintain a fine oil film covering which will protect it against "storage stain"; the amount of air trapped between the two oiled surfaces is negligible. Once the goods start to be handled the situation changes, as any movement of the coils from static position causes the coil to alternately change its shape slightly, thus causing the plating to flex and bend as explained earlier. While this is in progress the gap between the plating tightens and loosens, air is drawn in and is alternately expelled. The action, as explained, must be continuous on board the ship; therefore, if the dew point temperature or the ambient air rises above the temperature of the coil/coils the packing will start to sweat. This phenomenon is what is referred to as "cargo sweat" and often reaches such proportions as to cause free moisture to condense out upon the outer surface of the packing. This moisture can stream down the packing to become caught up in various metal projections and overlaps where it is directed inside to contact the steel and accumulates at the base of the coil to cause rusting damage. The introduction of so much moisture inside the coil raises the relative humidity of the air within the environs of the packing, thus causing the entire coil to sweat. Moisture in the form of globular drops forms, also contact with free moisture results in emulsification of the oil film, and rusting to some degree begins within a short period of time.

Case history

In connection with the above cause of damage developing, it is interesting to relate the details of a judgment which was given by the State of New York Supreme Court, in July 1965, concerning the voyage of two vessels which transported three different lots of packages of galvanised steel sheets from Rotterdam to New York. All packages originated from the same supplier, and in both cases the goods were shipped under bills of lading which were not qualified with regard to the apparent condition of the packages at time of shipment. With regard to both of these vessels, there was no cargo stowed in the same compartment as the packages that could be described as being incompatible with the packages of galvanised sheets.

At final destination, so it was stated in the court's findings, a sheet-by-sheet examination of the contents of all packages discharged from both vessels, disclosed that every sheet, without exception, was contaminated with a white or grey coloured rust deposit, stained, and wet to the touch. The goods could not be used for their originally intended purpose and were eventually disposed of as scrap.

Chemical analysis of the white rust and the packing paper proved that the white rust material contained sulphates but no chlorides, and was therefore not of salt water origin, and that the wrapping paper contained both sulphites and sulphates. The final conclusions of the court were:

> "The white rust contamination that made the galvanised steel sheets valueless for use in plaintiff's manufacturing operations resulted from a condensation created within the individual packages themselves, which was caused by an electro-chemical corrosive process, that took place within the packages because of the inherent nature of the contents."

Based on the above findings, the case appeared to be justifiably concluded as the judgment was in favour of the defendants who were the shipowners. The fact that every single sheet, in all packages, from both vessels was affected by white oxidation is strongly in favour of vice from within rather than damage developing from an external cause.

Effects of sea air

It is appropriate to point out that galvanised surfaces are more readily and severely corroded when salt particles are present in the air, or when they make contact with seawater. With regard to salt particles in the air, it is to be emphasised that the quantity of salt present in the air at sea is very small. In a ship's hold, even under conditions of forced ventilation, the amount of salt crystals in the air does not amount to very much. Nevertheless, analysis of white rust from galvanised surfaces which have made a voyage in a ship's hold, has shown faint traces of salt. These salt traces are responsible for intensifying the corrosion of zinc for two reasons.

(1) The salt attracts more moisture to the surface of the plating than would otherwise be the case were it not present.

(2) The moisture attracted on account of the presence of salt acts with the salt particles to form an electrolyte on the surface of the zinc, which together form electrolytic cells producing, in addition to white rust, an oxychloride of zinc. It is to be noted that zinc oxychloride is not hygroscopic, but zinc chloride is highly hygroscopic. The presence of salt particles does not therefore intensify the damage to the extent expected.

The faint traces of salt derived from the atmosphere are never likely to be confused with the large quantities of salt which would be present through contact with seawater. There would consequently be no confusion in distinguishing between damage arising from the two causes. Damage through contact with seawater also leads to the development of white rust, and also red rust, as the zinc coating, where contact is made, will be penetrated. Therefore, in addition to white rust, iron red rust, zinc oxychlorides and sea salts will be present. In order to determine whether or not seawater is involved, samples should be submitted for spectrographic analysis, whereby it can be determined beyond all doubt that the elements to their approximate values, always present in seawater, are involved (see SALINITY OF SEAWATER).

Silver nitrate applied to galvanised surfaces causes the test area to turn black, and for this reason is unreliable as a test for the presence of chlorides.

Possible causes of damage

Damage to galvanised steel sheeting can be caused by the following:

- —Poor preparation of the surface of the steel prior to applying the galvanised coating.
- —After pickling, washing of the plating may not remove all of the acid from the surface of the plate.
- —Water used for washing or cooling contains impurities.
- —The formation of white rust may develop if wetting from fresh water, rain and/or cargo sweat takes place. This may be particularly so if the goods reside in a still atmosphere where possibilities of evaporation of the moisture are very poor.
- —White rust will develop rapidly if the goods come into contact with seawater.

—Galvanised surfaces will deteriorate much more rapidly if they are permitted to reside in a marine atmosphere.

—If the material possesses a high temperature when it is packed in a humid atmosphere, moist air entrapped between the turns of plating, or sheets, may precipitate moisture when the material cools, with resulting deterioration of the zinc coating.

—Galvanised goods should not be stowed in the same compartment as fertilisers, or in a compartment which connects via some form of ducting, e.g. ventilation trunks, with another in which fertilisers are stowed. Fertilisers contain chemical ions which, when in contact with galvanised surfaces, and in humid ambient conditions, result in an aggressive medium for the promotion of the deterioration of zinc coatings.

—Transportation on board ship should be carefully considered, especially with regard to ventilation of the compartment where the goods are stowed—see section on Ventilation of steel cargoes.

—Plate 7 is of an import shipment of galvanised coils. The condition of part contents is shown in Plates 8 and 10. Plate 9 is of an apparently sound parcel awaiting shipment.

GALVANISED WIRE

This product is manufactured with hot rolled/cold drawn steel wire to which the zinc coating is applied in much the same way as the hot dip method of galvanising steel sheets. The wire is first annealed, then pickled, washed and fluxed, before being finally passed through a bath of molten zinc. Thereafter, the wire passes through wipers to regulate the thickness of the zinc coating. Such material is extensively used for fencing in the form of plain or barbed wire and also for the manufacture of wire rope, springs and nets, etc.

Galvanised wire is normally shipped in the form of loose coils which are on many occasions unpacked and unprotected. The principal points with regard to the deterioration of galvanised surfaces, as referred to under Galvanised steel, likewise apply to galvanised wire, and should be consulted in conjunction with this.

One manufacturer stored bright, new, unprotected coils of wire in a humidity controlled warehouse before shipment. Eventually, the goods were loaded into covered wagons and transported a short distance to the docks. In the quayside sheds the wire was placed upon new kraft paper and the entire lot was also completely covered with kraft paper. The wire was shipped on the following day into the tween decks of a vessel, which transported the goods to New Zealand where the wire was rejected by the receiver owing to its dull aspect, and was in due course sold. It was later discovered that the original receiver had repurchased the wire through a third party.

It is virtually impossible to ship bright, galvanised, unwrapped wire coils, and deliver them in the same bright condition at final destination. The actual oxidation process which takes place when galvanised surfaces are exposed to the atmosphere, causes a dull appearance to develop. Add to this the marine atmosphere experienced on a sea voyage and there is no doubt that the wire will be dull when delivered. Sea air circulated through a ship's hold contains a small quantity of salt particles which are also to some degree incompatible with galvanised surfaces and tend to advance the development of surface corrosion. Receivers contend that when marketing the goods, the dull appearance imparted to the galvanising—through a natural, unavoidable and even desirable oxidation process—prevents a satisfactory sales price being obtained against the market price of locally manufactured material which has a brighter and more appealing aspect. This is, of course, a commercial consideration because the dull condition of the galvanising does not amount to real damage.

Such goods should be kept dry at all times, and every endeavour should be made not to expose the material to moist still air. Direct transhipment into the sea carrier is desirable but not always possible. If pre-shipment quay storage cannot be avoided a well ventilated space should be provided. In such circumstances, the material must be kept clear of the quay floor by placing canvas tarpaulins beneath and over the top of the parcel or parcels awaiting shipment. Quayside storage prior to shipment should be kept to a minimum, as an accumulation of dust on the wire can absorb moisture and contribute to the evolvement of a dull aspect to the zinc coating. In one particular case, a nation of sparrows circulated in the roof of a quayside shed above a parcel of galvanised wire. An accumulation of their droppings, which was quite phenomenal, completely ruined the wire before it could be shipped.

As the main objective is to deliver the cargo in good order and condition it has to be handled with care so that scratch marks, scoring damage to the zinc coating and contact with dirt are avoided. Fibre rope slings should be used for hoisting or the wire placed upon trays. If fork-lift trucks have to be used, no sharp edges on the forks should be permitted. Quayside dirt holds moisture and is invariably impregnated with chemicals to some degree, hence the reason why loading trays are recommended.

Any apparent imperfections in galvanised wire should be taken seriously. Whether the wire is dull or bright in appearance must be considered. White powdery voluminous deposits in spots or patches, referred to as white rust, must be viewed as damage and likewise with regard to visible areas of red rust. Scratches, score marks, dirt and dust, also bending, twisting and/or kinking of windings are inadmissible. They must be regarded as damage or, in the case of dust and dirt contamination, as a condition which will eventually lead to the development of damage.

Moist wooden dunnage can damage galvanised surfaces and it should be ensured that any wooden dunnage used is dry (14%). This dunnage should be covered with a canvas tarpaulin or strong kraft paper. The transport of this type of wire can be much more successful if the goods are wrapped. Crepe paper wrapping against the wire, with an outer jute banding wrapper, offers good protection.

HANDLING STEEL CLAIMS

Introduction

Even though the surveyor/expert witness only intervenes partially in the handling of steel claims, his contribution is more often than not crucial to the success of such claims insofar as all the relevant facts are revealed and presented in a way which would enable a layman to understand the subject matter. To relate one experience which occurred in a North American court: an important claim which entailed considerable technical argument relating to the ventilation of a galvanised steel cargo, stowage, etc., was presided over by a judge who was judging his first ever maritime case; previously his speciality had been cases concerning fiscal law. A special effort was made by the expert witness to present his part of the case as simply as possible. Fortunately, the judge was extremely shrewd, highly intelligent and alert; his judgment of the case turned out to be very satisfactory.

In pursuing steel claims there are many documents which must be procured by cargo interests in order to evaluate their position and decide what further action is or is not worthwhile. Likewise, the respondent, usually the carrier, will also need to recover documents in order to prepare his defence. Using this information the lawyers will develop their views on the legal angles. The expert witness, drawing upon his past surveying experience, technical knowledge, superior knowledge, and factual evidence in the form of documents made available to him, will formulate a carefully considered opinion with regard to the actual merits of the case, to the extent that they concern the cause and extent of the damage which has been sustained by the cargo.

Depending on circumstances, cargo interests may consider proceeding against the manufacturer of the goods, the shipper, the ocean carrier or some other third party; in the latter mentioned case this is generally where there was evidence that the goods were not properly cared for after they were landed at the port of discharge. For obvious reasons, claims involving manufacturing defects and proven contact with seawater are less challenging than damage caused through contact with excess humidity, sweat water, etc., which all come under the heading of "fresh water damage".

The manufacturer

Where manufacturing defects are proven, whether evident or not at the time of shipment, there is a strong possibility of recovering against the mill. If, on the other hand, the claim concerns internal sweating of the material, it is not easy to prove that fresh water damage not apparent at the time of shipment, such as is often the case with galvanised steel sheet, took place before the goods were shipped. The defendant in such a situation might challenge the assessment of the loss if, as is sometimes the case, he or his representative was not invited to participate in the commodity survey. The reason why he is being attacked will undoubtedly reach him before any inspection of his mill can take place, and he will be well prepared if there is an inspection at a date which he will ensure is convenient to him. The most daunting issue will undoubtedly be proving to the manufacturer that there was no possibility of the damage occurring during the sea voyage, or after discharge when the goods were deposited in quayside stores, sometimes for months, before being called forward by the ultimate receiver. Internal sweating damage which is not outwardly apparent does, depending on circumstances, develop in stores prior to shipment. A lot of expertise is required on the part of the surveyor to recognise this damage and to successfully explain how it occurred in such a manner as to be convincing (see under the heading **Storage stain** on page 85).

Ship's hatch closing appliances

Much salt water damage to steel cargoes is caused by leakage of ship's hatches. Most P & I Associations insist on vessels undergoing a condition survey before club entry is taken up. Emphasis is given to the ship's hatches, and in many instances where hatches are found defective recommendations are made by the surveyor in order to restore these appliances to a condition whereby they will contribute to the ship being considered seaworthy. The value of all this activity depends very much upon the competence of the surveyor, the relevance of the recommendations made by him and the manner in which they were executed. Further to this, in most instances P & I Associations recommend that a hatch closing appliance survey be carried out before a cargo of steel is loaded. In pursuing a cargo claim against the carrier when leaking hatches are involved, the defendant, depending on circumstances, may in his defence rely upon survey reports and stress of heavy weather, and invoke his rights and immunities under the Carriage of Goods by Sea Act (COGSA). Where important claims are concerned, a claimant should consider endeavouring to obtain a copy of the *Condition Survey Reports* associated with club entry if these were of recent issue, and also a copy of the *Hatch Condition Survey Report* connected with the voyage concerned. It may very well be that a study of these documents by a competent expert can reveal technical abnormalities which bring into question the carrier's plea that due diligence was exercised to make the vessel seaworthy before commencement of the voyage. In the event of leaking hatches, where these are found in apparent good order and condition, often what might appear not to be a defect can be of major importance and affect seaworthiness.

Pre-shipment survey report

It is customary for the carrier to arrange for a pre-shipment survey to be held on steel cargoes. A copy of the pre-shipment surveyor's report should be obtained, since a properly written report will contain information with regard to:

—weather conditions prevailing during loading;
—details of any stoppages, disputes and stevedore damages;
—details of M/R and B/L per consignment of cargo;
—stowage position in the ship per consignment;
—where each parcel of cargo was stored prior to shipment;
—any remarks with regard to the apparent order and condition of the cargo at the time of or prior to shipment.

Statement of facts

These documents are prepared by the carrier's agent prior to departure of the vessel at both the loading and discharging ports. They provide, or should provide, a complete breakdown of the time the vessel spent in port from time of arrival to time of departure, giving times worked, idle time, times of any stoppages and the reason for such stoppages. These times, if found necessary, can be checked against the times and/or details contained in the pre-shipment survey report and especially the vessel's deck log book. Steel cargoes are so often damaged through the ship failing to cover hatches in sufficient time when rain is imminent. As a result of this the:

—ship's tank-tops become wet;
—cargo already in stow is rained upon;
—cargo brought to the ship's rail ready to be shipped sustains wetness;
—cargo already in stow can develop damage through rain entering the ship's holds, so raising the relative humidity within the cargo compartment and causing, in certain circumstances, fresh water moisture damage to the cargo.

All times and conditions can eventually be checked against the local climatological data provided by the weather stations.

With regard to closing ships' hatches in the event of rain, it is a fact that rarely are hatches closed in sufficient time to prevent rain entering the cargo holds. The reason for this is that the appropriate moment to decide to close the ship's hatches must be before rain starts to fall, in view of the fact that each set of hatches can take as long as 15 to 30 minutes to close depending on the type of hatches involved. The experience is that any ship master who makes a practice of taking this early action, obviously in the interests of caring for the cargo already on board his ship and cargo waiting to be loaded, would very soon be experiencing complaints from the charterers. If the master persisted, these complaints may develop into threats to put the ship off hire during those periods when work was stopped by the master when the weather was still dry. The ship master, in taking these decisions, of course runs the risk of stopping cargo operations when rain is threatening but no rain is experienced; in which case time lost is attributable to the master having incorrectly assumed that it was going to rain when in actual fact it did not. The charterers may well use these facts to their advantage. It is therefore not surprising that when the decision has to be taken to stop loading for rain some ship masters in a dilemma turn to their pre-shipment surveyor for guidance. Depending upon the character and experience of the surveyor, a decision may or may not be reached in time to prevent rain entering the cargo holds. In the circumstances it will be appreciated that when having to decide when to close for rain the ship master is often placed in an unenviable situation.

What usually happens is that the master waits until some rain is falling, which might be drizzle or light rain. Charterers may be of the opinion that such a degree of rain is not heavy enough to damage the cargo and will emphasise their belief that the rain will be of short duration, in which case they may pressurise the master into continuing to work cargo. Charterers might also offer the master a letter of indemnity holding him free of all consequences of any damage being sustained by the cargo on account of charterers requesting the ship to continue work during periods of rain. In most of these situations the captain decides to stop cargo being worked and he orders his crew to close the hatches. The number of crew members being available will very often only permit one hatch being closed at a time. The best time that can be made will be about 15 minutes per hatch; the worst, in the case where pontoons and tarpaulins are involved, can be up to 40 minutes per hatch. Therefore, where a five-hatch ship working three hatches is concerned, a simple calculation will give some idea of how long the open holds will be exposed to rain.

Certificate of notice of readiness to load

The reason for delay in acceptance of "Notice of Readiness" may mean that something was found wrong with the ship which prevented cargo from being loaded. Insufficiently dry cargo compartments, inoperable bilge lines, defective hatch covers, etc., would indicate lack of cargo worthiness and even unseaworthiness if whatever was involved was not fully and properly rectified before the ship started to load cargo. Depending upon what type of damage was involved and circumstances, this document could prove useful.

Stowage certificate

When loading, lashing and securing of the cargo is completed, some stevedoring companies present the master of the ship with a certificate of good stowage for him to sign. This document will state that he, the master of the ship, is satisfied with the manner in which the cargo is stowed, lashed and secured. In pursuing a claim for bad stowage such a document will only confirm what must already be evident; that in agreeing to sail his ship the master must have been satisfied that the method of stowage and securing was adequate for the intended voyage. Masters of ships carrying steel cargoes are not obliged to sign this document. In certain circumstances, any claimant may try to capitalise to some extent on this document by maintaining that it is written proof that the master exercised control over the operation of loading, stowing and securing of the cargo. The fact that the master agrees to sail his ship as things are, is sufficient proof that he has no complaints with regard to stowage, lashing and securing of the cargo.

Ventilation records

In proving that the cargo was properly cared for during the time it was on board the ship, it is customary to request from the carrier copies of the ventilation records for the voyage. *These records should begin at least when loading of the cargo commences and should not be terminated until the cargo has been completely discharged.* The information contained in such records for them to be complete, to be of any use in caring for the cargo during the voyage, must be:

—observations of wet and dry bulb temperatures taken inside and outside the cargo compartments;
—the temperature of the cargo;
—relative humidity and dew point temperatures and seawater temperatures.

This information should be recorded every four hours when the vessel is at sea. At the same time the records should clearly show how the ventilation was used in conjunction with the observations referred to.

Where moisture damage is concerned these records, or the absence of such records, invariably place the ship in a very vulnerable position. Unfortunately, sometimes inferior crews and what is considered to be inadequate crewing of ships, result in what is required to successfully transport steel cargoes through varying climatic conditions not being given the attention it requires. Frequently ventilation records cannot be produced. When such records have been conscientiously kept they very often reflect deficiencies in following the procedure that should be carried out to ensure the success of the exercise of ventilating or not ventilating the cargo. In such circumstances, the responsibility for most of what should have been done to protect the cargo, but was not done, ends up with the carrier who has to foot the bill. The wording of part of the first page of a recent U.S. court judgment confirms what has been stated above:

ASSOCIATED METALS & MINERALS CORP.

v.

M/V OLYMPIC MENTOR, ET AL.

United States District Court, Southern District of New York, December 20, 1995
93 Civ. 4330

Bills of Lading—142. Apparent Good Order, "Clean B/L"—1557. Ventilation and Protection—193. Inherent Vice, Decay—194. Insufficient Package, Breakage—1964. Sweat.

The cargo plaintiff in a steel coil sweat damage case met its burden of proving good condition at loading by evidence that wetting was from an external source and that the ship's personnel had seen sweat and condensation in the holds during the voyage even though no first hand testimony was provided as to pre-shipment packaging and handling. Evidence by the carrier that no claims were made by other consignees of steel cargo is not supportive of the inherent vice defence as there are many reasons why a claim may not be made. Shipper is not required to employ the most sophisticated water resistant packing materials so long as the packing is normal and typical for the cargo.

BILLS OF LADING—1212. Due Diligence—1213. Sea Worthiness—1557. Ventilation and Protection—1964. Sweat.

Carrier did not exercise due diligence to provide a sea worthy vessel to carry water sensitive steel coils where vessel did not contain equipment for measuring relative humidity and temperature within the holds and had no fans or mechanical ventilating devices to prevent sweating, but only a natural ventilation system.

BILLS OF LADING—1213. Sea Worthiness—1557. Ventilation and Protection—1964. Seat—CHARTER—156. Liability for damage to cargo.

Ship Owner may be held directly liable in personam to cargo owner for sweat damage to steel coil cargo due to the absence of forced air ventilation and equipment to monitor temperature and relative humidity in the holds where owner warrants in the Charter to provide the Charter with a seaworthy vessel "tight, staunch, strong and in every way fitted for ordinary cargo service".

The deck log book

It will be necessary to obtain a copy of the SCRAP deck log book or otherwise a copy of the pages relating to the voyage in question. This document should contain a complete history of the voyage. Details that must be contained therein, which are useful to the expert, are: courses steered, noon positions, engine revolutions (RPM), weather conditions encountered on voyage, any avoiding action taken during periods of adverse weather and the results of any cargo inspections. Further to this, stoppages for rain during both loading and discharging must be entered in the deck log book, also details of any unusual occurrences which might involve the loading, stowing, securing and eventual discharge of the cargo. Any log book which does not contain this information and other necessary information pertinent to the voyage in question, may be taken as an indication of lack of efficiency on the part of the vessel's staff and failure by them to conduct their shipboard activities in a professional manner in accordance with the normal practice of seamen. Much useful information can be gleaned from the pages of an efficiently kept log book. Much can be made from a log book which is not efficiently kept.

To relate an incident from experience: In one steel case it was necessary to obtain details of the ship's bunkers, i.e. to what temperature they were heated, in which tanks they were stored, and quantities per tank. Details of movement of ballast plus other engine room related information were required. In the circumstances it was decided to consult the chief engineer and he was asked to produce all necessary documentation relating to what was required. He only produced his log book which, when examined, had only the date entered on the top of each page and the chief engineer's signature on the bottom of each page. Obviously neither the chief engineer nor his log book was of any assistance whatsoever. In fact, a document such as the log book referred to, in the hands of the claimant, could prove damaging to the carrier's interests.

Bilge sounding records

Sometimes bilge soundings, which it is customary to record twice a day on board a well-run ship, are provided for by a section on each daily page of the deck log book. Generally, on chartered ships a record separate from the ship's log book is kept. Such a record can be most revealing in the event of sweat damage claims or where the ship's hatches have leaked, etc. In one case reviewed, the log book stated that Beaufort Scale force 12 winds were encountered with high seas and the entire fore ship was awash. In the bilge sounding records varying bilge level depths were reported over the heavy weather period. If the weather was so intense, no one could have gone onto the foredeck to take bilge soundings and therefore there was no harmonisation of the facts as presented.

Stevedores' outturn report

Everyone has their priorities in business and the priority of all stevedores, apart from the common priority of everyone to make money, is to make tonnage. In focusing on rapid discharge and removal of the goods from the ship's side, little attention is paid to the state of the cargo unless it is heavily mechanically damaged, or heavily rusted to such an extent that its condition could hardly miss being interpreted as defective. Add to this the fact that stevedores cannot be considered knowledgeable to any great degree with regard to some finer points concerning damage to cargo; so it is not surprising that their damage reports are unreliable. This is regrettable in some instances, since the document in question may be well-nigh crucial to the success or failure of a claim; it weighs heavily in favour of the carrier who has delivered material which was not in apparent good order and condition if the stevedores have not entered the appropriate remarks on their outturn report. There are instances in which these situations, where damage was known to be or must have been apparent at the time of discharge but was not mentioned in the outturn report, often come back to haunt the stevedores. In assessing the merits of a claim this must be one of the most important documents to obtain.

The cargo receipt

This is the receipt which is given by the receiver in exchange for his cargo when he receives the goods. The handling of this receipt is usually left to the trucker or some other person whose priority is receiving, loading and delivering to ultimate consignee. The receiver's representative may have put the carrier on notice with respect to landing cargo which was not in apparent good order and condition, but if the delivery/cargo receipt is signed clean, putting this together with a clean stevedores' outturn report may place the carrier in a position in which, in reality, he is not entitled to be.

N.B. There are some ports, usually where goods are being received directly onto the premises of an end user of the material, where a "cargo receipt" and an "outturn report" are not issued.

Ship's plans

An expert witness involved in cargo claims may not require all of these plans but he will in most instances need to obtain a copy of the "capacity plan" and the "general arrangement plan". Whether copies of the other plans are necessary will depend on the nature of the claim. The information provided in these plans varies from one builder to another but in general the information shown is as follows:

Capacity plan

This provides the following information:

—Deadweights, drafts and associated tonnages.
—Deadweight scale.
—Capacities of cargo holds.
—Details of tank capacities, e.g. bunkers, ballast and fresh water.
—Storage spaces.
—Sizes of hatch ways.
—Some plans show spot weight load of tank tops, tween deck and weather decks.

General arrangement plan

—This plan shows in detail the layout of everything on board the ship.
—Principal particulars.
—Details of the crew.

Piping plan

—As the name implies this plan gives the layout and details of all piping and pipe valve arrangements on board the ship.

Ventilation plan

—A layout of the ventilation system, including the ship's cargo holds.

Shell expansion

—An expanded view of a vessel's shell plating from which individual plates and side frame positions can be identified.

Surveyors and surveying

There are many cases involving fresh water damage to packaged steel which, as far as cargo interests are concerned, are seriously prejudiced from the moment the goods are landed. Where wrapped goods are concerned the condition of the wrappers need not necessarily be a reflection of the condition of the contents. Wrappers are often painted, so giving the unit a good appearance, whilst the contents may be affected by condensation. Fibreboard wrappers are used to pack coils, and galvanised packing if not too heavily oxidised may not raise suspicions at all.

The real problems arise when cargo landed with a clean stevedores' outturn report remains in a store, very often in the dock area, and is not called forward by the ultimate consignee for some considerable time. Eventually the goods are moved to the works where the packing is removed and it is found that the material is rust damaged. Climatological records may show that from the time the goods were landed up to the time that they were called forward for production, there had been considerable fluctuation in atmospheric temperatures. It is a fact that internal condensation in packed material can take place during transit without the wrappers being appreciably affected, or in any way affected at all. It is not difficult to prise open the lower area of the packing on items of cargo selected at random. If damage is found, the carrier can be invited to attend a survey for the purpose of jointly establishing the fact that damage exists. There are many cases on record where surveyors have accepted items of wrapped steel as being in sound condition, their opinions being based purely on the aspect of the wrappers. It is advisable to consider all wrapped steel cargo which has undergone a sea voyage to be suspect with regard to damage whether directly apparent or not. However, to what extent goods will have to be considered suspect by the surveyor will depend upon known factors and circumstances surrounding the voyage.

Where any steel claim is concerned, it would be of great assistance to the expert witness, also the claims negotiator, if surveyors would, in following discharge of the cargo, consider all items of wrapped steel suspect of containing damaged material. They should endeavour to determine whether cargo normally suspect in appearance, and also cargo where the packing is in apparent good condition, does or does not contain moisture-damaged material. Damage established on the quayside during or shortly after discharge would enable claimants to pursue claims based on proven factual evidence rather than circumstantial evidence as is often the case when damage is discovered some considerable time after discharge.

HEAT NUMBER

Heat or Heat No. is often seen in various documents covering consignments of steel. Heat refers to a batch smelting operation, a melting or a heat treatment cycle. Orders may be filled from more than one heat, although orders for some material uses may require one heat only.

HEAT TREATMENT

Annealing

The steel is heated to a temperature above the critical limit sufficient to permit all carbon to be absorbed into solid solution but still maintain the original grain size. The temperature is held for a time after which the steel is slowly cooled under control of the furnace. Following this procedure a desired structure in the metal can be produced.

Process of sub-critical annealing

For the purpose of achieving relief of internal stresses and re-crystallisation of the grain structure the steel is heated to about 600°C but not above the lower critical limit.

Normalising

This is similar to full annealing except that the steel is permitted to cool at room temperature. The cooling rate is fast enough to prevent complete precipitation of the carbon when passing from austenite through the upper and lower critical limits. As a result of this method of treatment the grain structure is improved, as are the toughness and hardness of the material.

HOT BLAST STOVE

These are situated adjacent to, and are as high as, the blast furnace itself. Each stove is fitted with a brickwork system heated to a high temperature by circulating hot gases, piped into the stove from the exhausts at the top of the blast furnace. In such circumstances, the stove is said to be "on gas" until such time that the brickwork reaches a predetermined temperature. Thereafter, fresh air is piped into the stove, now "on air", by a turbo blower heated to a high temperature through contact with the brickwork in the stove, later being blasted into the base of the furnace at temperatures between 800–1,200°C. Three of these stoves are involved, each being alternately "on gas" and "on air".

HOT ROLLED STEEL SHEETING

Manufacturing process

Ingots are placed in a soaking pit where they are reheated to a suitable temperature for further rolling into slabs, and then into steel sheeting in the form of a continuous strip. In some instances, slabs are taken from stock, in which case the slab is placed in a reheating furnace. The ingot or slab is brought to intense white heat to ensure that the temperature is uniform throughout this initial rolling. Temperature will be in excess of 1,000°C. During reheating, a layer of basic oxide forms on the slab and, in order to remove this, the slab passes through a scale breaker. The loosened scale is blown away by high powered jets of water. Onwards to the roughing train, where the slab passes through four to five sets of rolls in tandem which reduces the slab down to about one quarter of its original thickness, and at the same time squeezes the slab into becoming longer in length. Rolls, placed vertically, control the width of the slab as it reduces in thickness and lengthens into a continuous sheet of thin plating.

Finally, the hot plating, at a temperature of just over 800°C, passes to the finishing train where it passes through six or seven strands of rolls in tandem at great speed, and when the thickness of the plating is progressively reduced it emerges as a thin sheet of great length. Thereafter, the strip of sheeting, travelling at a speed of over 1,000 m per minute (see Plate 11) moves towards the coiling machine, before which it is sprayed with water to reduce it to a predetermined temperature. The entire operation is meticulously monitored by computer in order to ensure exact width and thickness of plating as well as a satisfactory cooling rate. The final product, in coil form, might consist of a plate thickness of about 1 mm to 5 mm, and 125 cm in width. Each coil usually weighs between about 7 to 15 tons depending on the length and thickness of the plating. Coils weighing up to 30 tons are not altogether uncommon.

Packing

Hot rolled steel coils are not usually wrapped for protection against contact with moisture and the development of rust. They are secured with a number of flat metal strapping bands transversely through the eye or core of the coil, and securing bands are fixed around the outer circumference of the coil. There are instances when hot rolled coils are pickled and then oiled (H.R.P.O.) for protection, after which they are wrapped in moisture-proofed kraft paper and completely enfolded in a metal envelope, all secured with flat metal strapping bands, usually three circumference bands and four placed transversally.

Pre-shipment storage

Unwrapped hot rolled steel coils are often stored in the open, uncovered and exposed to the elements. Therefore, it is not unusual for such material to be partly or completely rusty in appearance at the time when it is shipped. Wrapped hot rolled steel coils pickled and oiled (H.R.P.O.) should be kept dry, as any rust on the plating is inadmissible. H.R.P.O. coils should be given the same consideration as cold rolled coil (see Cold rolled steel sheeting).

Industrial uses

This type of steel may eventually be sheared into short lengths of plate and made up into bundles to be sold to the fabrication industry. On the other hand, it can be destined for re-rolling in order to produce cold rolled steel sheeting. See also **Hot rolled steel** on p. 80.

Handling damage

Often referred to as mechanical damage, this consists of physical defects to the plate edges caused during such manipulations as loading and unloading, when the side of the coil is permitted to strike some object. Heavy score marks across the plate edges may or may not be of importance depending upon the intended purposes for the material. If the plating is for re-rolling and the edges have been sheared to an ordered width, then deep incisions are not admissible; likewise, where tearage of the edges of the plating is concerned and depending on the depth of tearing. Buckling or bending of plate edges is of less importance unless the affected plating is actually turned over to form a fold or bent beyond the elasticity of the metal.

Usually there are rolls forward of the uncoiling machine which flatten out any bent edges. Another undesirable defect is telescoping and this applies especially to the centre core turns of plating (see Fig. 22(a)). Each unit should be coiled in such a manner that all plate edges are in line. If edges are projecting too far (telescoped), the possibility of damage developing during handling and in stow is considerably enhanced. In lifting coils, the use of chains and wires should be avoided. Only gear such as broad braided wire slings and "C" hooks should be used. The ideal type of lift truck used should be fitted with a circular bar prong if damage is to be avoided.

Rust

Both before shipment and after delivery, unwrapped hot rolled steel sheeting may spend a considerable period of time in open storage (see Plate 12(a) and (b)). The goods are exposed to rain and possibly a polluted atmosphere. Therefore, free moisture trickles down and across the edges of the plating, and seeps between the turns of sheeting to penetrate a limited distance whilst the coil is static. As rust is a natural phenomenon of steel, it is not surprising that most hot rolled products appear to be either rusty or partly rusty. A thin even film of surface rust on the plating, originally through contact with fresh water, is usually of no consequence. Internally fresh water rust development progresses slowly, in view of the fact that the surfaces of the plating are covered with a protective film of "mill scale", oxygen supply is depleted, and naturally once a rust condition becomes established the oxide film, also changing chemical reaction in the cells, tends to stifle the process. Before the plating can be used, the mill scale, or remnants thereof, and rust and extraneous matters which might be present, must be removed before the goods are cold rolled or provided with a protective coating.

(a). Plate edges affected by telescoping.

(b). Ovalised due to rough handling.

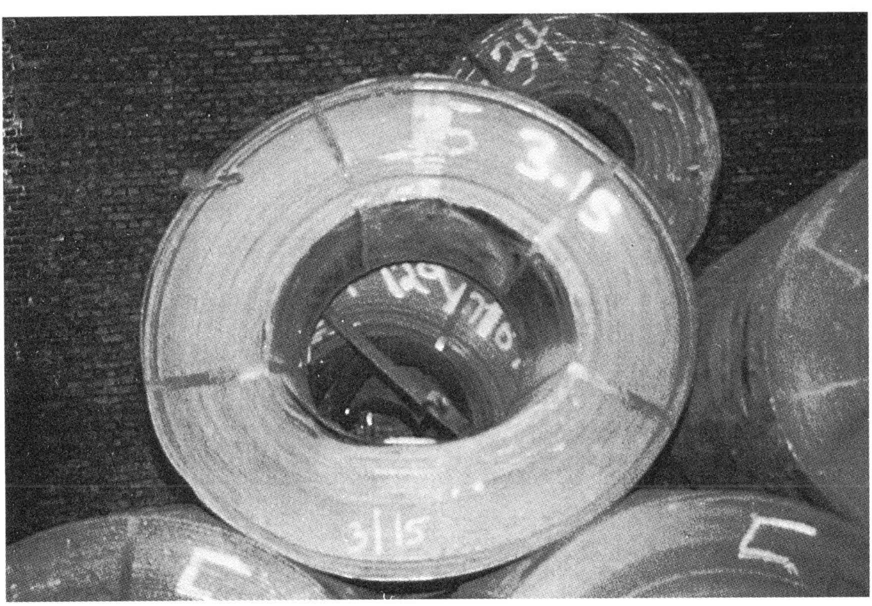

Fig. 22. Hot rolled coils.

This cleaning of the material is accomplished by subjecting the sheet to a pickling process, whereby the plating is passed through a hydrochloric acid bath.

Loading during rain

In consideration of open storage afforded to unwrapped hot rolled steel sheeting before shipment and after discharge (Plate 12(a) and (b)) such goods are often worked into, and out of, a ship during periods of rain. Naturally, consideration has to be given to other cargo already loaded, which might sustain wetness when coils are worked during rain, and which might also become damaged as a result of contact with a wet coil cargo. Further to this, even if cargo already on board can be protected from the rain when coils are being loaded during wet weather, a high level of humidity in the ship's holds, created by moisture which enters the cargo compartments with the coils, may create an atmosphere which is hostile and eventually damaging to other cargo. Consequently, the decision to work this material during periods of rain must be given the utmost consideration. If it is decided to load during rainy weather, it should be permitted only when very light rain is falling. This is on account of the fact that the open hatchway on board a ship serves as a large catchment area. Too much free moisture may collect on the tank top and bank up in the afterend of the compartment to such an extent that some coils might actually be partially submerged in water.

This water could become contaminated by chlorides from salt crystals left behind from previous seawater washings or rests of previous cargoes and/or impurities brought in with the cargo. If this occurs, those coils affected would sustain damage. Obviously, the ideal situation would be to receive dry cargo loaded only during dry weather. Nevertheless, owing to the fact that such cargo had sustained wetness prior to shipment, it would not be expected to suffer from further wettings during loading. Also, seeing that in many instances rust and wetness might be considered a normal pre-shipment condition for such material, which condition should not advance sufficiently to develop into damage during the average ocean voyage, ships' masters are pressed to receive cargo during rainy weather.

Contact with seawater

Seawater has a devastating effect upon steel products and rapidly causes a serious rust condition to develop. If action is taken before an advanced rust condition is established, the rust can be removed in such a way as to ensure that all chlorides and damaging salts are removed, so that the goods can be accepted as prime material (see Plate 13). The goods should be accepted as prime material after reconditioning but some receivers can be difficult, claiming that the goods are now reconditioned material and as such qualify for a depreciation.

On the other hand, if there is contact with chlorides prior to shipment, or earlier than a few days before the vessel's arrival at the port of discharge, or if contact with seawater occurs just prior to arrival of the vessel but the goods are left in store for some time before being processed further, there can be serious consequences. As previously stated, hot rolled steel sheeting is put through a pickling process to remove the mill scale and the rust. This is, in most cases, a normal part of events which takes place before the goods are ultimately used for whatever purpose they were intended. If the corrosive effects of salt water have become too firmly established, what should (after pickling) be an unblemished, clean dull grey surface to the plating, shows etch marks in various positions (see Plate 14(a), (b) and (c)). This situation can only be rectified by re-pickling the plating—two passes or more through the pickling bath—or subjecting the material to a slower pass through an acid bath. Depending on market values prevailing at the time, pickling may not be the best solution and other outlets must be found. Involvement in pickling will be considered when assessing the loss—the time factor possibly affecting the mill's production, chlorides reducing the effective life of the pickling bath, loss of material from the surface of the plating owing to excessive exposure to the acid in the bath. It is certain that if the chlorides/salts are not removed, the plating, where affected, will continuously deteriorate.

The importance of strapping bands

Strapping bands on the coils, at time of shipment, should be completely and sufficiently tight in order to afford maximum possibilities of delivering the goods still in a tightly wound condition. Coils which have slackened off, owing to broken securing bands, can upset the stability of the stow and cause problems in handling during discharge and afterwards. Difficulties can arise in the mill when uncoiling the material (as will be seen from Fig. 23) and claims may be forthcoming owing to disruption of a mill's set programme, loss of time affecting production and so forth.

It is worth mentioning that where the loose coils above were concerned, stones and grit penetrated between the slack turns of plating. The rolls, shown in Fig. 18, which were to be cold rolled, were reversed in an effort to tighten the turns of plating, in order to facilitate decoiling. As a result of this, the surface of the plating was seriously scored, so eventuating in an extension of the original claim.

Bundles

To suit customers' requirements, coils are often unwound, cut into plate lengths of about 2 m and stacked into bundles. Each bundle is secured with a number of criss-crossed flat metal strapping bands. Bundles of hot rolled steel sheets are usually unwrapped and unprotected against the development of rust. The goods will therefore probably have a rusty appearance at time of shipment.

For transportation by sea, all unpacked hot rolled mild steel plates and/or steel sheets, whether in bundle form or loose, should be given the same consideration as hot rolled steel sheeting in coils as already explained (see Fig. 24).

HOT WORKING

It might be said that hot working has the reverse effects of cold working. When steel at elevated temperatures is subjected to rolling and undergoes plastic deformation, the grain structure is broken down so that new and smaller grains are formed. The actual rolling process is not responsible for this restructuring of the grains, as it is the application of heat at suitable elevated temperatures that results in the desirable regrowth of the crystals.

Fig. 23. The effects of broken strapping bands—coils partly unwound.

Fig. 24. Bundles (unwrapped) of hot rolled mild steel sheets.

One of the ultimate aims in the manufacture of good quality steel is to produce a metal with the desirable properties which has the smallest possible grain size. Therefore, to counteract the unacceptable effects of cold working the material is subjected to heat treatment for the purpose of improving the grain structure.

It is appropriate to mention here, as an example, that ingots when cooling from a mass of molten metal develop varying grain structures. As the metal cools through its solidification temperature, varying speeds of cooling throughout the mass are involved with specific areas developing a grain size which varies from other areas within the mould. This heterogeneous crystallisation is rectified by eventual hot working when the ingot is hot rolled and re-rolled into slab and billet form.

INGOTS

These are huge heavy blocks of steel, trapezoidal in shape. They are intended for reheating and mechanical working to be reduced into semi-finished products such as slabs and blooms.

IRON

Iron is the most abundant of the elements and represents about 5% of the earth's crust. It is a white, magnetic, metallic element which is the most important of all the known metals. The uses for pure iron are few as its mechanical properties are unsuitable for most applications. Small amounts of other metals greatly modify its properties, and this applies particularly to carbon. The addition of a small amount of carbon produces steel. Pure iron melts at a temperature of 1,540°C.

IRON ORE

This is a term used for an iron bearing mineral which contains a sufficient quantity of iron to render it worthy of mining for commercial purposes. It is a chemical compound of iron and oxygen, it contains minor amounts of other elements and also unwanted matters such as rock, clay and sand. The following, owing to their high yield of iron, are the only minerals that are important:

Magnetite	(Fe_3O_4)	72% iron (black ore)
Haematite	(Fe_2O_3)	70% iron (red ore)
Limonite	($Fe_2O_3H_2O$)	60% iron (yellow brown ore)
Siderite	($FeCO_3$)	48% iron (light grey/brown)

N.B. Percentages of iron are approximate.

Some of the richest deposits of iron ore are in Sweden, Australia, Canada and South America, and are shipped to Europe by sea in very large bulk carrier ships each carrying anything up to 300,000 metric tons of the ore per voyage. The iron is won from the ore by heating it to a very high temperature in a blast furnace.

KILLED STEEL

Liquid steel is subjected to a deoxidising process by adding aluminium and ferrosilicon. As a result of this, there is no porosity in the ingot but such metals are distinguished by slight piping and peaking at the top of the ingot.

The killing of the steel imparts to it high strength characteristics, higher resistance to impact and enhanced welding properties. Al-killed steel is tougher than Si-killed steel at a low temperature and vice versa.

LADLE CAR

This is a large vessel on wheels which is used to transport molten iron from the blast furnace to the department involved in the steel-making operation.

LETTERS OF CREDIT

Some problems in negotiating letters of credit

A letter of credit (L/C) is an undertaking by a bank, in the form of a contract, to pay money to a beneficiary in as far as the documents required are provided in the correct manner and in due time. One of these documents can be a bill of lading, defining the shipment of cargo to be transported between two or more ports.

When goods are shipped under a CIF contract, the seller must pay the ocean freight and the buyer will require a clean negotiable bill of lading. In the case of an FOB contract the buyer will take care of all details re payment of the ocean freight. According to the International Chamber of Commerce a clean bill of lading is defined as follows:

> "A clean shipping document is one which bears no superimposed clause or notation, which expressly declares a defective condition of the goods and/or the packaging. Banks will refuse shipping documents bearing superimposed clauses or notations unless the credit expressly states clauses or notations which may be accepted."

If the bill of lading is qualified by the insertion of a superimposed clause, this will mean that, in all probability, the condition of the goods will not reflect the value of the cargo and selling conditions as declared in the original invoice. Such a situation raises serious problems for the seller and the following example might make this clearer.

A manufacturer in Europe, who has sold goods to a purchaser in Japan, appoints a forwarding agent (who may be considered as the shipper) to arrange transport of the goods. In the contract between the seller and forwarder there can be a clause which states that the forwarder shall arrange matters in such a manner that, within a specified period of time (usually a few weeks) after the goods have been shipped on board the ocean carrier, he, the seller, shall receive from the advising bank the proceeds of sale in full and this only after negotiating of the L/C providing a.o. the presentation of the required shipping documents. In the event of the forwarder defaulting on this conditions, interest will be paid by the forwarder to the seller and a stipulated rate of interest will apply.

The exchange of money between buyer and seller is usually arranged through an irrevocable letter of credit, either confirmed or not. This means that the bank of the seller negotiates with the bank of the buyer which will temporarily hold the bill of lading. In the event of the bill of lading not being signed clean, credit will be withheld from the forwarding agent, if such is stipulated in the contract between manufacturer/seller and forwarding agent.

It could take longer than six weeks before the cargo is landed in Japan. If claims are entered, surveys will be held, and in due course a survey report will be produced which specifies to what extent, if any, the cargo has depreciated in value. Not only will the forwarder have to pay interest to the seller on the funds which he has been unable to produce, he may in due course be required to make up the difference between the original sound value and the depreciated value of the material as determined by the surveyors. Further to this, and if stipulated in the said forwarding contract, the credit facilities extended by the bank may be urgently required to enable him indirectly to finance other shipments. In such circumstances, the forwarder will be operating with an acute capital limitation, and failure to negotiate the B/Ls for one shipment might eventuate in his business being terminated.

LIABILITY AND SEAWORTHINESS

Both of the abovementioned areas of opinion belong within the domain of the marine lawyer but, at the same time, it is necessary that the surveyor is conscious of the value of the information he is providing and therefore his decision as to what to write down and how to present it is of importance. An experienced surveyor and especially an expert witness, may be sufficiently knowledgeable as to confidently permit himself to put forward limited categorical statements based on past experience.

There can be varied interpretations as to what constitutes "lack of due diligence", "negligence" and "recklessness" arising out of the master's/owner's obligation towards the carriage and care of the cargo, and the following may be of help in determining how these terms are viewed. In the case of one vessel, an air pipe to a double bottom ballast tank was fractured at an accessible position half height in the cargo hold. Before cargo is loaded, a competent deck officer, or member of the owner's staff on board the vessel should inspect the cargo hold. It would be expected that a notation be made in the deck log book to the effect that the cargo space was examined and reported as it was found to be. In the case referred to, the chief officer did examine the cargo hold but made no notation regarding this in the deck log book. When the vessel was discharging cargo the double bottom tank in question was ballasted and water flowed through the defective part of the pipe and damaged the cargo. The chief officer testified that he did examine the pipe, but did not observe the defect, which according to photographs provided was readily visible. In the circumstances it was found that the chief officer was "lacking of due diligence" and the owner was found responsible for the damage and lost the case.

In the same instance, had the fracture in the pipe been situated on the outboard side of the pipe, i.e. facing the shell plating, it would have been considered an inaccessible defect, not readily detectible by a reasonable inspection of the pipe when viewed from the level of the tank top by a person exercising sufficient due diligence. In this instance, the owner should escape liability. If it could be proved that in the first instance, whoever was delegated to inspect the cargo hold did observe the defect but failed to report it, the respondent may escape liability by claiming negligence of the person who inspected the hold. On the other hand, had a member of the owners staff, e.g the owner's superintendent, inspected the pipe and no repair was carried out for whatever reason, the owner would in all probability find it difficult to escape liability

According to the Hague-Visby Rules, Article III, paragraph 1a, "The carrier shall be bound before and at the beginning of the voyage to exercise due diligence to (a) Make the ship seaworthy". Where steel hatches are concerned, in the event of leakage and damage to cargo, the master/carrier would be obliged to prove that everrything humanly possible was done to ensure that the hatches were suitably and well maintained, in order that there was no doubt that every possible precaution had been taken to ensure that leakage of the hatches could only occur under extenuating circumstance, e.g. stress of heavy weather.

In another case, a full cargo of steel beams was being loaded and the master's attention, via the chief officer, was drawn to the fact that some of the dunnage was breaking under the weight of the cargo as it was being loaded, and stability of the stow was affected. Further to this, in two hatches, not fully completed, the securing was considered inadequate. Discussions were held with the stevedores regarding employing stronger dunnage, but loading continued uninterrupted and nothing was done to rectify the situation. During the voyage the vessel experienced winds of beaufort force 12 and the cargo broke loose in three hatches.

During discharge, the beams were found to be warped, twisted and bent. In one hatch the cargo had become a tangled mass that had to be virtually torn out of the stow. The view was taken that when the vessel sailed she was unseaworthy as the cargo was not properly secured, nor stowed in a manner that could be considered safe for the intended voyage. Considering the worst scenario that could be encountered, had the vessel experienced an engine breakdown during the heavy weather, drifted beam-on to the sea and developed synchronized rolling with the seas, together with a shift of cargo, she may very well have been lost. It therefore seems that some legal minds use the term reckless when loss of life may be involved. In the case in question the master was described as being negligently reckless. However, in spite of everything, the owner escaped liability on a legal technicality.

LIMESTONE

Iron ore contains earthy materials and trace elements which are unwanted in the production of pure iron. Limestone is introduced, under the great heat of the furnace, which combines with the impurities to form a liquid slag. Limestone acts as a flux which scavenges the charge of parts considered to be extraneous to iron contained in the ore.

LOCK FORMING QUALITY

A quality attached to galvanised sheet, which means that the sheet can be bent 180° with the galvanising remaining intact.

LUDER'S LINES

A surface defect sometimes encountered during the forming or deep pressing of mild-steel sheet. It consists of a series of dull lines that appear on the surface of the metal and are due to non-uniform distortion. These appear on the metal as soon as deformation starts and are first noticeable on the parts that are least deformed. As the deformation proceeds, the lines appear in increasing numbers. They can be seen as light depressions or slight ridges, according to whether the metal has been subjected to tensile or compressive forces. When large grain material is subjected to considerable plastic flow (elongation), wide bands cutting across many grains may destroy the surface flatness and spoil the appearance of the sheet. The lines are also referred to as "flow lines" and "stretcher strains".

MALLEABILITY

The ability of a metal to deform permanently under compression without rupture, e.g. angle iron and various constructional shapes achieved by bending, rolling or pressing.

MANGANESE

Manganese is always present in commercial steel as it improves the quality of the material. It is essential to steel making, as it is used as a deoxidiser and desulphuriser in the steelmaking process. Acting as a scavenging agent it has a cleansing effect which improves the durability and forgeability of the finished product.

MATE'S RECEIPTS AND BILLS OF LADING

In English law the carriage of under-deck cargoes is governed by the Carriage of Goods by Sea Act 1992, which is based on the outcome of an international convention held for the purpose of creating a set of rules relating to bills of lading. A set of rules was agreed in 1921 in The Hague (hence the name Hague Rules) and a revised version was embodied in a convention signed in Brussels on 25 August 1924. These Rules were amended in certain respects by a protocol signed in Brussels on 23 February 1968, and the amended set of rules is referred to as the Hague-Visby Rules.

Whether the rules are in force in a particular territory depends upon whether that state has signed and ratified the convention (or acceded to it as the case may be). The parties may give effect to the convention by giving it the force of law, or by including the rules in appropriate form in their national legislation. The current situation is that, although many of the leading maritime nations are parties to the Hague-Visby Rules, those parties to the Hague Rules that have not adopted the 1968 Visby Amendment are still subject to the original set of the Hague Rules. There are some differences between the two sets of rules and these mainly concern:

(a) Limitation of liability, including special provisions relating to containerisation (Art. IV).
(b) The application of the Rules (Art. X).
(c) The "Himalaya problem" whereby the limitation provisions could be circumvented by suing servants or agents of the carrier in tort (Art. IV bis).
(d) Statements in the bill of lading being conclusive once it has been transferred to a third party acting in good faith (Art. III rule 4 as amended by Visby).
(e) Amendments to the time bar, including in relation to indemnity actions (Art. III, rule 6 bis).

A new convention sponsored by the United Nations at Hamburg in March 1978 drafted a new set of rules known as the "Hamburg Rules" which became effective from 1 November 1992. Most of the world's leading maritime nations have ratified the Hague Rules, the Hague-Visby Rules and a few have ratified the Hamburg Rules. See the Appendix for the text of the rules.

The Rules

The difference between the Rules (see Appendix), which would be of a direct and immediate concern to the master, would be that in the Hague/Hague Visby Rules (Art. 3), the Master is not obliged to render the vessel seaworthy; he must prove that he exercised due diligence to make the vessel seaworthy. They also state that other requirements must be met with regard to the manning and equipment of the vessel and the carrier's responsibility in the care of the cargo. The Hamburg Rules mention none of this, but in context the wording of the paragraphs are framed in such a manner as to indicate the period of his responsibility for the goods, i.e. when he is in charge of the goods (Art. 4). During this period, should the goods sustain damage the reasons given for the said damage will be judged, and liability directed or apportioned, according to the merits of the case. As an example, where seaworthiness is concerned, if any defect were known, would a prudent owner have required that it should be made good before sending his ship to sea had he known of it? If he would, the ship is not seaworthy. The difference is that, where in the Hague and Hague-Visby Rules the Master is only obliged to prove the exercise of due diligence to make the vessel seaworthy, under the Hamburg Rules there is no escape, either all measures humanly possible were exercised to render the vessel seaworthy or they were not.

Further to the above, as under the Hague/Hague-Visby Rules, the interpretation of the term "commencement of the voyage" means that the voyage commences at the first port of loading. Under the Hamburg Rules, the term "voyage" is interpreted as a single bill of lading voyage, regardless of the number of stops the ship may make at ports along the way. Therefore, every new voyage would demand the same stringent seaworthy conditions as referred to above. It is considered that every shipmaster should be fully conversant with all of these Rules in order for him to fully understand his lawful responsibilities with regard to the carriage of cargo by seagoing vessels.

The bill of lading

An original bill of lading is a document of title to whatever goods it refers to; it is a form of receipt for the goods received on board the ship and it is accepted as a negotiable document throughout most of the banking systems of the world. A bill of lading can also be looked upon as evidence of a contract between the carrier and the shipper, even though it is only signed by one of the contracting parties. One of the important functions of the bill of lading is that it serves to protect what might be referred to as the "innocent third party" involved—the receiver who finally becomes the bearer of the bill of lading.

The receiver is not expected to be an authority on shipping or to have any knowledge of the carriage of goods by sea. On the contrary, he is merely a person who has ordered goods from another country, and his knowledge of the entire shipping venture is not expected to extend further than the fact that the bill of lading, which he holds, is *prima facie* evidence that his goods where shipped on a given date, on a named vessel, in an indicated apparent order and condition. If his goods arrive in a damaged condition, he would be expected to enter a claim either against the carrier or against the cargo underwriters (who, upon settlement, would become subrogated to the rights of the bearer of the bill of lading).

If the condition of the cargo, by the time of its reception, has changed to such an extent that the receiver has grounds for entering a claim, the carrier would seek to exercise his rights and immunities (see HR IV). This means that in due course the carrier will have to prove to the receiver (or subrogated cargo interests) that he exercised due diligence to provide not only a seaworthy but also a cargo-worthy ship at the commencement of the voyage, and that he did properly and carefully load, handle, stow, carry, keep, care for and discharge the goods carried (see HR III 1 and 2). In other words the carrier will have to prove that (outside of anything absolutely beyond his control) he operated a sound system in order to protect and carefully carry the cargo with which he was entrusted. Whether due diligence, proper care etc., have been exercised is a matter of fact in each case.

Whilst the Hague and/or Hague-Visby Rules clearly and concisely set out the legal provisions for the carriage of goods by sea under a bill of lading, the situation becomes somewhat ambiguous because of differences of interpretation in conjunction with the laws prevailing in various countries. As an example, there are the conditions acceptable as constituting heavy weather at sea. In the past there has, in Great Britain, been a tendency for wind force 8 and above, of the Beaufort Scale, to be accepted as heavy weather, whereas, in some other countries, anything under wind force 10 is not accepted as constituting heavy weather conditions. From experience it seems that, in general, weather conditions in claims settling in the developed world are looked upon as being: Beaufort wind force 7–8 boisterous weather; 9–11 heavy weather; 12 and upwards unusual and exceptional weather.

Mate's receipts

In some ports a mate's receipt is not used, because the master puts any exceptions relevant to the apparent condition of the cargo directly into the material bill of lading at the time of signing this document. In larger ports using more modern methods, the system might be that the shipper buys two sets of "unified documents" (two or three originals and a number of non-negotiable copies) and the other set consists of the mate's receipt, the loading permit, the tally sheets, etc. The shipper completes these documents (which are virtually carbon-copies of one another) and so is deemed to have guaranteed their accuracy at time of shipment (see HR III 5). The shipper then presents these sets to the agent, who keeps the bill of lading but sends the other set to the stevedores at the loading berth. In due course, the cargo is loaded, whereupon the chief tally clerk presents the mate's receipt to the ship's command with (where appropriate) his advice as to advisable exceptions, if any. The master or chief officer may, or may not, insert exceptions when signing the mate's receipt and the latter then goes to the agent's office. Where necessary, the bill of lading is amended and then that document is signed and is at the disposal of the shipper against payment of freight and any other material charges.

The uninitiated might well ask why, in the face of so much paper work, increase it by introducing a mate's receipt. If the bills of lading and manifests (per port of discharge, plus the "Dangerous Goods" manifest) had to be dealt with by the master after the completion of loading, the vessel would obviously be delayed, whereas by use of the method outlined the vessel can sail as soon as its cargo has been stowed, lashed and secured, and the vessel battened-down. The mate's receipt is thus a saver of a considerable amount of time.

As we have seen, nevertheless, the mate's receipt should be regarded as nothing more than a receipt for the cargo loaded aboard the vessel in the apparent condition mentioned in the receipt, its use being to facilitate true and correct issue of the relevant bill of lading. The mate's receipt is *prima facie* evidence of quantity and condition of goods received. Notwithstanding the above, considered to be the situation following the normal course of events, trade custom may establish that mate's receipts are treated as documents of title in the same way as bills of lading. There is no internationally documented or judicial obligation for mate's receipts to be issued or signed.

In some instances the vessel's chief officer is asked to sign an exceptions list and this could be considered to be a "consolidated mate's receipt" to facilitate true and correct issue of the relevant bills of lading. It is the signature of the master (or someone specifically authorised to sign "For the Master") on the bill of lading which validates that document.

Delivery of damaged cargo

In theory, the master delivers the cargo to whosoever shall present the bill of lading but, in practice, the ship gives the cargo to the customs authorities at destination, and in due course the bearer of the bill of lading collects the cargo from the customs shed when authorised to do so. If the goods are damaged prior to being received, ex Customs, by the bearer of the bill of lading, he should take written exception or he may call for a joint survey. Removal of the goods into the custody of the person entitled to delivery thereof without holding joint survey or previously at time of delivery giving written notice of loss or damage, constitutes *prima facie* evidence of delivery by the carrier of the goods as described in the bill of lading (see HR III 6), whereas when exception is taken, it is considered there is *prima facie* evidence of possible involvement of the carrier's liability. In the latter case the carrier will seek to nullify his liability by invoking his "rights and immunities" (see HR IV) but it is self-evident that the recording of pre-shipment damage by a suitable exception in the bill of lading, prior to its issue, is the carrier's best defence. This is emphasised by the fact that in the United States of America, there are cases on record of steel being delivered "clean" then travelling inland to be sold and resold. The final buyer may complain about the condition of his purchase and enter a claim against the seller, and that claim can travel back down the line of merchants and end up being brought against the ship several months after the goods were delivered. Such actions can be successful, to the detriment and cost of the carrier. The defence of the carrier is very much easier where there is a clause actually in the bill of lading referring to the pre-shipment condition of the goods if they are presented for shipment in a wet, rusty and/or otherwise damaged condition.

Clausing of mate's receipts and bills of lading

Clauses to be inserted in mate's receipts and bills of lading should be carefully considered and used with discretion. Unnecessary and/or meaningless clauses, and those which are not strictly confined to the apparent condition of the goods, should be avoided. The carrier or the master or agent of the carrier is required to issue to the shipper a bill of lading showing the apparent order and condition of the goods (HR III 3 and 3(c)). The purpose of a pre-shipment inspection of the cargo, therefore, is to determine whether or not the apparent condition of the goods conforms with the description of that cargo as shown in the relevant bill of lading. If, for example, the bill of lading refers to a number of steel profiles without any qualifying remarks relating to their condition, and if upon inspection it is seen that they are made up into bundles which are unwrapped and thus unprotected and/or affected by rust, the master should mention these facts in the bill of lading.

Shippers prefer to receive a "clean" bill of lading. This means that in such a document the original description of the goods, as supplied by the shipper, remains unaltered in any manner or form. A claused bill of lading cannot be readily negotiated and payment cannot be obtained, without delay, against the material letter of credit (see under Letters of credit).

Useless and dangerous clauses

In view of the fact that shippers provide cargo for ships to carry, it would appear logical that their wishes should be viewed with a certain measure of respect. This, together with what has been said earlier, raises the question of unnecessary clauses in bills of lading. For instance, a clause in a bill of lading such as "Goods Uncovered on the Open Quay" does not refer to the condition of the cargo and therefore is an unnecessary and unwarranted insertion in a bill of lading. Similarly, a remark such as "All Cargo Loaded with Chain Slings" not only serves no useful purpose but is also self-defeating and damaging to the carrier's interests in that the carrier is bound to "... carefully load ..." etc. (HR III 2). There is also the clause "Quantity and Quality Unknown" which hangs on the lips of so many masters. HR IV 2 clearly states that "Neither the carrier nor the ship shall be responsible for loss or damage arising or resulting from ... quality ... of the goods" and, as the master of the vessel does not possess the necessary knowledge to judge whether or not poor quality is involved, it becomes evident that a clause embracing the word "quality" as its main theme should not have any place in a bill of lading.

The first part of that clause (referring to "quantity being unknown" gives rise to ambiguity. It is normal commercial practice to acknowledge, by signing a mate's receipt or a bill of lading, that goods have been received into the carrier's custody and that a certain quantity is involved (the more so in that the freight is based on that quantity). In such circumstances, it is both contradictory and illogical that the master should sign for goods received as shown in the bill of lading, but at the same time, clause the document to the effect that he does not know what quantity he has on board the vessel. There are other types of clause which also really are of a damning nature to the carrier's interests and which, from time to time, are entered in mate's receipts or bills of lading. The following are examples.

Example 1. "Cargo Loaded During Rain"

HR III 2 provides that the carrier shall properly and carefully load. It follows that, during the loading operation, the goods are in the care of the carrier, and if the master allows them to become wetted by rain he is failing to carry out a legal obligation. He should stop the loading and close the hatches during periods of precipitation.

Example 2. "Damaged by Stevedores"

HR III 2 provides that the carrier shall properly and carefully load. The stevedores are normally the servants of the carrier. Once again the carrier admits failure to carry out a legal obligation in that the stevedores have not been directed to load properly and carefully.

The above holds true even in those contracts of carriage where there is a "free in" clause, because the latter only changes the agreement as to which party shall pay for the loading. Under "liner terms" it is the carrier who pays, whereas under "free in" the loading costs are paid by the cargo interests. In both cases, however, the liability arising out of negligent, careless and/or improper loading by the stevedores rests firmly with the carrier, not merely under the quoted provision of HR III 2 but also under those of HR III 8. This does not mean that the carrier has no redress against the stevedores, but it does mean that the carrier cannot shrug off the consignees by inviting them to address their claims directly to the stevedores or to any other "third party".

Stevedore damage

From all the foregoing it becomes evident that a "stevedore damage" clause has no place whatsoever in either a mate's receipt or a bill of lading because any and every such clause can be described as an "own goal" or "grievous wound" self-inflicted by the carrier.

Nevertheless, in the event of stevedore damage, the carrier should put the stevedores on notice immediately and in writing and if deemed necessary, to establish the cause and extent of the damage, the carrier should call for a joint survey. As we have seen, when the cargo claim comes in, the claimant cannot refer the claimant to the stevedores (HR III 8) but the carrier's notice to the stevedores and/or the joint survey will considerably enhance the chances of the carrier obtaining a recovery from the stevedores or from their insurance. Failure to put the stevedores on notice, immediately and in writing, leaves the carrier with virtually no chance of obtaining redress.

The loading operation

Situations will arise when cargo sustains damage, whereby the master of a ship is obliged to decide whether the damage referred to should be entered as a superimposed clause upon the relevant mate's receipt and bill of lading (pre-shipment damage), or be designated as "stevedore damage" contracted during the actual loading operation or by the stevedores on board the vessel after the goods have been loaded. In many instances the division between the two categories of damage, i.e. before loading and after loading, is clearly defined, but quite often differences of opinion can arise where damage is sustained in the vicinity of the ship's side. As far as the master of the ship is concerned, damage sustained by cargo prior to loading is pre-shipment damage, damage sustained after the goods have been shipped, which includes during the actual loading operation and any handling damages contracted after this, is stevedore damage.

In view of the above, the question will be asked as to when the goods are deemed to be under the control and care of the master of the ship. At some point during the loading operation the responsibility for the cargo passes from shore-based interests to the ship when the master takes over the responsibility for the cargo. In many instances the master's control over the goods, and his responsibility for them, is considered to be effective from the moment the loading operation commences. This is deemed to be as and from the moment the goods are hooked onto the gear which will hoist the cargo onto the ship, irrespective of whether this may involve a shore crane or ship's gear.

Although in many instances the above will be applicable, it has to be borne in mind that the extent to which the carrier has to undertake the loading of the vessel may depend not only upon different systems of law but upon the custom and practice of the port. Further to this, contracts or arrangements may exist between the various parties involved in the shipping venture which may very well affect actual applications of what is stated in the Hague Rules, Article III, and the Hague-Visby Rules, Article III (see Appendix).

Letters of indemnity

A bill of lading is a negotiable document and, almost without exception, the shipper requires a "clean" document in order to be able to negotiate it and thus obtain payment for the goods. The law, however, requires the master on demand of the shipper to describe the quantity and the apparent order and condition of the goods (see HR III 3). Problems therefore arise when the goods are presented for shipment, e.g. when loading commences, when there is a discrepancy between the quantity and/or the apparent condition of the goods as opposed to what is stated on the relevant mate's receipt and/or bill of lading.

If the master clauses the mate's receipt and/or the bill of lading the carrier may lose that shipper as a client, whereas if he does not clause the documents in line with the noted discrepancy, he becomes party to a fraud in signing the documents "clean". The practice has arisen of issuing a "clean" bill of lading against the carrier's acceptance, from the shipper, of a letter of indemnity in which the shipper undertakes to hold the carrier free and harmless from all consequences which may arise from the fact that the bill of lading is issued "clean".

The issue of a "clean" bill of lading against the acceptance of a letter of indemnity puts the carrier at enormous risk. The practice in itself is illegal and consequences could develop from this fact in itself. The shipper may not honour the letter of indemnity, which leaves the carrier with no alternative but to foot the bill. Further to this, the position of the protecting and indemnity associations is that such practices are not recognised and participants' P & I cover may very well become prejudiced.

Pre-shipment surveys

Steel products are more highly susceptible to damage resulting in claims than most other products carried in seaborne trades. In many instances loading operations commence immediately upon the vessel's arrival, so that there is no time for the ship's personnel to inspect the cargo prior to it being loaded. It is also not unusual that the complement of a vessel has never previously had to follow the loading of a steel cargo, thus putting it at a considerable disadvantage with regard to the formulation of adequate damage and/or rust clauses when discussing such defects with shippers specialising in such cargoes. Further to this the ship's officers probably will not be familiar with the systems of marking and identifying the various parcels of cargo which are involved, because such systems vary from port to port. In consequence of the foregoing it has become customary for shipowners, managers and/or charterers to appoint a competent local surveyor, experienced in pre-shipment survey work, to inspect steel cargoes prior to their shipment and during loading.

The surveyor's duties include the formulation of suitable descriptive clauses relating to any visible rust on the goods, physical defects (handling damage other than that caused by the stevedores), structural abnormalities that might be construed by third parties as defects which develop after shipment, contamination by extraneous materials such as identifiable and/or non-identifiable powders, grease, moisture, chemicals, etc. The surveyor should be sufficiently competent to have his terms of reference extended, if necessary, to permit his intervention with reference to stowage problems and any other subjects giving rise to disputes which can develop during the course of loading.

Descriptive clauses used have been discussed earlier in this book (see pages 60–72). Recently, pressure has been brought to bear upon the various P & I Associations to scrap the suggested P & I Club clauses, as mentioned in the 28

February 1964 International Group of P & I Clubs circular letter headed "Letters of Indemnity" (see p. 63). Exporters have been pressing the industry to use the coded photographs that are used between buyer and seller, which are based on the Swedish Standards. Many surveyors, generally under the influence of shippers, are showing interest in the use of these clauses and the American Rust Standards Guide has gained some popularity. Thc time-honoured method of correcting any deficiencies in the description of goods as referred to on the face of a bill of lading, is by the ship master inserting superimposed clauses in that document. It is recommended that this procedure is not interfered with by using coded figures and letters referring to photographs, which will in any case not be attached to the relevant bill of lading and, for a large shipment, can involve many hundreds of bills of lading. The various propositions are discussed further below.

In the 1964 P & I Club circular it is stated, "In no case should any reference be made to the degree of rust such as 'atmospherically' or 'superficially'." In fact, anything other than that there is rust on the goods without qualification is undesirable. This is on account of the fact that the master of a ship is not an expert on merchandise and therefore is unqualified to determine more than what he sees, which is rust, i.e. the apparent condition of the goods. It is generally accepted that the master cannot truthfully and with confidence state otherwise.

ISO 8501-1 : 1988. There are something like 24 photographs to choose from. The colours and examples are not good and this is one reason why they would not lend themselves to pre-shipment survey work.

American Rust Standards Guide

The American Rust Standards Guide refers to "Light Surface Rust", "Heavy Surface Rust", "Heavy Rust" and "Badly Damaged". Such decisions cannot be made on the basis of photographs; this is only possible from a detailed inspection by an experienced steel surveyor who would be obliged to examine the surface of the steel after cleaning away the removable rust. Neither the master of a ship nor his officers would be competent to make such decisions with any degree of accuracy. Further to this, the coded method/photographs fixes the condition of the steel at one specific point in time, whereas there are many varying conditions and aspects of the rust condition within the scope of any of the standard P & I Association rust clauses from "Rust Stained" through to "Rusty", thus making allowance for the change in colour of the rust expected during the course of an ocean voyage and exposure of the steel to a marine atmosphere.

It is accepted that the P & I Club circular of 1964 is outdated and that some of the clauses are in many respects duplicated, and it has been found that some of the clauses are never used and therefore superfluous to needs. On the other hand, the clauses most frequently used have stood the test of time and most surveyors know through experience how to apply them. However, the list has been carefully studied and compared with requests for change from a number of areas around the world. Some P & I Associations have already informed their members of the following list, which explains the meaning of each clause and how it can be determined.

"Rust Stained"

When 75% or more of the surface area of the cargo, or item of cargo, is covered with a fine, light-coloured film of rust, which when removed by wire brushing, scraping or wiping reveals a smooth, bright metallic surface.

"Partly Rust Stained"

When under 75% of the visible surface of the cargo, or item of cargo, is covered with a fine, light-coloured film of rust as described for "Rust Stained" above.

"Rusty"

When 75% or more of the surface area of the cargo, or item of cargo, is covered with a dark-brown-coloured rust formation, which when removed by scraping or wire brushing reveals an uneven, dull surface still rusty in appearance to some degree.

"Partly Rusty"

When less than 75% of the surface of the cargo, or item of cargo, is covered with a dark-brown-coloured rust formation as described for "Rusty" above.

"Rust Streaked"; "Evidence of contact with free moisture—drip down rust streaks"

These clauses, or similar clauses, will apply mainly to the packing of wrapped material but may also be applied to hot rolled products when found appropriate. They are very important, as drifting free moisture can penetrate the overlaps of packing and contact the contents, so causing damage direct by contact and creating internal sweating of many types of steel products.

N.B. It will be found that the following rust conditions will on many occasions be used in conjunction with the general rust conditions as above indicated.

"Rust Spots Apparent"

Localised very slight penetration of rust in the form of rusty spots and especially in areas where the mill scale coverage is broken or fragmented. The spots are not prominently bulbous in appearance and when cleaned away leave a smooth steel surface.

"Rust Spotted"

Localised penetration of rust through areas of broken mill scale. Bulbous, revealing an uneven surface when removed by wire brushing.

Apart from the above-mentioned rust clauses the following clauses may also be applied where appropriate.

"Rust with Pitting"

Deep-brown-coloured rust which, often with rust scale formation, when removed by wire brushing or scraping reveals pitting of the surface of the steel.

"Covered with Snow / Ice"

Visible surfaces partly or totally covered with snow and/or ice.

"Wet before Shipment"

Partly or totally wet surfaces apparent before shipment but which may only become apparent when moved for shipment, e.g. lifts of structural steel, when lifted water runs out from inside. When lifting hot rolled coils, flexing of the windings often squeezes out water from between the turns of plating; in which case, although initially apparently dry, the goods are wet before shipment.

"Areas of steel surfaces reacting to silver nitrate solution tests"

Suspect areas should be tested. If they react the test indicates that in all probability chlorides or salt water contamination is involved.

Stained by an unidentifiable powder

These powders are often hygroscopic and contain chemicals which can be aggressive to steel surfaces. The colour of the powder should be mentioned. Samples of the powder should be obtained and kept for future reference.

Oil / Grease Stained / Spotted

During warm weather oil and grease drops from the cranes onto the steel prior to and during shipment. Depending on circumstances the defect may have to be considered a stevedore damage and not pre-shipment.

N.B. See also page 244, "Steel cargo pre-shipment survey report".

Charterers' and shipowners' interests

Insofar as the loading and carriage of cargo is concerned, the interests of the charterers and the shipowners should be parallel. Nevertheless, conflicts of interest can arise in disputes relating to stowage and/or the securing of the goods.

Unfortunately, in most instances, such disputes cannot readily be resolved without the intervention of one or more surveyors. Another source of trouble can be the welding of pad-eyes to the structure of the ship (in the holds of the vessel and for the account of the charterers) where molten metal can fall down between the overlaps of the packing of wrapped steel and thus set alight the paper and/or jute interior packing. Invariably such fires are doused with water when, in fact, the cargo involved should be kept absolutely dry. Further conflicts of interest may arise from the condition of the hatch-closing appliances, the complete watertightness of which is essential in the preservation of steel cargoes.

Steel carried on deck

Contracts for the carriage of deck cargoes are not covered by the Hague or Hague-Visby Rules except by virtue of a clause paramount in relation to a non-negotiable document (see COGSA 1992, section 1). Whenever it is intended to carry steel on deck the carrier should make a separate written agreement with the shipper which (irrespective of any printed clause on the reverse side of the bill of lading to the effect, for example, that "The carrier has an option to ship the cargo on deck") must be handwritten on the face of the bill of lading.

If the above is not done, the carrier has no acceptable evidence, *vis-à-vis* the third party bearer for value of the bill of lading, that a contract to ship on deck actually existed, and in the not unlikely event of subsequent claims, it would be held that:

—the unclaused, or improperly claused, bill of lading is evidence of a contract to carry the goods under deck
—by shipping the goods on deck the carrier is in breach of contract
—the carrier forfeits the limitations of liability which otherwise would protect him.

The consequential cost to the carrier thus could be enormous as the claims would be indefensible.

Situations do arise, however, where structural steel pieces and even galvanised materials are shipped uncovered and unprotected on the vessel's weather deck fully exposed to the elements. With the already suggested agreement handwritten into the bill of lading, the master may believe that he is fully covered in all respects and that he has no further duty to the deck cargo. The fact remains that in spite of the agreement to carry on deck the goods remain in the care of the master whilst they are on board the ship. Under common law the master may be held liable for damage sustained by the goods if it could be proved that by the exercise of some reasonable avoiding action, which at the material time was realised but ignored, such damage could have been prevented.

Much of what has been said herein above is based upon situations which have arisen during the loading of steel cargoes. It should, however, be abundantly clear that application of the wording of the Hague or Hague-Visby Rules cannot be too categorically predicted. The application is greatly influenced by the customs and practices of the port and the interpretation in law of the particular territory. For obvious reasons it is recommended that steel products are not carried on the open weather deck, always under-deck.

For the text of the Hague Rules, the Hague-Visby Rules and the Hamburg Rules, see the Appendix.

MERCHANT IRON

This type of material is small in cross section when compared with structural steel, and shipped in lengths of 20 to 40 feet where angles, flats, beams and rounds are concerned. Bars of about 5 cm and below are usually 20 feet in length. Such goods are made up into bundles secured with wire or metal strapping bands.

The use of these goods is considerably diverse. The flats and angles may be used in the construction of racks, for instance, in stores, whereas rounds, flats, square bars and half rounds can be employed in the construction of railings and iron stairs. Rounds, octagonals and hexagons might serve in the manufacture of washers and nuts which screw onto the ends of bolts. These are but some of the many uses to which the goods are put.

This is one type of hot rolled material which, although in many instances it is stored uncovered prior to shipment, should be protected from developing too much rust. Experience has proved that these goods can be susceptible to claims through rust which has developed through contact with fresh water. Prior to shipment, if the goods cannot be stored under cover, they should at least be covered with tarpaulins in order to keep rust development to a minimum, and prevent an accumulation of free moisture within the bundles.

When compared with structural steel, merchant iron is, for obvious reasons, much more fragile and therefore must be handled with care. Projecting ends on bundles, where the ends are not flush with each other, are easily bent and damaged. The goods are also very prone to damage if the correct slings are not used. In stowing these goods, and especially when they are overstowed with other cargo, the possibility of crushing damage should be kept in mind.

MILD STEEL

Mild steel, often referred to as low carbon steel, is iron which has been strengthened by the introduction of 0.15 to 0.25% carbon. Relative to most other steels, it can be made cheaply and rapidly without impairing the main desirable features, which are strength and ductility. Compared with the manufacture of other steels, it is simple to fabricate and requires no added use of heat treatment.

This type of low carbon steel is in evidence all around us, being in the bodies of the cars which we drive, in the bridges which cross rivers, also in the frameworks of large buildings and the hulls of ships—to mention but a few outlets for this material. Mild steel forms about 85% of all steel manufactured.

Most of the steels comprising cargoes carried in seaborne trades are mild steels in the form of hot and cold rolled steel sheeting, steel plates, the basic plating for tin plate and galvanised steel sheeting and also structural steels and merchant iron.

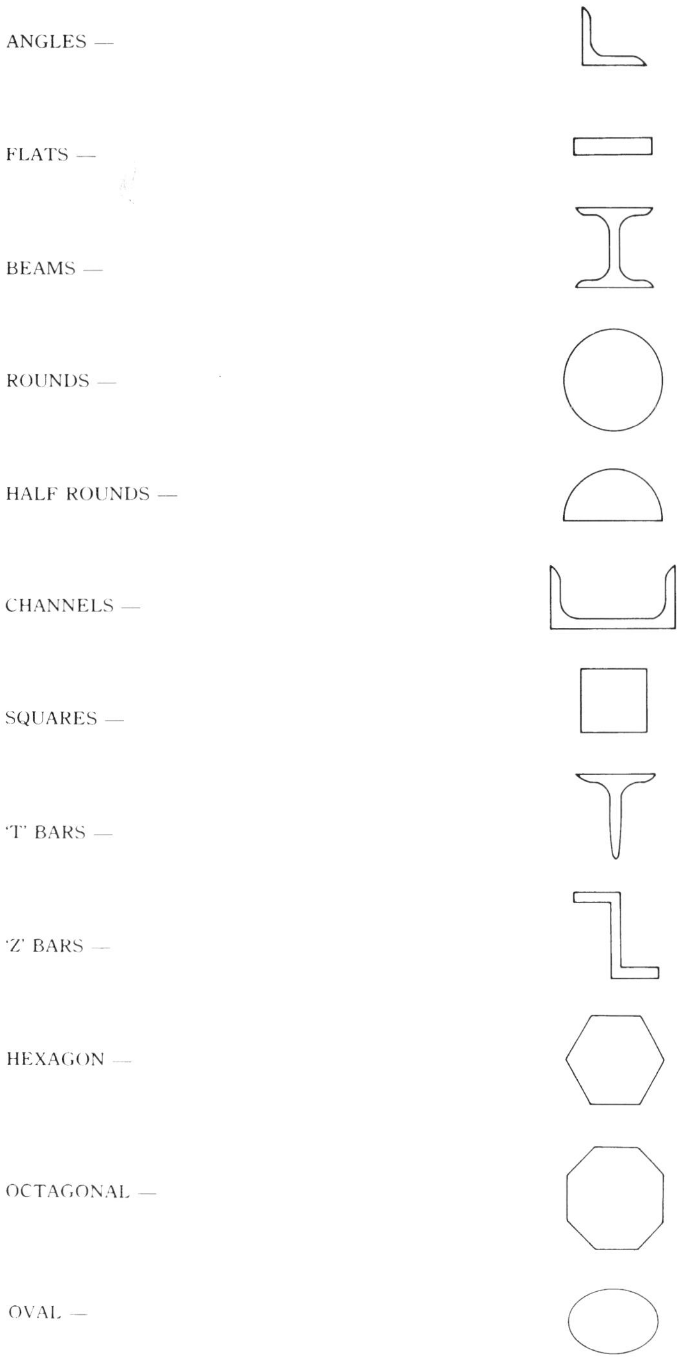

Fig. 25. Merchant iron profiles.

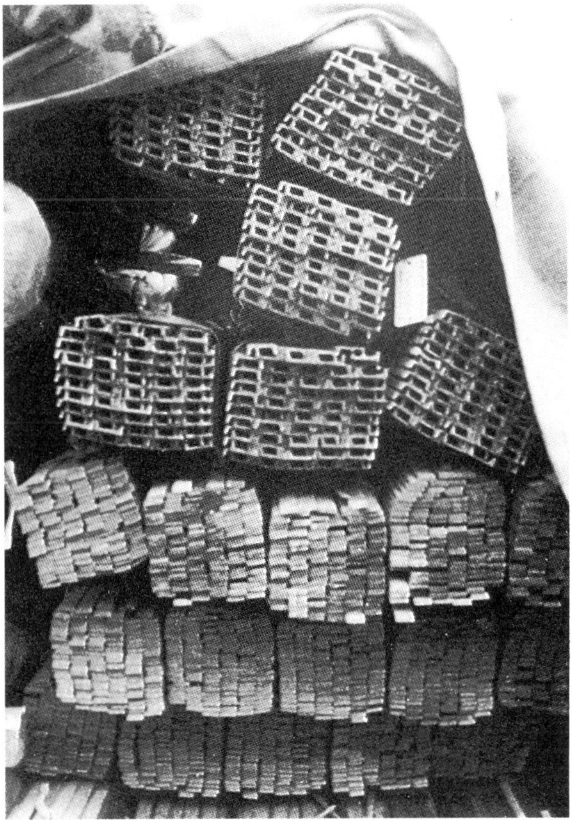

Fig. 26. Bundles of flats and furring channels.

Those steels which have carbon content in excess of 0.25% belong to the range of medium and high carbon steels where greater strength and hardness, at the expense of ductility, are required. Such items manufactured from these other ranges of steels are files, knives, chains, weapons, machinery parts, etc.

MILL DEFECTS

Hot Rolled Coils

Jagged edges, sometimes referred to as "Saw Tooth", are usually, as the above expression implies, a series of jagged edges having developed in the rolling process. Oscillations during rolling and winding can cause problems with decoiling and especially slitting.

Cold Rolled Coils

Lack of sufficient oiling allows rust to form on internal parts of the steel sheet away from the edges. Dents in the surface of the sheet (see CAMBER and LUDER'S

LINES). Reel creases, caused by the original processing machinery gripping the coil too tightly during winding.

Hot Rolled Wire Rod

Too much pressure applied by the rolls can cause the wire to become oval-shaped or out-of-round. Overfill or piping is hollowness, which causes the rod to break when being drawn. Fins, on one side or both, cause difficulties in processing. The above defects have to be cropped out of the wire.

MILL SCALE

During the manufacture of hot rolled steel products, which commences at temperatures in excess of 1,000°C, iron at the surface of the metal combines with oxygen in the ambient air to form basic iron oxide. Scale forms in a rather thick layer upon the surface of the material, and it is this scale which imparts to new steel a characteristic blueness in appearance.

The scale usually develops in three layers: FeO (Wustite), Fe_3O_4 (Magnetite) and Fe_2O_3 (Hematite). This is purely academic, but it is interesting to note that there are iron ores of the same name and chemical composition. This indicates that mill scale is the result of corrosion at elevated temperatures, when the iron in the steel combines with oxygen in the atmosphere so reconverting to it natural state—iron oxide.

Mill scale is not part of the material, it is only an attachment which will eventually become detached even if not removed before protective coatings are applied, which might in isolated instances be the case. Intact mill scale is protective to the steel surface beneath but it is very brittle. In any form of handling or manipulation it cracks and fragments so that the exposed areas quickly rust (see under Surface preparation before coating).

MINI-MILLS

This is a method of producing steel which bypasses the necessity for using blast furnaces, iron ore, coking ovens, etc. For obvious reasons in recent years this method of steel production has become very popular and consists of feeding scrap metal into an electric arc furnace; some mills have DRI facilities. Naturally, as more and more of these mills are set up around the world, the demand for and the price of scrap increase. At such a time, if costs of raw materials fall and integrated mills increase their productivity and efficiency, as has been the situation recently, mini-mill production may become less attractive.

NICKEL

Nickel is mined as an ore. It is metallic element which imparts to alloys special properties such as hardness, resistance to corrosion and greater strength. It has a wide application in industry and creates qualities in steel that permit a diversity of uses. Nickel is well known for its contribution to the manufacture of stainless steels.

PACKING

The packing of steel products was devised and visualised to suit the needs of factories, production mills and customers, probably without any thought being given to carriage by sea. The handling equipment in the mills and factories is designed to handle only the particular material on the premises, whereas the handling gear of stevedoring companies, also ship's gear, is designed to handle a diversity of merchandise. In mills, factories and steel traders' stores there is no great or pressing need to make tonnage when handling the goods, which is the reverse of the situation where stevedores are concerned. Further to this, during loading and discharging operations the goods are often exposed to inclement weather conditions. In consideration of these facts, it is not surprising that the packing provided for the protection of steel products carried by sea sometimes seems to fall short of satisfying the requirements demanded for an ocean voyage.

Where steel products are concerned, those unit lifts are wrapped which need extra protection against physical damage which might be contracted during handling. Such goods are naturally more sensitive, in many respects, than those which are shipped in an unwrapped condition. Apart from damage which might be sustained in handling, often referred to as mechanical damage, the packing is there to protect the material from contact with the atmosphere.

The packing of steel products carried in seaborne trades is mainly confined to cylindrical coils of strip steel sheeting and oblong packages containing steel plates. The actual material is in general confined to cold rolled steel sheeting, coated steel sheeting (galvanised, aluminised, painted), electrical steel, tin plate and hot rolled steel which is pickled and oiled. The diagrams (Fig. 27(a), (b), (c)) show the most popular types of packing which are in use and customarily have been used for decades. Variations on these types of packing were resorted to from time to time, by using plastic sheeting, fibre board, extra metal edge protectors, etc., as an additional precaution against the goods sustaining damage during sea transport. Nonetheless, in spite of what might be referred to as isolated endeavours to improve the packing, and provide additional protection to the material, the situation eventually reverts to the customary and time-honoured form of packing as indicated herein.

Packing costs money. It is not part of the goods but only there to protect them and must generally be regarded as an unwelcome, but unavoidable, necessity. Nevertheless, necessary it is, and in order to satisfy its function its requirements will be kept to a minimum. Present day packing could be regarded as a compromise between all factors involved, unwittingly arrived at through a process of trial and error in consideration of initial costs, market requirements, degree of protection offered by the packing and eventually its capabilities to withstand the rigours of an ocean voyage.

Goods are sold by weight which can, in some circumstances, include the weight of the packing. With such high density material, freight is more often than not

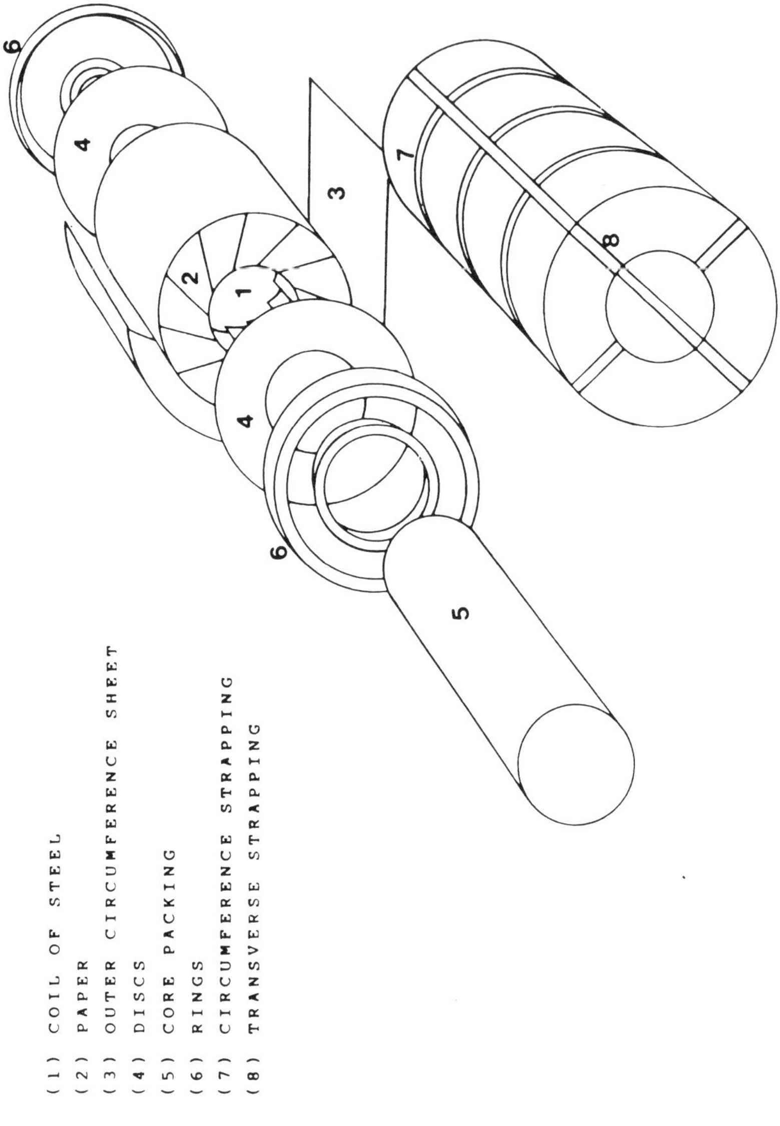

Fig. 27(a). Packing steel coils—method 1.

(1) COIL OF STEEL
(2) PAPER APPLIED
(3) DISCS PLACED ON ENDS
(4) OUTER CIRCUMFERENCES PACKING APPLIED
(5) CORE PACKING APPLIED
(6) SECURING STRAPS CIRCUMFERENCE
(7) TRANSVERSE STRAPPING

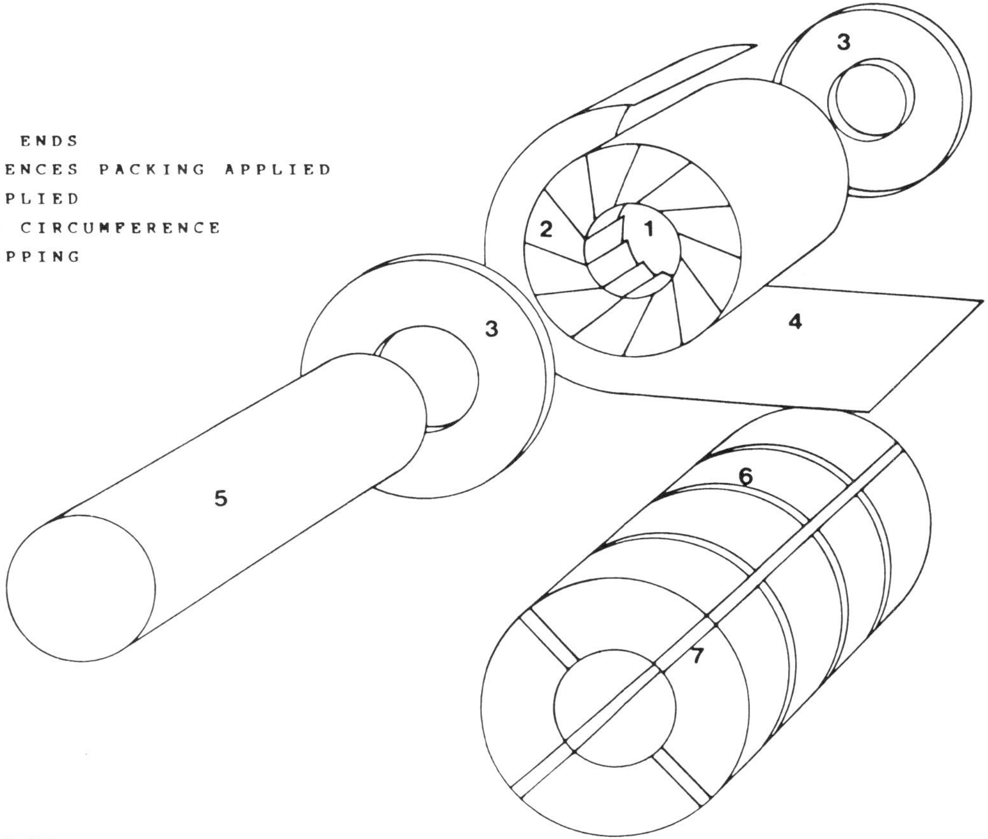

Fig. 27(b). Packing steel coils—method 2.

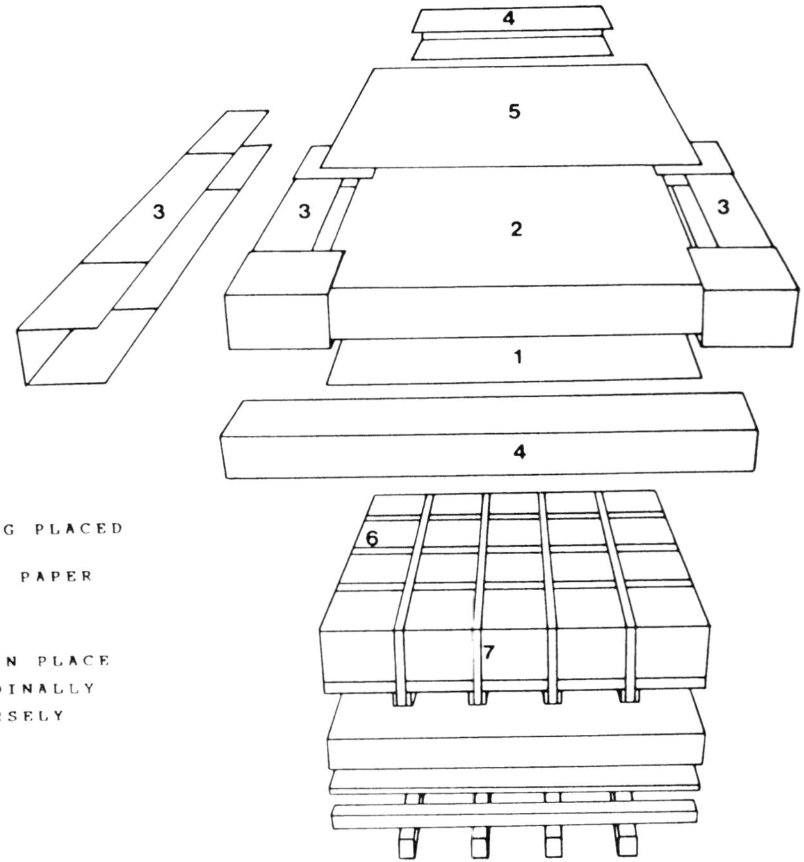

Fig. 27(c). Packing steel plates.

calculated on weight, therefore the weight of the packing must be kept to a minimum. As the diagrams show, the method of packing is basically similar for both coils and packages. The actual material is first wrapped in a single sheet of kraft paper, lined with an adhering film of plastic, the overlaps of which are not usually sealed. The unit is then fitted with an outer metal wrapper composed of waste steel sheets. Thereafter, flat metal strapping bands are applied around the circumference of the unit and transversely through the centre core.

Examining the two types of packing for steel coils it will be seen that one type has a metal disc fitted on the end, with protective rings in the form of angle irons surrounding the inner and outer circumference edges (see Fig. 27(a) and Plate 16). The purpose of this arrangement is twofold, one being to offer protection to the edges of the coil, the other to fit a plain disc on the end, not specifically manufactured to fit one size of coil, any shortage of diameter in the end disc being taken care of by the overlap of the protective rings.

In the case of the other type of coil packing (see Fig. 27(b) and Plate 15) the end disc is fabricated to fit the end of a specific size of coil; it cannot be used for any coil of another diameter. Apart from the fact that this type of packing is more expensive than the other, there is also less protection offered against edge damage occurring. Nevertheless, it has other redeeming features and is much more to be recommended for use with steel sheeting in coil carried by sea.

In the shipping industry it is often the practice to refer to steel plates of the approximate dimensions 1,000/2,000 mm, when stacked and secured together but unwrapped, as a "bundle". Similar material, which is wrapped, is distinguished from the unwrapped goods by calling it a "package". Therefore we have the terms "bundles of steel sheets" (unwrapped) and "packages of steel sheets" (wrapped).

The method of packing packages of steel sheets is shown in Figs. 15 and 27(c). The stack of steel sheets is first wrapped in kraft paper which is lined with a plastic film on one side—the overlaps of this paper may, or may not, be sealed. On the ends of the stack of sheets, and also to the long sides, are applied steel channel type protectors, whereafter a protective steel sheet is fitted to the top of the package as well as below. All is then placed upon a wooden pallet and secured as shown in Fig. 27(c) by the application of a number of flat metal strapping bands. It is of considerable importance that the packing is close fitting to the stack of sheets, also that the top sheet extends exactly to the perimeter edges of the package. The wood forming the pallet or skid must be sufficiently hard and strong to withstand a weight of much more than two tons and upwards without compressing or breaking. Further to this, the timber used should have a low moisture content. To avoid damage from lifting gear, the longitudinal outermost bearers must occupy a position on the extreme outside edges of the package.

Apart from the type of packing shown in Figs. 15 and 27, there is a rather large variation on this method with another arrangement, whereby the side packing and the top sheet is dispensed with and replaced with a specially manufactured steel hood. This is fitted over the top and down the sides of the stack of sheets, which are always wrapped in plasticised kraft paper, all being finally secured with metal strapping bands as already explained. The two types of packing might be considered, in their arrangement, i.e. paper packing, metal covers and strapping bands, analogous to the packing of the coils.

In consideration of the strength and quality of the materials employed, for them to be fully effective the edges of the steel sheeting comprising the material must be level, i.e. ends flush with each other—coils should be tightly wound.

Weak packing

The term must mean that the aspect of the packing would suggest that it is insufficiently robust to withstand the rigours of an ocean voyage. Suspicions would be raised by a preponderance of broken bands and apparent disjointing of the packing, together with perhaps some form of movement of the contents visible, partly visible or not visible at all. The terms “inadequacy” and “insufficiency” of packing could apply as they would in the case of lack of packing or packing of adequate strength but arranged or applied in such a manner as not to afford maximum protection to the actual goods.

A ship master is not considered to be an expert on merchandise, nor is he usually sufficiently knowledgeable to understand the technical intricacies of packing of goods. Nevertheless, there are certain signs of weak packaging apparent, as explained earlier, which would suggest to a ship master that the packing was weak, insufficient or inadequate. If, on the other hand, the master engages a surveyor to survey the cargo on his behalf, it is customary and generally accepted, and expected, that such a surveyor would be an expert who is fully conversant with all aspects of steel cargoes carried in seaborne trades. It is therefore considered that with such assistance available to him the master could no longer claim to be ignorant of the technical aspects of packaging of steel products offered for shipment.

In many instances when cargo leaves the packing area it is picked up and stacked somewhere on the premises of the mill or factory. Later, it is again picked up and placed on a road or rail wagon, transported to the docks/port area where it is offloaded into the quayside warehouse, picked up again and taken to the ship’s side and finally hoisted on board the ship where it is nearly always handled into the stow by a forklift truck. This means that from the time the packing is applied, the goods are handled six times before they finally come to rest stowed in the ship’s hold. If apparent weak packing is involved, it usually evidences itself by the time it is being loaded into the ship. Deficiencies of packing of a more technical nature would only be detectable by the expert, e.g. the surveyor.

Paper packing

The resistance to permeability of the wrapping paper by moisture vapour will depend upon temperature and relative humidity of the ambient air. In general it can be said that resistance is good. However, if the overlaps are not sealed with similar moisture-resistant paper there will be no protection against an upsurge of relative humidity beneath the paper brought about by increased vapour pressure of the ambient air. This type of packing offers no resistance to contact by free moisture. Some of the latest paper packing being advertised to wrap steel is what is referred to as being in the class of Volatile Corrosion Inhibitors (VCI). The VCI chemicals are incorporated into the inner film layer of the paper and diffuse into the atmosphere surrounding the coil, allegedly preventing the onset of corrosion.

From experience, it has been found that the paper cannot cope with excessive increases in relative humidity and cannot prevent damage if free moisture contacts the paper. The paper is used with steel coils, packages and wrapped wire rod.

Metal packing

The type of steel waster sheeting used varies in thickness from one factory to another. In general, it is expected that the thickness of sheet used would be about 0.5 mm.

Strapping bands

There are no international standards dictating what dimensions, tensile strength or metal composition should be used specifically for securing coils, packages or bundles of steel sheeting. We can, however, fall back upon normal practices in the industry and past experience, which dictate that the bands be of sufficient strength to withstand normal handling and the usual stresses imposed and normally encountered upon an ocean voyage. Of major importance is the securing of the ends of the bands, which should be crimp sealed. Particulars of one of the worldwide manufacturers of steel strapping, used in the steel industry for many years, with coils up to about 15 tons and packages, are as follows:

Dimensions of straps:	1.1/4 × 0.31 (32 × 0.8 mm)
Breaking strength:	2.472 kg/cm^2
Tensile strength:	+ / – 95 kg/cm^2
Elongation:	10%

The crimp seals of strapping bands can cause damage to the material when pressure is brought upon them, such as when the coils turn in stow or are pressed in stow. Correct positioning of the circumference seal is shown in Fig. 29(a) which is as near to the top of the coil as possible. Figure 29(b) indicates how the seal can damage the plating, which damage is transmitted through to many turns of plating. When securing coils the lashings should be applied in such a manner as to prevent turning in stow (see section on Stowage).

PALLETED CARGO

It is not unusual for various types of cargo in coil form to be placed and secured upon wooden pallets, e.g. tinplate, aluminum, galvanised steel, etc. It is essential that the moisture content of the pallet is not higher than 14%, and in order to achieve this quite often kiln-dried timber is used. If the moisture content of the pallet is too high, and there is no continuous circulation of air around the unit, the air adjacent to the pallet will take up moisture from the wood and will develop a high relative humidity. How far above the pallet this reconditioned air extends will depend on how high into the base of the coil the moist air penetrates. If the penetrated air has a dew point temperature higher than the temperature of the plating, condensation will take place directly upon the material, with ensuing damage developing.

Another cause of this damage developing concerned a pallet which was first

Damage sustained by a coil of aluminium which had been stowed upon a wooden pallet. It shows the effects of moisture penetration from the pallet.

Fig. 28. Palleted cargo.

covered with a plastic foam pad as a shield against the moisture in the pallet. The coil was enclosed in a plastic bag placed over the top of the coil, with the open end folded beneath the coil and the foam pad. The pallet was shipped in a container, the inside of which sweated profusely after it was landed at the port of destination. Sweat water falling onto the plastic cover/bag streamed down onto the foam pad, penetrated beneath the coil to the open end of the bag and caused the same type of damage. Had it been possible to terminate the ends of the bag beneath the foam pad, the damage could have been avoided. Therefore, in view of the above, the packing of coils shipped upon wooden pallets must be carefully considered. The case referred to was considered to be a case of "inadequacy of packing" (see Fig. 28).

PATTERNED PLATING

Such material is shipped either in unprotected coils or bundles. It is hot rolled material and should be considered in exactly the same way as hot rolled steel sheeting. The only difference is that the surface on one side of the plate, or both sides, has a "rhombus pattern" or is "stippled", sometimes referred to as "tear drop plating". Such goods are: floor plating, which can be seen in many public buildings, fire escapes, stairs, in factories as floor plating on catwalks built around machinery, etc. If used outside, the steel will be given a protective coating but inside buildings, where the relative humidity is never too high, no protective coating may be applied.

Fig. 29(a). Suitable positioning of crimp seals when coils in stow preferably on top.

Fig. 29(b). Damage caused by crimp seals to coil plating.

PHOSPHORUS

This is a non-metallic element, the presence of which is undesirable in steel. A too high phosphorus content leads to the development of a coarse grain and brittleness. To avoid poor quality material developing from this cause, the phosphorus content is generally kept at about 0.05%.

PIG IRON

This is the product of the blast furnace and is the molten iron extracted from iron ore which is tapped off at the base of the furnace. It is the basic raw material which is used to make steel and cast iron, and contains about 4% carbon, also small quantities of silicon, sulphur, phosphorus and manganese.

To produce one ton of pig iron it is necessary to process, in the blast furnace, something like two tons of iron ore, one half ton of limestone, one ton of coke plus many tons of air.

The name is derived from the fact that the molten iron is run off into channels which have branches, and in earlier days the outlay of this system, as viewed from above, could, by some stretch of the imagination, resemble new born piglets suckling the sow, hence the name pig iron.

Pig iron is in a suitable form for handling and storage purposes. It is not unusual for loose pigs of iron to be transported by sea, and in view of the fact that the goods have to be re-melted to the molten state for the purpose of refining, no harm of a significant nature can be sustained by the material during normal transport conditions.

PIPES AND TUBES

This type of material comes into two categories with regard to its manipulation and carriage by sea: (1) large diameter pipes, shipped in single pieces; (2) small diameter pipes and tubes, shipped in bundles.

Large diameter pipes (unprotected)

Such pipes are usually manufactured from low carbon mild steel which may, if the thickness of the plate forming the pipe is not too great, originate from a hot rolled steel coil, slit into strips (referred to as skelp) to form the width of the plate intended to be fabricated into a pipe. The long edges of the sheet, or plate, are brought together and welded to each other so forming a pipe. Such pipes usually have a visible weld running the full length of the pipe.

The type of pipe line mentioned above is manufactured in all lengths to order; 30–60 foot lengths are not unusual. Another method of manufacturing such pipes is by spiral welding when a flat strip of plate is wound to form a hollow spiral tube. Adjacent edges are welded together and the pipe is formed with a spiralling weld visible along its length.

These pipes are, in general, used for the conveyance of gas or water. They have to be welded together and in consequence of this the end edges of each individual pipe is bevelled (Fig. 30). The type of pipe referred to is the most common form of piping carried by sea. Full ship loads, plus a deck load, are not uncommon. The following are the types of defects which will provoke claims.

Damages to the bevelling on the end of a pipe may eventuate in the defective end having to be cut off and rebevelled. Such damage is usually in the form of a nick or score mark which has depth. Shippers often specify that any pre-shipment remarks, regarding this type of damage, should clearly state the depth of the defect. It is indicated that incisions in excess of 3 mm will result in reconditioning being necessary. On the other hand, minor incisions, score or chafe marks should always be considered as a defect which may or may not result in eventual loss to the end user of the goods.

Denting on the end of the pipe can result in the end being out of the round to such an extent that it cannot be welded to the end of another pipe. The length of pipe which is so affected may have to be cut off and rebevelled.

An appreciable dent on the body of the pipe may pull the ends off square to such an extent that the pipe is rendered useless and has to be scrapped.

These types of pipes are usually shipped without any protection against the development of rust or mechanical damage. As the goods are transported from the mill to the port of embarkation on uncovered rail wagons or in open lighters and they may also spend considerable time stacked uncovered in the open, it is not unusual for this type of piping to be rusty in appearance at the time of shipment.

Very careful handling is necessary to avoid damage. The method of slinging each pipe is to fit a hook over the area of the bevelling at each end, so enabling the pipe to be lifted by wires threaded through an eye on each hook. To avoid damage, the

Fig. 30. Bevel ended pipes.

inside of the hook, where contact is made with the pipe, must be lined with copper or fitted with a type of heavy duty plastic. These protective liners need to be regularly inspected for wear.

On board the ship, end and body damages to pipes have often been attributed to contact with end bulkhead stiffeners and contact with ship side frames. Such damages can be avoided by careful handling, proper dunnaging and good stowage.

Depending upon the thickness of the metal forming the pipe, it will usually be found that the permissible height of stowage will exceed the depth of hold on board the average bulk carrier. Nonetheless, confirmation of stacking limits from the shippers should be insisted upon. Exceeding the stacking limits may result in deformation of pipes in the lower parts of the stow.

Large Diameter pipes (protected)

Mild steel and cast iron pipes are protected by the application of special coatings which may be fusion bonded epoxy coatings, polyethylene coating, neoprene, bituminous materials, coal-tar and asphalt mastics. These coatings are applied on the understanding that they will remain intact. During manipulation, owing to the cumbersome nature of large pipes and their weight, the protective coatings often sustain some degree of damage. In many instances, shippers have personnel standing by at time of loading into the ship, for the purpose of effecting repairs to the damaged areas of the coating.

The most difficult part of the transport operation is carriage by sea, owing to the pipes having to be manipulated into and out of the very confined spaces within a ship's hold. If unsuccessful pre-shipment handling of the pipes is so apparent as to be evidenced by the fact that protective coating repairs have to be executed alongside the ship's rail, it is logical to predict that the ship will likewise deliver pipes with defective coatings.

Stowage

The aim, in the stowage of single pipes, should be to keep the stowage level and properly orientated as shown in Fig. 31. The stowage shown in this diagram involves large diameter pipes, and it will be seen that large quantities of timber have to be used as packing between the outer pipes and the ship's side. The reason for this is fairly obvious, as without such an arrangement the stowage could not be kept level and such pipes cannot be haphazardly dumped into the wings.

The smaller the diameter of the pipes the less need there will be to use wooden packing and situations will arise whereby it can be dispensed with entirely. Some packing or the building of wooden stools will in most instances be called for in those holds where the ship's sides are less irregular in shape, such as the case might be in holds one and five in a five-hold ship. It will be appreciated that, in compartments where the ship's sides narrow down at one end the pipe stowage cannot be splayed out—all pipes must be kept strictly in a fore and aft direction (see Figs. 32–33).

At the commencement of loading the usual procedure is to lay double flat (6″/1″) dunnage athwartships upon the vessel's tank top plating. This dunnage will be spaced at intervals, measured in a fore and aft direction of about 3 m. The laying of the dunnage will terminate at about 1 m from the end of the pipe stowage. This

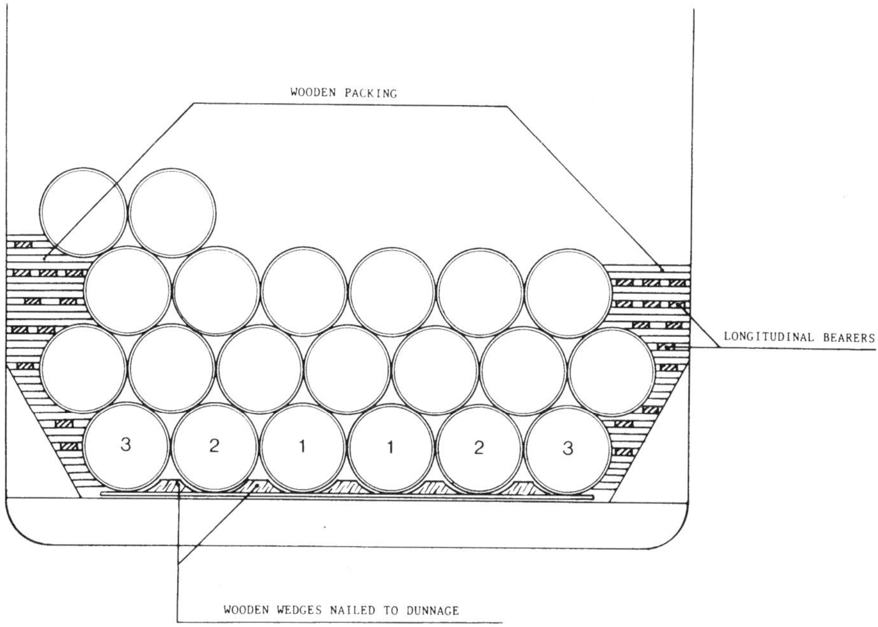

Fig. 31. Stowage of large diameter pipes. Ship's hold in cross section and partly loaded.

dunnaging will also be extended up the slope of the lower wing tank side plating. In those instances where the tank top plating is very level, not undulating or excessively dented, single flat dunnage should be used. Flat dunnage should also be placed between the pipes and other component parts of the vessel's structure.

It is imperative that the pipe stowage is kept strictly in a fore and aft direction. Pipes should not be stowed athwartships. Very serious deformation damage can result from stowing pipes in such a manner that they cross (see Fig. 35).

The above covers general directions for loading single pipes which are uncoated. Where coated pipes are concerned special instructions may be issued with regard to the stowage in order to avoid damage being sustained by special coatings. This can be best demonstrated by the following, which is relevant to a full shipload of specially coated pipes.

The shipment consisted of single pipes of which the bell end measured 137 cm and the small end 127 cm. There were also double lifts (one pipe inserted inside a larger pipe). Each lift varied from 3.2. to 11.38 metric tons. Shippers' instructions, received via the charterers, were as follows:

—Cast steel pipes—cement lined and seal coated.
—Outer coating zinc cabolac epoxy.
—There is to be no contact with dunnage or with the vessel's structure. Polypack will be used to avoid this.
—Lifting hooks must be protected.
—Each lift to be slung with a bridle so that lifting wires are near to the vertical.
—Each row of pipes must be protected with polypack.

Fig. 32. Cast piples—inner cement coated.

Fig. 33. Special shipment of coated pipes.

Fig. 34. Packing between pipes and hopper tank.

—Lashing wires, where used, not to come into contact with the pipes. Sections of rubber hose will be supplied for this.
—Coating damage will be repaired on the quay before shipment.
—6/13 cm wooden dunnage to be laid athwartships upon the vessel's tank top—seven lines.
—Between each pipe a double wedge is to be fitted.

Large stools or wooden packing, as shown in Fig. 31, also in Fig. 34, were built between the ship's side and the pipe stowage.

Stowage on deck

Pipes are often stowed on the deck of the ship. An ideal type of vessel for this is one specially designed to carry timber cargoes. These ships, strengthened to carry logs and timber, have provisions to fit steel uprights at intervals of about 3 m, ranged along the outer extremities of the weather deck to port and to starboard. On board such a vessel the pipes can be stacked, fore and aft, right across the ship, in a solid block between the uprights. In attempting such a stowage on board an ordinary

Fig. 35. An example of how not to stow steel pipes.

bulk carrier or conventional cargo carrier, attention has to be paid to the strength of the bulwarks, as there are cases on record where bulwarks have carried away under the weight. Apart from the support provided by the uprights, the block of pipes has to be lashed in a suitable manner by which the stack will not be afforded the opportunity to roll beneath the lashings. One of the popular methods of stowing pipes on deck is to build the stack in the form of a pyramid. Figure 36 shows such a stowage and a recommended method of securing for stowage upon the ship's hatches. Such pyramid type of stowage can be extended out to the ship's sides (see Figs. 36 and 37).

Small diameter pipes

Welded pipes and tubes

A strip is cut from a hot rolled coil to the width required to form the walls of the pipe, the edges are planed to a "V" shape after which the strip is curled to the form of the pipe as it passes through rolls. This brings together the two prepared edges, which are welded.

After fairing of the welding and clearing the inside of the pipe by forcing a plug through it, the pipe is passed through an annealing furnace, thereafter being stretch reduced to size. The pipes are then cut to length and put through an ultrasonic test. Depending on customers' requirements the pipes may be given a protective coating, left with plain ends or threaded ends, or fitted with a sleeve, as the case may be.

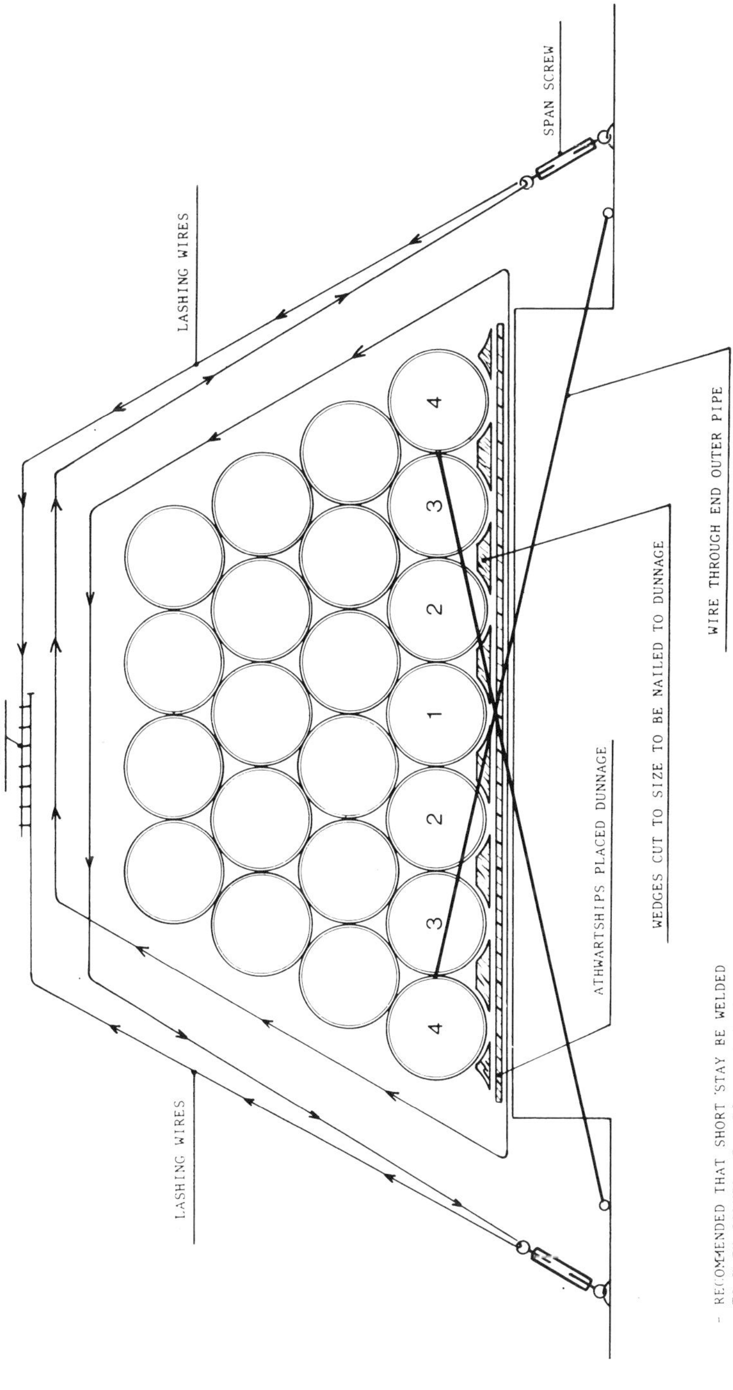

- RECOMMENDED THAT SHORT STAY BE WELDED TO EACH CORNER OF HATCH IN WAY OF OUTER LOWEST PIPE EACH SIDE.
- NUMBER OF LASHINGS IN RELATION TO WEIGHT OF PIPE STACK.

Fig. 36. Pipes—pyramid type stowage on hatches.

Fig. 37. Deck cargo of pipes. Part stowage of large white pipes eventually lost overboard.

Seamless pipes

Sometimes referred to as "solid drawn pipes", this is a type of piping or tubing which is preferred for all purposes where internal pressure is high and service severe. The seamless pipe is achieved in three stages as described below.

A round steel billet is removed from the furnace and the end is pierced in a rotary piercing mill, Thereafter, a mandrel is forced through the billet to transform it into a hollow cylinder. The cylinder is then put through a series of rollers so that it is elongated into a shell. There then takes place a final rolling, after reheating, to form the required size of tube. Again, depending on customers' requirements, the pipes may be given a protective coating, left with plain ends or threaded ends as the case may be.

These small diameter pipes and tubes are used in the manufacture of bicycle frames, furniture for offices and homes, gardening tools and equipment, vehicle shock absorbers and exhaust pipes, boiler and condenser tubes, frameworks for the roofs of sheds, to mention but a few. Popular sizes range from 8 mm outside diameter, wall thickness 1 mm, weight per running metre 0.173 kg to 105 mm outside diameter, wall thickness 2.5 mm, weight per running metre 6.316 kg.

Small diameter pipes and/or tubes are shipped in unwrapped/unprotected bundles and the quality of the bundling is important. The pipes/tubes have to be symmetrically orientated in the bundles and rigidly secured, if damage is to be avoided (see Figs. 38 and 39). If not properly bundled slackness will develop, the stow will become unstable and pipes will be crushed and dented. In handling, pipe ends commence to protrude from the ends of the slack bundles and become bent in handling and in stowing, which is not facilitated by the protruding pipes. The bundles should not be allowed to become wet from rain, cargo sweat or ship sweat. The moisture can be held in suspension within the bundles where individual pipes butt up against each other and heavy rusting and pitting may develop which can be damaging to thin walled pipes. Galvanised pipes, whether chromated or not, should never be allowed to come into contact with free moisture, as a line of heavy white rust will develop inside the bundles where adjacent pipes rest against one another. Bundles are sometimes wrapped in plastic—however, the same precautions apply.

Hollow steel bars

This is a type of pipe created by forcing a hole through the centre of a round bar in much the same way as described for solid drawn pipes. Supplied usually in lengths of up to 10 m, they are used for structural and machinery applications and also in the oil industry. Sizes can be from about 32 mm outer diameter, bore 20 mm, weight 3.84 kg per running metre to about 200 mm outer diameter, bore 160 mm, weight 92.25 kg running metre. These goods are, more often than not, shipped loose but sometimes in small unprotected bundles. They are considerably stronger than the average pipe of comparable size; nevertheless, they can sustain damage in handling as explained for pipes and tubes.

Small diameter pipes, which might be distinguished by the fact that they are not shipped loose but in bundles, will usually have a protective coating applied. They may be greased for protection, varnished, painted or galvanised. Owing to their thin wall structure they will not withstand too much rust development before they become defective. Further to this, rainwater, usually with a high pH in industrial areas, remains as free moisture within the bundles and can cause serious damage. Also, rust formation on small pipes can provoke claims in respect of commercial depreciation. In view of the above, small diameter pipes should always be kept dry. For carriage by sea galvanised tubes should be given a passivating dip, usually in a di-chromate solution.

With regard to the stowage of small diameter pipes in bundles, to avoid damage they should not be overstowed. If this cannot be avoided for some reason or other, it should be remembered that they will not withstand too much weight. Where such goods form the brow of a stow, the top two tiers of bundles can be block lashed together with wires as a precaution against movement.

The packaging of small diameter pipes requires special consideration and it is recommended that when possible these pipes are bundled in such a manner that their ends form a hexagonal shape as shown in Fig. 38, being well strapped up with tensioned flat metal strapping bands. Packed in this manner they are, in most instances, capable of withstanding the rigours imposed upon them during a sea voyage. The pipes shown in Fig. 39 are poorly bundled and strapped with wires which are hand tightened; such bundling invariably promotes bending, and denting promotes the disintegration of bundles, which might otherwise have been avoided.

PLATES

Mild hot rolled steel plates of various sizes are used extensively in tank construction, ship building and in the fabrication industry. The larger sizes are shipped in single pieces and in most instances stored outside, unprotected against the elements, prior to shipment. Therefore, they are generally loaded in a rusty condition (see Plate 17).

Stacks of these plates when viewed from the side often look wavy owing to wrong placing of dunnage. Prior to shipment the stacks are not under pressure, as they might be in the hold of a ship, so that the apparent deformations will probably only be temporary. Sometimes in handling, a plate will develop a transverse kink which must be considered as damage as the defect is permanent and will require a re-rolling operation to reflatten it. In handling large plates special gear is required in the form of hooks coupled to chains as shown in Fig. 40.

Long plates are usually stowed athwartships in the hold of the ship, but care must be taken to use plenty of suitable dunnage placed in line through the stow; this being particularly required when other cargo is to be loaded on top. One incident involved a coil stowage on top of plates, which is not to be recommended—through lack of 'and improper placing of' dunnage, the entire shipment of plates was deformed to a serious extent.

Small dimensional hot rolled steel plates are shipped in unprotected bundles which are secured with metal strapping bands. The sizes of these bundles vary, but dimensions of plates around 1 to 2 m by 1 m are not uncommon. Such goods usually

Fig. 38. Small diameter pipes: acceptable bundling for sea transport.

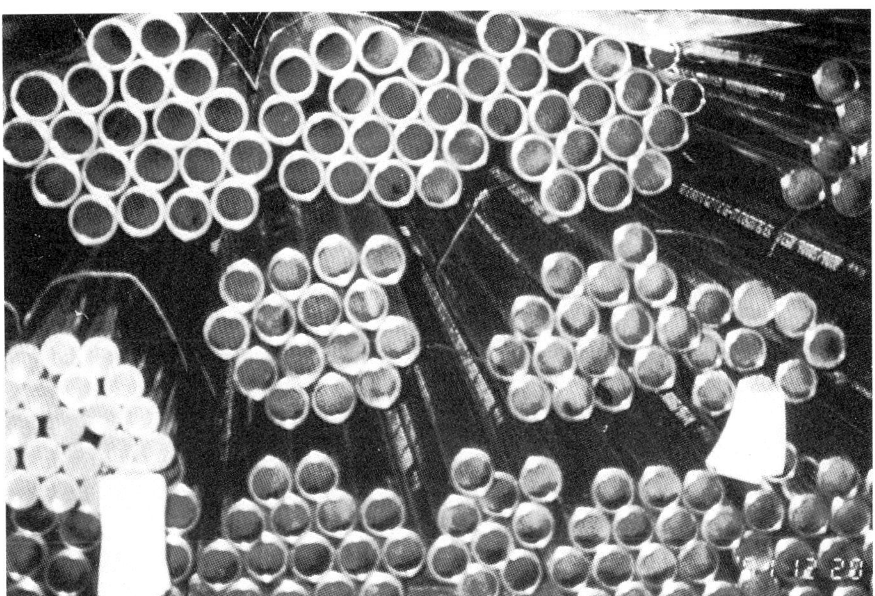

Fig. 39. Slack bundling of pipes leads to damage.

Fig. 40. Special gear for handling plates.

appear to be in a rusty condition when shipped (see Figs. 24, 40 and 41 for the type of material in question).

Care must be taken in slinging the type of long plates shown in Fig. 40. The type of clamp often used is a single plated clamp with a vertical extension to the ring part attached to the lifting chains. Further to this, they can be of a home-made variety manufactured by the stevedores. The use of such clamps must be avoided as they are responsible for accidents, involving loss of the load, which bends and vibrates if the lifting gear is not complete. Figure 40 is a good example of how the goods should be slung; note that the plates in the lift are level and not bending. Further to this, the transverse section of the chains, between the clamps, is fitted with a distancing chain, at each end of the lift, from the centre of the transverse chain, longitudinally, to the end of the plates, where it is hooked over the end. In this way the side clamps are always adjusted to the right position along the sides of the lift and the plates are kept level. In Fig. 40 the type of double clamp recommended can be observed; single plate clamps should never be used.

PRIME MATERIAL

Such material would be on specification as per the standards applied with regard to sizes, tolerances and surface condition, i.e. the quality of the actual surface. "Prime condition" means that the goods are, in appearance, as they left the production line, free of any rust and in conformity with the contract between buyer and seller. This contract may, in the case of hot rolled material, permit some surface rust (see Plate 13).

Fig. 41. Bundles of hot rolled steel plates.

QUENCHING

From a temperature just above the upper critical level, the steel is rapidly cooled by subjecting it to complete immersion in either water, oil or iced brine. The change from the austenite face-centred lattice to the body-centred lattice structure is depressed and all carbon remains in solid solution. The type of steel resulting from this treatment involves an unstable distorted face-centre lattice structure. Such material is referred to as "martensite", which is extremely brittle and hard. In order to improve the quality of the material, which in its stated condition is too brittle, it is reheated to just below the lower critical level and then allowed to cool. This treatment is called "tempering" which relieves the stresses and improves the structure of the material.

REINFORCING BARS

Long round bars of steel, usually 20 or 40 feet in length, with a diameter of up to approximately 20 mm. They are manufactured with a deformed surface, the patterns of which are various. Such goods are shipped in unprotected bundles, which are usually secured with pieces of wire rod wrapped round the bundle at intervals and with the ends twisted together. The number of pieces per bundle will depend on the diameter of the bars, but some idea can be obtained from the following:

Diameter of bar	No of bundles	Length of bar	Pieces in bundle	Theoretical weight
10 mm	835	8 metres	420	1,571.126 metric tons
13 mm	822	8 metres	240	1,570.349 metric tons

When the goods leave the production line, they are, as is all hot rolled material, covered with a layer of mill scale and blue-coloured in appearance. The mill scale is extremely brittle, and is not part of the material but only a temporary attachment, a large proportion of which will eventually become detached and fall off. As the bundles are not rigid, every time they are handled they bend, and the mill scale fractures and breaks off in places—usually to an extent commensurate with the size of the bar, i.e. length and cross-sectional area. In those places where the mill scale becomes detached, thus exposing the basic metal to the atmosphere, a film of rust rapidly forms.

Such material, more often than not, is stored in the open prior to shipment and exposed to the elements, and as a result of this the goods when shipped are invariably rusty or partly rusty in appearance. Further to this, prior to shipment the bundles may appear dry but once hoisted to be transported to the ship's side water may be seen pouring out of them.

Reinforcing bars are used extensively in the construction industry for the reinforcement of concrete. They can be seen lying around building sites for lengthy periods before being buried in concrete, where a satisfactory bond between the bar and the concrete may not be achieved if there is too much mill scale remaining on the surface of the bar or bars. Usually mill scale presents no problem, since with the scale being lost in handling and a proportion lifted off by rust, the rusted portion of the bar is invariably greater than that with the mill scale still attached. Not only would it be unnecessary but it would also not be a viable proposition to subject such material to a process of surface preparation before use; loose scale is usually removed by wire brushing. Even though a liberal amount of rust is acceptable, pitting, loose rust scale and/or voluminous corrosion of the material may qualify as damage. Contamination by grease and powders can provoke claims. Various international standards apply but requirements are similar to what is stated in British Standard BS 8110 as follows:

> "7.4 Surface Condition
> Reinforcement should not be surrounded by concrete unless it is free from mud, oil, paint, retarders, loose rust, loose mill scale, snow, ice, grease or any other substance which can be shown to affect adversely the steel or concrete chemically, or reduce the bond. Normal handling prior to embedment in the concrete is usually sufficient for the removal of loose rust and scale from reinforcement."

As the bundles are not rigid, are long compared with their cross-sectional area, loosely secured and unwrapped and the actual bars are rather thin, there is a tendency for bars to become bent in handling, which also can result in claims.

It will sometimes be seen that a consignment of re-bars is with "CARES" approval. The "CARES" organisation describe themselves as follows:

> "Cares is an independent, non-profit making certification body which was established in 1983 to provide the construction industry with confidence to

users, purchasers and specifiers of reinforcing and prestressing steels through a regime of regulation, testing and inspection. It operates for the benefit of the construction industry offering certification schemes to companies that produce materials, components or offer services to the reinforced concrete industry. Clients can specify 'CARES' approved companies and products with confidence that they will comply with the relevant product standards and without the need for verification testing by the purchaser or contractor."

In actual fact "CARES" acts in the same way and for the same purpose, as a quality control company (see pages 178–179). "CARES" approved material is usually embossed with their marks as proof of their approval of the product; this fact may be mentioned on the face of the Bill of lading. As already explained, "mill scale" is an attachment to the bars, not part of the material; it partly fragments in handling and becomes detached. Further to this, rusting of the bar assists in lifting the "mill scale" and together they usually render the bars sufficiently free from "mill scale" to be directly buried in concrete. Usually, a bar can be used, as it is, with 30% of the mill scale intact. The strength and performance of reinforced concrete depends on a good bond between the steel and the concrete. This can only be achieved if the steel is in good condition and free from contamination by grease, oil, mud, excessive rust (especially if it is flaking rust) and loose "mill scale". When being loaded into a seagoing vessel during warm weather grease from the wire sheaves at the head of the crane jibs can fall onto the cargo, as can also remnants of previous cargoes.

On the building site there is much discussion regarding rust and mill scale, but in most instances wire brushing of the bars, to the engineer's satisfaction, is sufficient to solve these problems. As far as sea transport is concerned, contamination with sea water, usually owing to leakage of transverse hatch joints, results in claims being entered against the vessel. The fact is that when surveyors test with silver nitrate, they often refer to a positive reaction as proof that the goods are in a damaged condition. Such an off-the-cuff description is unacceptable, unless it can be determined that the bars are heavily corroded with flaking rust scales, or pitted to such a degree that the tensile strength of the bar is affected, which can only be established by laboratory analysis. In order to prevent the chloride contamination from damaging the steel it will have to be removed by the use of fresh water high-pressure hoses, whereby the bundles can be thoroughly doused. Water from the city mains cannot be used as it normally is contaminated with fluoride. The only other viable proposition is for the goods to be heavily rained upon.

Seawater from leaking hatches usually causes prominent localised rusty strips across the bundles. In such circumstances, if damage is involved, it is very much confined and only affects a small percentage area of bars in each bundle. Bending and cutting of the bars is sometimes carried out on-site, but most suppliers to building contractors perform cutting and bending services. Various forms of bending and cutting are required, as is cutting to various lengths. Consequently, there is always the opportunity to manipulate cutting and bending, with the associated cuttings put to scrap, containing those small areas of each bar that are chloride contaminated. If this can be done no loss through chloride contamination will be caused.

Pre-stressing wire for concrete, see WIRE RODS—FOR PRE-STRESSING.

RESIDUES OF PREVIOUS CARGO

Before loading steel products, it is absolutely essential that all rests of previous cargo are removed from the cargo compartments. Dust from most of the commodities carried by ships can be damaging to steel surfaces on account of the fact that in the event of the steel being wet or sustaining wetness during the voyage, a situation which need not necessarily result in damage to the steel cargo constitutes a potential danger, because dust harbours moisture and does not readily dry out. Pits in the surface of the material can develop. Most of the bulk chemicals, e.g. fertilisers, absorb moisture from the atmosphere and these types of deposits can be particularly damaging.

Coal with a high sulphur content, when wet, can be extremely damaging to steel surfaces. One case in particular attracted much attention when wet coal was loaded into a ship. It was claimed that the sulphur content of the coal caused pitting of the ship's tank-top plating up to depths of 20 mm, when the thickness of the plating itself was only 24 mm.

Sulphur is another commodity which, whether in a wet or apparently dry condition, can devastate steel surfaces. In a dry condition, the sulphur dust permeates behind rust scale where it remains unnoticed. In due course, it sets up strong corrosive cells which degrade the metal. When sulphur is wet the situation is even worse, and this was demonstrated in the case of a vessel that loaded damp sulphur, which was so aggressive to the metalwork of the ship during a single voyage, that deterioration of the tank-tops resulted in them having to be renewed.

There is on record the case of a vessel which, prior to commencement of loading steel, carried a cargo of phosphate rock, sulphur, muriate of potash, ammonium sulphate and cement in bulk. The vessel was not properly cleaned out and large amounts of dust from deposits lodged on projecting parts of the ship's structure, and also considerable amounts of dust which were left behind, contacted the cargo of hot rolled steel sheeting. It rained during the course of loading, and failure to close the hatches in sufficient time resulted in the cargo being wet when the vessel sailed. The ship was equipped only with low level natural-draft ventilators, the tops of which terminated at the same height. A considerable amount of sweating of the ship's metalwork and the cargo occurred during the course of the voyage, partly as a result of the additional moisture which entered the cargo compartments when the cargo was loaded. The consignees entered a massive claim for rust and pitting damage to the steel, which was eventually attributed to contact with rests of cargo carried on the previous voyage. The offending materials involved turned the otherwise harmless fresh water into a powerful electrolyte and the resulting attack upon the steel was very aggressive.

In view of what has been stated above, it will undoubtedly be appreciated that more or less all rests of previous cargo, irrespective of their nature, can in certain circumstances cause damage to steel surfaces.

RIMMED STEEL

When a low carbon steel is cast into ingot form, deoxidation is controlled in such a manner as to produce an outer area at the surface of the ingot which is almost free from carbon or impurities. The outer area of the skin of the ingot takes on a smooth clean outer surface and there is no pipe formation in the top of the ingot. Rimmed steels is non-deoxidised, and is used when a fine surface finish is required, such as is the case in the manufacture of cold rolled steel sheeting.

Low carbon rimming steels (0.07) to 0.15% carbon content) possess extremely good pressing and deep-drawing qualities. For this reason they are extensively employed in the production of wire rods, nails, rivets, re-bars, cold rolled strip plating for car bodies, tin plate tubes and pipes, etc.

ROLLS IN TANDEM

Series of rolling mills are arranged in line, each set of rolls being referred to as a stand. The rolls in each stand are usually placed in an order which is referred to as four high and this means that the item of steel being rolled passes through the centre, two rolls above and two below. The purpose of the upper and lower rolls—back up rolls—is to exert sufficient pressure upon the two centre rolls, which are in effect the work rolls, to prevent excessive distortion or bending of these two centre rolls.

ROUGHING TRAIN

In the early stages of breaking down the cross-sectional area of semi-finished products, such as billets, blooms and slabs, to prepare them for further rolling in other mills and turn them into finished products, they are passed through a rolling unit known as a roughing mill. More than one set of rolls is involved, usually four strands, four high—four rolls per stand. This is the first stage in the process of hot rolling.

ROUNDS

Round bars are shipped in bundles which are usually unprotected and secured with wire bands or flat metal straps. They are, in most instances, stored outside in the open, unprotected from inclement weather conditions prior to shipment. Because of this, rust development is apparent before and during loading. The diameter of the individual bars varies considerably from one shipment to another.

These goods have a wide range of uses. They can be shipped as round bars (reinforcing bars) for structural work, in billet and bloom form for the eventual fabrication of seamless pipes or even for the manufacture of shafting, to mention but a few end uses. Large heavy round bars may be shipped in loose condition. In bundle form the tightness of the securing of individual bundles is of importance with regard to the loading operation, stowage and eventual delivery of the goods still in the form of bundles (see Fig. 42).

Fig. 42. Rounds—reinforcing bars in wired bundles.

Bundles which are not even-ended, individual bars projecting, are susceptible to bending damage and loosely secured bundles can present problems in stowage and in handling.

SALE OF DAMAGED STEEL

Occasions do arise when damaged, or even undamaged, steel has to be offered for sale. In such circumstances, the surveyor responsible for selling the material will usually make out a sales circular. This circular will give particulars of the goods to be sold, the conditions of sale and other particulars pertinent to the sale. Care must be taken in the wording of the circular and the presentation of the information provided in order that all will proceed smoothly. Details of the damage need not be mentioned, as any prospective buyer will have the opportunity to examine the cargo, thereby deciding for himself as to its condition, extent of the damage and further uses in which it can be employed. It is of great importance that permission to sell is obtained from the owner of the goods in writing. The following is a form of circular sales letter which has been successfully used over many years.

SALE OF DAMAGED GOODS

Goods in transit—Origin (give country of origin)—Stored at (give address of place of storage)—Price offered to be per kg./per kg. ton, per lot based on the advertised weight as mentioned hereunder:

Description of the Goods: Type of material for sale, e.g. steel coils, beams, etc. (full and accurate description necessary).
Quantity—Number of pieces.
Dimensions (millimetres).
Weight (metric kgs. or kg. tons).
Standard employed, indicating specific grade of steel involved.

The goods can be inspected between the hours of 0800/1600 hours from Monday to Friday, both days inclusive, upon production of this circular at the place of storage as above mentioned where (give name) should be contacted.

Without prejudice as to any unforeseen or non apparent adverse developments affecting the state, condition, or quality of the above mentioned goods, we shall proceed to dispose/sell the damaged material, for whoever it may concern, by public sale and by submission of sealed offers to be opened in our office on 26th —— at —— hours.

The offers should be submitted as requested and in accordance with our general conditions, which will be recognised and accepted by the eventual purchaser of the goods, i.e. the goods are accepted in the state/condition in which they are found "as they are and where they are" without any written reserves whatsoever, such as results of analysis of samples, etc.

All information, including technical details, concerning the goods is provided in good faith but without responsibility as to the correctness of such information. Any costs incurred by any potential or eventual purchaser of the goods in checking any details advertised in this circular will not, in any event, be for the account of the seller/sellers. On completion of the sale no complaints whatsoever are acceptable owing to the fact that the material was always available for inspection prior to the sale taking place.

Any offers, under sealed envelope, must be in our possession not later than 26th —— 1999 at —— hrs. The envelope should be addressed to us mentioning "Sale of Damaged Cargo. Ref. No. —— ." The eventual successful applicant for the goods will effect payment to us by cheque made out in the name of (the owner of the goods who gave permission to sell), in our office, without any deductions whatsoever, and effect payment on or before 27th —— at —— hours.

The purchaser of the goods must take reception of the material not later than 28th —— 1999 by —— hours. Any customs formalities, import licenses, clearances, etc., will be attended to by the purchaser and he will also accept the responsibility for any necessary or imposed taxes, levies, tariffs, etc., of whatever nature, for his account. After the above mentioned removal time and date, the goods remain on the quay (or in store as the case may be) for the account of the buyer who thereafter accepts all risks and costs.

The sellers reserve the right to sell all or part of the goods to the highest bidder, or at any time to spontaneously withdraw goods from sale, without giving reason and without being held responsible for this or any related action taken.

Surveyors who dispose of damaged cargo by sale must be careful to ensure that the market is properly tested. If it can be proved that an insufficient number of prospective buyers were approached, surveyors stand to be criticised for not pursuing the sale vigorously enough in order to mitigate the loss. Where large amounts of damaged cargo are involved foreign dealers should be given an "opportunity to bid". Often, where various different types of cargo are concerned, e.g. wire rod, coils etc. (and even with one large amount of cargo of the same type, e.g. Cold rolled steel coils), the best results are achieved by offering the cargo for sale in smaller blocks and/or as per type of cargo with the emphasis towards smaller blocks of cargo.

SALINITY OF SEAWATER

The chemical elements in seawater originate mainly from the earth's crust and find their way into the seas and oceans of the world from rivers, submarine volcanic activity, glacial break up and from the earth's atmosphere. It can in fact be said that seawater is a dilute solution of almost everything, the main constituents being present in nearly constant quantities in all seawater everywhere.

The salinity of seawater can be defined as the total amount of solid material in grams in 1 kg of seawater when all the carbonate has been converted to oxide, the bromide and iodine replaced by chloride, and all organic matter completely oxidised. In simple terms salinity represents the total amount of dissolved substances in seawater. The total amount of solid material in seawater is as follows:

Substance	Symbol	Grams per kg	Percentage
Chloride	Cl^-	19.353	55.0724
Sodium	Na^+	10.76	30.6195
Sulfate	SO_4^{-2}	2.712	7.7175
Magnesium	Mg^{+2}	1.294	3.6823
Calcium	Ca^{+2}	0.413	1.1753
Potassium	K^+	0.387	1.1013
Bicarbonate	$HCO3-CO2^{-2}$	0.142	0.4041
Bromide	Br^-	0.067	0.1907
Strontium	St^{+2}	0.008	0.0228
Boron	B(OH)3 B(OH)4	0.004	0.0114
Fluoride	F^-	0.001	0.0027
		35.141	100.0000

Marine and cargo surveyors all over the world, in order to determine the origin of rust on steel products use a solution of silver nitrate. A quantity of this solution applied to a rusty surface will turn milky white if chlorides are involved, when the pattern of rusting on the cargo corresponds with possible, or obvious, sources of leakage of water through the vessel's structure into the ship's hold; in most instances this is considered sufficient evidence that the cause of the damage is contact with seawater.

A qualitative analysis, such as a silver nitrate test, is not conclusive evidence that seawater is involved. At best, such a test should only be considered as an indicator to the possibility of salt or seawater wetting and cannot be, without further substantiation, considered a definite proof of such a wetting.

A better determination can be made by what is generally referred to as a "quantitative analysis" such as a "spectrographic analysis". As indicated above, seawater while varying somewhat in total solids is, for all practical purposes, uniform throughout the world. The composition of the solids is the same, and the compounds forming the solids appear in the same ratios of total solids throughout the world.

In determining whether or not a wetting is the result of salt water or seawater, the analyst conducting the "spectrographic analysis" should search for all the chemical compounds indicated in the above list. He should also be determine whether or not they appear in the proper ratios as shown.

Questions are often raised with regard to the amount of sea salt in the air at sea, what quantity enters the hold of a ship, and whether the amount which might enter the hold causes or contributes to the rusting of steel cargoes. There is less sea salt in the air during fair weather than there would be during boisterous weather when spray is being carried by the wind from the breaking crests of the waves. The amount of salt which enters a ship's hold will vary with the state of the weather, irrespective of whether or not the ship's ventilators are open and the method of ventilation employed, i.e. mechanical or natural ventilation. If no ventilation is provided there will naturally be no salt-laden air entering the cargo compartments.

The quantities of salt crystals in the air inside a vessel's hold are infinitesimally small, but by means of present-day methods of analysis the solids in a sample could be separated into the various solid constituents, but not quantified. Steel in contact with such minute quantities of chloride and sodium—merely traces of NaCl—would not sustain any damage on account of this over any period of time. Such contamination is considered less harmful than impurities normally found in the atmosphere. Were such minute contacts detrimental to steel cargoes then masses of claims would have been entered on most voyages attributable, or partly attributable, to this one cause. In such circumstances, such a contributing factor to the detriment of steel cargoes carried by sea should have taken a prominent position in the minds of all claims adjusters, but it has not.

When a ship enters port and hatches are opened it is customary in many ports for surveyors to proceed onto the steel cargo and test surface areas with a silver nitrate solution. This exercise may be carried out because it is the habit of the surveying fraternity in that particular port to do so where steel products are concerned. Unusual configurations of rust, or just the colour of the rust, may raise their suspicions as to the possible involvement of chlorides. Another reason may be the fact that all or part of the visible cargo is wet. When the cargo is dry, or wet from pure condensation water, there will be no reaction to silver nitrate tests. This means that if salt crystals are present on the material through contact with sea air, the amounts are so small that silver nitrate solutions are insufficiently sensitive to react. On the other hand, if samples of the steel are properly and carefully selected the presence of chlorides, in very small amounts, may be detected by laboratory analysis. Pure seawater reacts strongly to silver nitrate solution, which turns white and gives the appearance of curdled milk. The reaction to diluted salt water is less severe and such a reaction is referred to as a brackish reaction. As the dilution progresses the term "slightly brackish" is used and at this stage the reactive colour of the liquid is what might be referred to as a misty steely blue. Such a reaction is detectable if the test is carried out in distilled water washing, but is virtually undetectable if the test is carried out directly upon the surface of the steel. Because of this, it is evident that testing and sampling must be carried out correctly if the true situation with regard to the presence of chlorides is to be determined. Rust scrapings must be removed directly into a suitable receptacle and soaked in distilled water before applying silver nitrate solution to the liquid. These samples, i.e. rust scrapings, must not come into contact with the hands of the sampler, and this applies especially if the samples are for laboratory analysis, because various solids such as sodium and chlorides, etc., are exuded through the pores of human skin. Taking into consideration the delicate nature of the analysis where solids can be defined, but not quantitively in view of the minuteness of the amounts involved, contact with the human hand will upset the reliability of such analysis.

Often the question of the presence of traces of chlorides is raised in maritime disputes, where, due to their presence, cargo interests claim seawater contamination is involved. In one such case an analysis of sound wrapping paper produced a chloride content of 0.062%, which is not surprising as tap water contains chlorides and water from the city mains is used in the paper-making industry. Carefully selected samples of rusty and allegedly contaminated plating were selected jointly by the surveyors and the analysis produced a chloride presence of 0.006% and less. It is incumbent upon the cargo interest's surveyor to determine the nature of the damage, and in these circumstances the surveyor will submit samples to a laboratory for a spectrographic analysis. If contact with seawater is involved all solids proving this will readily be identifiable, as situations concerning seawater leave behind more than just traces of the constituent parts. When dealing with traces of chlorides, the appropriate analysis invariably fails to produce results proving the presence of the complete range of seawater solids. In the case above, the conclusion was reached that condensation water had leached out chlorides from the packing paper, which accounted for the lesser traces of chlorides found in the rust samples.

SALT WATER/SEAWATER CONTAMINATION

Steel cargoes coming into contact with seawater during the voyage or sustaining salt water contamination prior to shipment are not uncommon. Such contamination to cold rolled products ruins that part of the material it contacts, as does fresh water (see **Cold rolled steel coils**, p. 29). Similar contamination of hot rolled products, e.g. structural steel, hot rolled steel coils and plates, etc., although it eventually leads to damage developing, if given appropriate attention within a reasonable period of time, will not lead to damage developing. Where fresh water contact with hot rolled steel is concerned, in the short to medium term it is usually of no consequence. Salt water can cause severe rusting with pitting and accelerate an already established rust condition.

How long after contact with sea/salt water damage can be expected to develop, is a question often asked. This is a question which cannot be answered precisely, because quality of ambient air, relative humidity, oxygen supply and atmospheric temperatures are controlling factors which vary considerably from country to country and location to location. From experience it can be stated that in general, where hot rolled steel is concerned, from first contact, given ideal storage conditions, action must be taken within four to six weeks if damage is to be avoided. The best and safest policy to adopt is immediate action. Within a reasonable period of time the effect of contact with salt/seawater is contamination and not damage. The latter can be avoided by subjecting the material to a decontamination process discussed as follows.

Acid pickling

In production this is a process used mainly with hot rolled strip steel in coils and hot rolled plates. It is used for removing scale and extraneous matter from the surface of metal by immersing it in a suitable chemical reagent—sulphuric acid or hydrochloric acid—which will attack the scale and remove the unwanted contaminants. Protection of the steel *substrate* is about 90% achieved through introducing inhibitors to the pickling solution. After pickling, the material has to be washed with water. This can be quite costly when individual plates have to handled separately. The method is used mainly with steel in coil, which is put through a continuous strip pickler, and sometimes with plates and heavy sections.

If hot rolled strip in coils is involved and in production destined to be cold rolled, it would in any case have to undergo the process of pickling. If this can be done within a reasonable time after discharge from the ship, any claim arising from the salt/sea/chloride contamination should be minimal compared with what is involved if the pickling costs become part of the claim.

Shot blasting

This is a process which is referred to on p. 171 and consists of bombarding the metal surface with an abrasive shot. It is used to clean steel plates and structural material, such as contaminated beams, angle flats, etc. There are companies which have established and publish yearly tariffs for this type of work.

Fresh water washing

This may or may not be the cheapest method of dealing with sea/salt water contamination. Fresh water can be obtained from the water mains but in most instances it cannot be used to decontaminate steel since this water is usually contaminated with chlorides, a fact often overlooked by surveyors. The water should be tested with silver nitrate before being used. The only alternative would be to transport water from an inland lake or use uncontaminated river water.

There is always the possibility that such work is not permitted on the dockside and it is likewise possible that it cannot be done on the premises of the receiver. Each plate will have to be hosed down and labour employed to run a yard boom over the already rusted areas, especially around the edges of plates. The costs of transport, handling equipment and labour can develop into a sizeable claim. Depending on the circumstances, the costs of one method of cleaning should be compared with another before a final decision is taken about how the reconditioning work might be done.

Even with hot-rolled coils, prolonged exposure to heavy rain can wash away all the chlorides. Claims for salt/seawater contamination should be viewed with scepticism, since in nearly every instance hot rolled steel surfaces have to be given a protective coating, before the application of which the steel has to be shot blasted or pickled.

Invariably receivers will maintain that they have no customers for the goods and that it is destined to go into store as stock material. Another excuse might be that the material does not fit into the present production schedules and will therefore be stored for some time. This may or may not be true; however, it is necessary for surveyors to pursue a method of approach to such a situation which will reveal how warranted the claim actually is.

SCALE BREAKER

During the process of heating up ingots or slabs to raise them to a temperature necessary for rolling, the iron in the steel at the surface of the unit combines with

This coil of cold rolled steel sheeting (conventionally packed) remained for 14 days submerged in sea water. The strip was oiled, tightly wound and there was very little penetration of sea water or rust and no pitting. Hot rolled coils (unpacked), on board the same vessel, were more heavily damaged by rust. This was typical of the entire shipments. both types realised a salvage value of around 50% of the sound value.

Fig. 43(a) and (b). Steel coils submerged in sea water.

the oxygen in the atmosphere to form a layer of basic iron oxide in the form of scale. This scale has to be removed, as it might otherwise be rolled into the surface of the metal. Removal of the scale is accomplished by subjecting the item to a light passed through a single set of rolls—situated between the roughing and finishing trains—to crack the scale which is then blown away by jets of steam and water (see under Mill scale).

In the actual removal of mill scale from semi-finished products, there is a process of passing hot-rolled material through a machine whereby it is subjected to a number of reverse bends. The process shatters the mill scale and also suppresses any further tendency to kink when being manipulated. Figure 44 shows such an arrangement for breaking the scale from wire rod destined for the manufacture of road mesh.

SCRAP

As far as steel making is concerned, scrap is composed of materials returned within the steel plant for recycling, industrial scrap returned from customers, and also scrap from the population in the form of cars and household goods. The quality of the scrap is important, as is size, cleanliness, rust and chemical composition. Carried by sea, scrap can be in the form of bales. Some shipments consist of small loose pieces up to large pieces weighing more than a ton each. There are plenty of cases on record of tank tops being penetrated during loading.

There are two distinct methods of steel production in use today. The basic oxygen furnace takes molten iron from a blast furnace and refines it to steel. Scrap—old steel—is mixed with the hot metal immediately prior to the steel-melting

Fig. 44. Scale breaking arrangement used with wire rod.

operation. The other method of steel making is even more dependent on the scrap industry. The electric furnace uses scrap as the major—often the only—feed stock. So virtually all the steel we use today is made, to a greater or lesser extent, from scrap. The United Kingdom specification for iron and steel scrap, for the manufacture of iron and steel, sets out scrap specifications. The introduction to the specifications is the general conditions which are applicable to all grades of scrap and they comprise:

A. Safety. This refers to dangerous objects in the scrap such as pressurised gas, fuel, shells, bombs, and material of an inflammable nature.
B. Cleanness. Whereby it is stated that all grades shall be free of dirt, non-ferrous metals or foreign materials of any kind, excessive rust and corrosion.
C. Residual and other alloys. This means that the scrap shall be free from alloys. However, inclusions of any negligible amounts are permitted.
D. Grading. Here it is stated that no delivery shall contain a mixture of grades. However, this is not intended to preclude the accidental inclusion of negligible amounts of other than the advised grade.

Finally there is O.A., which is old heavy steel scrap, and O.B., which is also old heavy steel scrap.

From here we come on to the grades, of which there are many. However, to give some idea of what is involved, we quote from one grade, 3B, which is fragmented scrap, and here it is stated as follows:

> "Old light iron and steel scrap fragmented into pieces almost all not exceeding 200 mm in any direction. Must be commercially free from dirt, non-ferrous metals and foreign material, and exclude grindings, swarf, turnings and borings. Must conform to the following specification: Density—0.80 tonne per metre cube minimum, Tin (Sn) content—0.03 per cent maximum, Copper (Cu) content—0.25 per cent maximum."

One of the methods of producing shredded steel scrap is as follows.

The shredded steel is made in a large electric-powered griding machine. Metal is fed into a conveyor to a mill of rotating hammers and material is crushed through a steel grating approximately 3 inches square. The process creates high heat which removes oil/liquids. The crushed/shredded material then goes over a magnetic and radioactive separator—to a forced air cyclone to remove fluff and impurities—and is then cooled along a flat conveyor where manual labour picks out any remaining impurities. The product is made of approximately 80% auto-bodies (engine block usually removed) and 20% light steel refrigerators (light material less than 18th of an inch, all compressors removed).

With regard to the carriage of steel scrap by sea, there are certain types of scrap such as baled scrap which can be carried with impunity even though some types of scrap, such as shredded/fragmented steel scrap, will heat on the surface if it sustains wetness, from either being wet at time of shipment or becoming wet through contact with sweat water or leaking hatches. The heating can of course become more serious in the event of seawater contacting the goods. Other material such as cuttings and iron/steel swarf is liable to self heating and to ignite spontaneously. This heating is certainly more frequent in those cargoes which are in a particularly finely divided form, loaded in a wet condition and contaminated with such minerals as unsaturated cutting oil, oily rags and other combustible matter. The cause of heating of these cargoes is due to the high surface area to mass ratio, which creates massive oxidation surfaces, friction between items comprising the scrap due to the working of the vessel in a seaway and vibration from the engines, extraneous inflammable matters permeating the bulk and of course moisture. Seawater contacting such a cargo can result in the cargo becoming white hot with conflagration. Such cargoes have been treated with inhibitors to prevent excessive oxidation taking place but the effectiveness of this is questionable, especially when seawater is involved.

The IMO publication *Code of Safe Practice for Solid Bulk Cargoes*, 1994 edition, states under Appendix B. UN No. 2793 IMO Class 4.2—Ferrous Metal, borings, shavings, turnings, or cuttings, in a form liable to self heating, iron swarf, steel swarf:

> "These materials are liable to self heating and to ignite spontaneously, particularly in a finely divided form, wet or contaminated with such materials as unsaturated cutting oil, oily rags and other combustible matter."

Under the heading of special requirements the article points out that prior to loading the temperature of the material should not exceed 55°C, also that the surface temperature of the cargo should be taken prior to, during and after loading and daily throughout the voyage. It then goes on to point out the dangers of going into the cargo spaces while the vessel is at sea and that if entry is found necessary then special breathing apparatus should be used. It also points out that if the surface temperature exceeds 90°C, during loading, further loading should cease and should not recommence until the temperature has fallen below 85°C. The ship should not depart unless the temperature is below 65°C and has shown a steady or downward trend in temperature for at least eight hours. It is important to note that, during loading, it is advised that the material be compacted in the cargo spaces as frequently as is practicable with a bulldozer or by other means. Also, after loading, the material should be trimmed to eliminate peaks and should be compacted.

In section 3 it is pointed out that, whilst at sea, any rise in surface temperature of the material indicates a self-heating reaction problem. If the temperature rises to 80°C a potential fire situation is developing and the ship should make for the nearest port. They go on to say that water should not be used at sea; early application of an inert gas to a smouldering situation may be effective. In port, copious quantities of water may be used but due consideration should be given to stability. The United States exports a considerable amount of ferrous metal borings, shavings, turnings or cuttings. Serious problems have been experienced with these cargoes, such as heating, even to the point where vessels have been in serious trouble while at sea. Included is a copy of the US Coast Guard Regulations 148.04–13. It will be seen that these regulations are similar to those of the IMO; however, they are somewhat more comprehensive and provide other bits of useful information.

The above deals with what is considered to be the potentially dangerous scrap cargoes. There are other scrap cargoes which are fined down to such a degree that there is a somewhat high surface area to mass ratio and which, if they sustain wetness and/or are contaminated by some extraneous materials to a higher than permitted degree, will develop heat. A combination of these two factors could classify the cargo as being potentially dangerous. However, if the contamination is within acceptable specified limits the amount of heat generated on the surface of the stow, or within the bulk, is not in any way alarming but must be monitored. Fragmented scrap may form such a cargo but the general consensus of opinion, based on experience, is that fragmented/shredded steel scrap in itself is not self igniting and is not considered a potential fire hazard.

Prior to the loading of scrap cargoes such as those mentioned above, the vessel's cargo compartments must be efficiently washed out with fresh water in order to get rid of any salt crystals left from previous salt water washing; in this respect the underside of the steel hatch closing appliances must be given special attention. Any wood or other combustible materials are to be removed and the holds must be clean. It is of great importance that the weather deck hatches are watertight and remain watertight. The steel hatch closing appliances must be in top condition. Before the cargo is loaded it is suggested that the hatches be ultrasonically tested and hose tested to confirm watertight integrity. Sealing tape should be applied across the seams of all joints expressly for the purpose of rendering the compartments as airtight as possible. Similar provisions should be made with regard to any ventilation system.

Provisions must be made to monitor the temperature of the cargo. It is not recommended that bilge-sounding arrangements be used for this purpose because often bilge-sounding pipes are too close to the ship's sides and may be affected by sea temperatures. The wells often project into heated fuel oil double-bottom tanks, so that temperature readings may not be reliable. Provision should be made to record the surface temperature of the material in a number of positions, also the temperature within the bulk by means of distance reading thermometers.

In the process of heating of such cargoes, whether autoxidation is involved and/or oxidation caused through an electrochemical process, heat, moisture and oxygen are conducive to the cargo eventually developing more heat. Steel scrap is an inert cargo and does not require ventilation unless atmospheric temperatures are falling, when the incidence of ship sweat may be involved. Incorrect ventilation when atmospheric temperatures are rising may lead to cargo sweat on the surface of the bulk; this must be avoided by properly sealing off the compartments and making them as airtight as possible. If possible, i.e. if there is a choice, compartments with heated double-bottom tanks should be avoided.

In those instances when a vessel has to divert because cargo temperatures appear to be getting out of control CO_2 gas should not be used, since in contact with heated iron it can be reduced to carbon monoxide, which is toxic and flammable. Inert gas may have to be used eventually and nitrogen gas is recommended.

When vessels are directed to load cargoes that are liable to heat and create problems on voyage, prior to arrival at the loading port it would be advisable for the ship's master to consider having a competent surveyor appointed to ensure that maximum precautions are taken to ensure the success of the voyage. Temperatures of the cargo recorded during the voyage should be obtained by means of distant reading thermometers.

American Regulations

§ 148.04–13 Ferrous metal borings, shavings, turnings, or cuttings (excluding stainless steel)

(a) This section applies to the stowage and transportation in bulk of hazardous materials described as ferrous metal borings, shavings, turnings, or cuttings on board vessels (excluding stainless steel). However, unmanned barges on which the article is stowed for or transported on a voyage entirely on the navigable waters of the United States are exempt from the requirements of this section. Ferrous metal borings, shavings, turnings, or cuttings (excluding stainless steel) must not be stowed and transported in bulk unless the following conditions are met:

(1) [Reserved]

(2) All wooden sweat battens, dunnage and debris must be removed from the hold before the article is loaded.

(3) During loading and transporting, the bilge of each hold in which the article is stowed or it is to be stowed must be as dry as practicable.

(4) During loading, the article must be compacted in the hold as frequently as practicable with a bulldozer or means that provide equivalent surface compaction. Upon completion of loading, the article must be trimmed to eliminate peaks or mounds and compacted.

(5) Other cargo must not be loaded in a hold containing the article if:

(i) The cargo to be loaded in the same hold with the article is another hazardous material as defined in this part or a combustible material;

(ii) The loading of the article is not completed first; and

(iii) The temperature of the article in the hold is above 130°F or has increased within eight hours before loading of the other cargo.

(6) During loading, the temperature of the article in the pile being loaded must be less than 130°F.

(7) Upon completion of loading, the vessel may not leave the port unless:

(i) The temperature of each article in each hold is less than 150°F and, if the temperature of the article in a hold has been more than 150°F during loading, the temperature of each article has shown a downward trend below 150°F for at least eight hours after completion of loading of the hold; or

(ii) The vessel intends to sail directly to another port that is no further than twelve hours sailing time for the vessel concerned, for the purpose of loading more of the article in bulk or to completely off-load the article, and the temperature of the article is less than 190°F and has shown a downward trend for at least eight hours after completion of loading.

(b) For the purposes of each temperature requirement of this section, the temperature of the article is the highest temperature taken between eight and fourteen inches below the surface at ten-foot intervals over its length and width.

(c) The master or person in charge of a vessel that is loading or transporting the article must ensure that the temperature of the article is taken:

(1) Before loading;

(2) During loading, in each hold and in the pile being loaded at least every twenty-four hours and, if the temperature is rising, as often as necessary to ensure the conditions in this section are met; and

(3) After loading, in each hold at least every twenty-four hours.

(d) During loading, if the temperature of the article in a hold is 200°F or higher, the master or person in charge of the vessel must notify the Coast Guard Captain of the Port and suspend loading until the temperature of the article is less than 190°F.

(e) After loading:

(1) If the temperature of the article is 150°F or above, the master or person in charge must notify the Captain of the Port and ensure that the vessel remains in the port area until the conditions of paragraph (a)(7)(i) of this section are met; or

(2) In the case of a short-duration voyage to which paragraph (a)(7)(ii) of this section applies, if the temperature of the article in a hold is 190°F or above, the master or person in charge must notify the Captain of the Port and ensure that the vessel remains in the port area until the conditions of paragraph (a)(7)(ii) of this section are met.

(f) Except for shipments of the article in bulk which leave the port of loading under the conditions specified in paragraph (a)(7)(ii) of this section, after the vessel leaves the port, if the temperature of the article in the hold rises above 149°F, the master must notify the nearest Coast Guard Captain of the Port as soon as possible of:

(1) The name, nationality, and position of the vessel;
(2) The most recent temperature taken;
(3) The length of time that the temperature has been above 149°F and the rate of rise, if any;
(4) The port where the article was loaded and the destination of the article;
(5) The last port of call of the vessel and its next port of call;
(6) What action has been taken; and
(7) Whether any other cargo is endangered.

(g) To meet the conditions of this section, the master of a vessel that is transporting the article must ensure that each temperature taken is recorded.

SEMI-FINISHED PRODUCTS

These are blooms, billets, slabs, sheet (hot rolled) and wire rods, etc., which require further work treatment; they are not a finished product.

SEMI-KILLED STEEL

This is a steel, the characteristics of which are shared between killed and rimmed steel. Deoxidising agents are added in small quantities. This type of steel is used in constructional work, plates, etc. The ingot is distinguishable by piping at the top (see under Killed steel).

SHEET PILINGS

These goods are usually shipped in lengths of 20, 40 and 60 feet; moderately longer lengths are not uncommon. Depending upon the thickness of the metal involved, the approximate weight per running metre varies between about 40–60 kg. The type of goods involved can be seen in Figs. 45 and 46.

These piles are driven vertically into the ground and locked together by means of the keying devices provided on the outer edges of the flanges. In this way the piles can be arranged to form a barrier where the retention of earth work is required.

The construction of the sheet piles provides considerable strength; however, the keying arrangement on the flanges is susceptible to damage, and this should be kept in mind during handling operations—braided wire slings are best used and slings should not be overloaded.

Up to about seven piles are stacked together to form an unsecured bundle, and it has been found that more than seven piles per bundle may result in the top pile cracking. Adequately sized strong dunnage has to be inserted per layer of maximum seven piles to keep the bundles apart, bind the stow together, and facilitate slinging during discharge.

Sheet pilings are sometimes coated for protection, but are generally not so

Fig. 45. Sheet pilings stacked ashore awaiting shipment.

treated, in which case they will be affected by rust at time of shipment. Pre-shipment storage is usually in the open.

SHORT DELIVERY OF CARGO

Short delivery of cargo has plagued the industry for many years and undoubtedly will continue to do so. There are bulk carrier full loads of steel which are covered by perhaps up to 1,000 bills of lading and in the itinerary of the ship there may be more than one loading port and more than one discharge port. The carrier has to take the necessary precautions to avoid discharging cargo, or part consignments, at the wrong port(s). There was a time when the carrier took much care to ensure that the various parcels of cargo were well separated in the ship. This was done by laying specially prepared strips of burlap or burlap cloth to segregate the cargo for one port from another. Even specially prepared tapes with port names printed on them have been used. The fact remains that it is incumbent upon the carrier to ensure proper and efficient separation of cargo by port and to carefully monitor the discharge in order to avoid delivering any cargo at the wrong port.

As far as the carrier is concerned the weight of steel cargo is usually of no concern to him. He has no reasonable means at his disposal of checking the weight of individual parcels of steel cargo and cannot therefore be held responsible for any discrepancies in weight. However, having signed a bill of lading referring to a quantity of cargo expressed in unit numbers, he must deliver the same number of pieces or units at the designated port of discharge. Failure to do this will place upon him the burden of proving why there was a short delivery. If there is not feasible

Fig. 46. Sheet pilings stowed on deck.

explanation and the master is unable to exercise any lawful rights or immunities, e.g. loss of cargo through proven and accepted perils of the sea, the carrier will be obliged to make good the value of any proven loss arising from short delivery of cargo.

The quantity of cargo loaded can be checked by appointing a firm of Sworn Weighers and Measurers who will tally the cargo into the ship. Liner companies with breakbulk ships would make such arrangements. Where chartered ships are concerned there is often a clause in the governing charterparty whereby it is stipulated something to the effect that charterers will arrange for the cargo to be tallied into the ship. Some ship masters insist on bills of lading being claused "as per shore tally", thus indicating that the tally was not carried out on board the ship. Such a clause does not in any way relieve the carrier from delivering the correct quantity of cargo.

Steel coils are usually electronically weighed, but weights may not always be reliable, as weighing machines need to be calibrated periodically. Where short delivery is concerned steel coils are not a great problem. The problems arise with unwrapped hot rolled steel, such as single steel beams and channels which are grouped into unsecured lifts. There are then bundles of small scantling material, see pp. 123 and 125, which are secured with either wires or flat metal strapping bands. The carrier would not be liable for the number of pieces in intact delivered bundles, but would be liable for the official number of unit lifts comprising any consignment for which the ship's master has signed a bill of lading. Other steel cargoes regularly giving rise to shortage claims are billets, blooms, plates in secured bundles, or single plates grouped or lifted singly, reinforcing bars, road mesh, pipes, etc.

Small scantling material is packed into unwrapped bundles in the mill where it is manufactured, and it is there that the number of pieces per bundle is determined. These bundles and also the larger items of cargo, e.g. large beams, channels and blooms, etc., are transported to the docks where the labelling and/or paint marking is checked. The various parcels are identified, and the number of units or lifts are checked and measured for volume by the tally clerk. During the course of loading the tally clerks stand by, in the vicinity of the various hatches, either on board or ashore at the ship's side and check the quantities taken on board as per the predetermined amounts placed in each sling or lift of cargo. The same procedure is, or should be, followed at the time of discharge. Basically this is what does take place; however, procedures can vary from port to port and from country to country. It is of course incumbent upon the carrier to supervise the loading and the discharge and to ensure that all cargo for any given port is discharged at that port.

With regard to the weight of hot rolled structural material, this weight is determined by calculation. By example, if by measurement a given unit section volume is calculated, this volume when multiplied by the internationally accepted specific gravity of steel (7.85) will give the weight of the piece. It is then simple arithmetic to apply this to the running length of the material. A steel plate measuring 8 mm × 600 mm × 6,000 mm would on the basis of the calculation have a weight of 22.6 metric tons.

There may never have been any reason to believe that partial shortages of steel cargo have been caused by barratry of the master or crew, or caused by the carrier stealing cargo during transit. Shortages generally result from:

—initially poor marking of the goods;
—mistakes in tallying;
—all round inefficient organisation;
—failure on the part of the carrier to properly and clearly segregate the goods in the ship. As a result of this some cargo for port "B" may be discharged at port "A" and vice versa;
—in extreme cases theft prior to loading or after discharge.

Generally speaking, the most common cause of shortage in numbers and of weight can be put down to human error. Bad marking of the various consignments must be at the top of the list.

Where bundles of hot rolled material are concerned, metal or plastic labels attached to the securing wires are the usual method of identification; these are sometimes torn off or damaged. Usually bundles and other items/lifts of cargo are paint marked in strip fashion, e.g. two black stripes or brown stripes. These are bad colours since they are hard to identify during discharge after dark. Bright colours are better but all colours fade through being weathered. The longer the goods are left exposed to the weather the more faint any paint marks become, thus causing difficulties in identification of the cargo at the port of discharge.

In general, the correct quantities are usually loaded; this is especially so where all of the cargo is assembled at one berth for shipment; when the berth is empty; or there is no steel left on the quay, and everything has been loaded into the ship. It is somewhat more difficult when cargo is being shipped direct from barges, coasters, rail and road wagons or direct from shore storage; all of this activity can take place at the same time. There was one prominent manufacturer and exporter of hot rolled steel beams who was fully equipped with state-of-the-art computerisation for the handling, pre-shipment storage, counting and documentation procedures. The entire set-up was a complete disaster. The company in question had a long history of mixing cargo from individual bills of lading as well as loading incorrect quantities.

Hot rolled products are more often than not given open storage prior to shipment. They are exposed to the elements, bundles are wet inside from rain water and they may even be visibly wet before shipment or loaded during rain. The cargo remains wet in the ship and will invariably be further wetted from ship sweat brought about by the wet condition of the cargo. Over a period of three to four weeks in such an environment the paint marks if not destroyed become seriously faded and can render the material unidentifiable at time of discharge.

With regard to the quantity of cargo loaded, often the ship's master will want to know how he can check the number of pieces mentioned on the shipping documents for which he will be responsible. He will be informed that the cargo is being tallied into the ship by an independent tally firm. If a chartered vessel is involved the master may feel that he cannot trust the tally, as it is being carried out on behalf of the shipper or the charterers. Of course, he is quite entitled to ask to examine the tally sheets before signing the bills of lading. The ship owner may be able to appoint a company to tally the cargo on his behalf but usually this is considered to be a waste of money. Two tally men at one hatch, one for the owner and one appointed by whoever has appointed the stevedores, will make sure that their joint tally agrees. Further to this, one will carry on while the other answers the call of nature, or whatever, and the exercise in general is quite likely to be unreliable.

There have been instances where the crew has been used to tally cargo. This is something which is beyond today's crew member. A bulk carrier loading a full cargo will most probably be working three hatches continuously from 0600 hrs through to 2200 hrs. Night gangs may even be employed, so that cargo is loaded round the clock. Even if the crew managed to carry out some form of tally which did not agree with the tally of the official tally clerks, the ship's tally would not be accepted against a shore tally conducted by an independent tally company. Tallying ships' cargoes is a profession and the ship's crew tally would be looked upon as the work of amateurs as opposed to that of professionals.

Finally, there are companies, carriers and cargo interests who appoint cargo surveyors to check quantities of cargo being either shipped or landed. In certain circumstances this might be possible and of some value where small quantities of cargo are concerned. But again the surveyor is not equipped to tally and follow the discharge of a full shipment of steel; further to this, under the rules of the port he may not be permitted to officially tally cargo, which must by carried out by certified tally clerks attached to firms of sworn weighers and measurers, who are often members of a union.

Recently a rip-proof label has been introduced to the market and advertised as specialist labels for the steel industry. For these labels it is claimed further:

—Suitable for Hostile Environments.
—Extremely Hard Wearing, Tear and Abrasion Resistant.
—Good Barcode Readability, Does Not Attract Dirt.
—Excellent Thermal Transfer Printing Properties.
—Plain or Pre-Printed in up to 4 Colours.

This system may be the future in marking and identifying cargo.

SHOT BLASTING

The use of metal abrasives for cleaning the surface of steel in preparation for the application of protective coatings. Rust, mill scale and any other unwanted matter is removed by bombarding the surface of the material with metal filings or specially manufactured shot. Various grades of abrasive are used to produce specific surface quality, after which a blast primer is applied. Where rust is concerned, if possible, pickling is preferable as blast treatment may not always eliminate entirely all corrosive salts.

SINTER

This is a method of using up the fines from ore and coke. Iron ore dust and lumps below a certain size, and also similar sized material from coke, cannot be utilised as they are, because the strong blast of air would blow them through the top of the furnace. Apart from this, sintering minimises production by achieving a minimum of coke consumption per ton of iron produced.

The ore is crushed and screened after which the ore fines are mixed with coke fines and limestone. This mixture is then roasted into an agglomerate which takes the form of lumps of a clinker-like material. In the process of sintering the volatile matter is driven off by the heat.

Apart from maximising production, the use of sintering lends itself to efficient smelting. Sinter can be as high as 50% of the furnace charge, and some furnace charges consist entirely of sinter and coke together with the ore.

SKIN PASS

A term used to describe a light cold rolling of heat-treated sheet or strip. The operation improves surface finish and suppresses the tendency to kinks, flats and stretcher strains on further manipulation. The rolling involves only a slight reduction of the sheet, etc.

SLABS

After the steel is made to a predetermined specification, it is poured from the converter vessel into an ingot mould. These ingots, if not remaining in the ingot form, are taken to a soaking pit where they are held at a high temperature until they are transferred to the rolling mill (blooming mill) where they are rolled into a semi-finished block of metal referred to as a slab (see Fig. 47).

The finished slab has dimensions of about 250 mm thick × 2,500 mm wide and is produced in various lengths. These slabs will eventually be either directly or indirectly rolled into thin plate, to ultimately form hot rolled steel coils. Slabs are transported by sea and are unwrapped and unprotected against rust. They are stored in the open, unprotected against the elements and usually fairly well covered with a light film of rust through contact with the atmosphere and weather conditions in general.

Fresh-water rust, which is uniform and of recent development, is of no consequence. Each slab weighs many tons and in a ship's hold, providing the top tier is well chocked off in the gaps between individual slabs, should not shift or cause any difficulty. Owing to the weight involved, these goods need to be stored in the bottom of the ship. Dunnage beneath the lower tier must be placed directly above the transverse floors to prevent spot overloading of the tank-top plating.

The slabs shown in Fig. 48 each weigh eight tons. Dunnage was placed between each horizontal tier but did not project from one block to the other in an interlocking

Fig. 47. Slabs awaiting shipment.

Fig. 48. Slabs—commencement of stowage in No. 5 Hold.

fashion. The slabs were stowed seven high and secured in blocks as indicated, each block being secured with metal strapping bands. Apart from the dunnage between the horizontal tiers, and the strapping bands referred to, no other securing was used. A number of voyages were made to clear this order, apparently without difficulties arising.

Such a method of stowage is not recommended. A more reliable and seamanlike stowage is obtained by stowing the slabs athwartships. In such circumstances, the slabs can be winged out to the slope plating of the side tanks. Slabs in the various tiers will overlap and the stow will be locked into a solid block. Slabs occupying the top stowage position can be wedged with timber where gaps appear.

The fact is that there are many variations on stowage and even new types of stowage coming into existence, all of which are promoted by a drive to save money. When steel slabs are unitised, and not handled singly, obviously there is a saving. When these heavy blocks of cargo can be lowered directly into stow, e.g. hatch square stowage, there is no need to use fork lift trucks and a further saving is made. This can go on with shortening up on dunnage, reducing wires, strapping bands, etc. This is not really stowage in the true sense of the word, as the cargo is just landed in the hatch square. In due course such stowage, after many voyages have been made, shows up in a port, where experienced, independent professional surveyors are concerned, with large gaps left between the termination of the stowage and the sloping sides of the lower wing tanks. The eventual conclusion is that such stowage arrangements are decidedly dangerous.

In its defence, those who favour that method of stowage always point to the many voyages where such arrangements have been employed without anything untoward occurring. Granted, disasters with slab shipments are rare, but when those cargoes shift at sea, as they have been known to, the results may be ill-fated. A victim of a steel slab cargo shift was the Italian motor vessel *Tito Campanella* where the cargo was stowed in a fashion referred to as the Swedish Oxelosund System. This was apparently devised to cut down on handling by stowing slabs in large fore and aft positioned blocks, so leaving gaps in the wings as already referred to. The top slabs were in some instances winged out to the ship's sides, by placing them athwartships. The vessel was lost with all hands and the stowage was blamed for the incident.

Another vessel loaded in the same manner in a Mediterranean port was bound for the USA. The ship was off the Portuguese coast in force 5 weather conditions when a deep rumbling was heard within the holds of the ship, which immediately developed a heavy list to starboard. Fortunately, it was possible to get the ship into a nearby port and restow the cargo. Had this shift of cargo occurred in mid-Atlantic the end result might have been quite different. One of the latest of these types of stowages to become a controversial subject of discussion originates from the U.S.A. (California stowage). Such stowage might be acceptable on ships which have box shaped holds, but on ships fitted with lower wing tanks gaps have to be left in the wings; such an arrangement is not acceptable to many surveyors. Slabs stowed in a fore and aft direction may partly pile up on one side of the hold, and have been known to; a complete pile up would undoubtedly be more serious.

Supporters of the types of stowages referred to are quick to designate or establish such arrangements as being customary in the trade. On the other hand, how reliable is such a label when what it signifies is undoubtedly flawed? Those who defend these methods of stowage can put forward some convincing supporting arguments in the form of numerous successful voyage histories, all of which would be of no value to them in the event of a ship being lost, because such documentation would be conclusive evidence that all measures humanly possible had not been taken to prevent the cargo shifting.

SLAG

In iron production, slag is the name given to the residue which floats on the surface of the molten metal in the furnace. It is composed of the impurities separated out from the ore to produce iron and this is achieved by the introduction of a flux to the furnace charge in the form of limestone.

The slag in molten form is poured off through the slag notch at the bottom of the furnace and when cooled it is cinder-like in appearance. Although from the point of view of iron making, slag is a residual waste material, it is used in the building industry for the manufacture of insulation material and as a fertiliser, etc. Such goods are usually transported in jute bags and are odourless but very dusty.

SLAG NOTCH

Sometimes referred to as slag hole. It is a hole in the side of the furnace through which the slag can be run off. When not in use the hole is plugged. As the slag floats on the top of the molten metal it is so arranged that the slag notch is situated at this level.

SPACE LATTICE

Metals are composed of atoms which arrange themselves in regular geometric patterns known as space lattices. Where iron and steel are concerned there are two distinctive patterns.

Within each grain comprising the crystalline structure, the lattice structure of the atoms may be a body-centred cube having one atom positioned at each of the eight corners, and one atom at the centre.

The body centred cubic pattern may be changed to the face-centred cubic pattern, and vice versa. When face-centred, an atom is positioned at the centre of each of the six faces of the sides forming the cube (see Fig. 49).

STAINLESS STEEL

Stainless steel cannot be defined as a single specific material as there are many types of stainless steel. Modern stainless steel is principally an alloy of iron and chromium but at the same time embraces a group of corrosion resistant steels with a minimum of 8% chromium and varying additions of nickel, molybdenum, titanium and niobium.

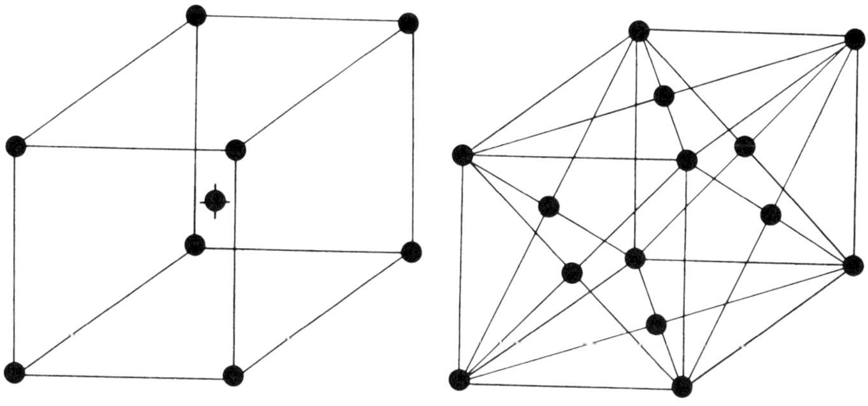

Fig. 49. Lattice structures.

The characteristics of stainless steel are that it is resistant to attack by the atmosphere and chemical reagents, impervious to oxidation and scaling, and also maintains its strength at elevated temperatures. It is believed that its resistance to corrosive attack is provided by a film of chrome oxide at the surface of the alloy which is so fine that it is undetectable. If this surface film is removed, it immediately re-forms through oxidation with the atmosphere. The three main types of stainless steel are as follows.

Martensitic

This type may have a chromium content of between 11 and 18% and varying proportions of carbon. Martensitic stainless steel is magnetic in character and may be hardened by quenching but has poor formability properties and is therefore not particularly ductile. It is mainly used in the cutlery industry.

Ferritic

This type may have a chromium content of between 16 and 30% and a carbon content of 0.2% or less. It has magnetic properties, is susceptible to taking strains and its uses are more or less confined to the automobile industry.

The following table shows the chemistry of some grades of Martensitic and Ferritic stainless steels:

Steel Type	*C* % *Max.*	*Mn* % *Max.*	*P* % *Max.*	*S* % *Max.*	*Si* % *Max.*	*Cr* %	*Ni* % *Max.*	*Mo* %
M 409**	0.03	1.00	0.040	0.020	1.00	10.50–11.70	0.50	–
M 410	0.15	1.00	0.040	0.030	1.00	11.50–13.50	0.75	–
M 410S	0.08	1.00	0.040	0.030	1.00	11.50–13.50	0.60	–
F 420*	***	1.00	0.040	0.030	1.00	12.00–14.00	–	–
F 430	0.12	1.00	0.040	0.030	1.00	16.00–18.00	0.75	–
F 434	0.12	1.00	0.040	0.030	1.00	16.00–18.00	–	0.75–1.25
F 439	0.07	1.00	0.040	0.030	1.00	17.00–19.00	0.50	–

Austenitic

This type of stainless steel normally has a chromium content of between 18 and 20% and a nickel presence of between 8 and 18%. The material is non-magnetic and is characterised by its toughness combined with ductility and also its failure to be hardened by heat treatment, but it does harden rapidly when cold worked.

The 18% chrome and 8% nickel austenitic type of stainless steel (8/18) is one of the more important types and is used in a wide range of applications. It is also more resistant to corrosion than the martensitic and ferritic types of stainless steels. Its anti-corrosive properties are enhanced by additional nickel and further improved by smaller additions of molybdenum and copper.

Austinitic stainless steels have the grades which give the perfectly highly polished surface finish. The following table shows the chemistry of some grades of this material:

Plate 1. Internal parts of a cold rolled steel coil damaged through penetration of cargo sweat

Plate 2. Packing affected by cargo sweat. Note drip mark configuration of stains

Plate 3. Collection of moisture in lower part of protection ring penetrates packing

Plate 4. "Rust stained" round bars

Plate 5. Steel in a "rusty" condition

Plate 6. Electrical steel sheeting in packages. Note the heavy cargo sweat rusted condition of the cover of the package beneath. The steel hood type cover preserved the silicon coated electrical steel sheets

Plate 7. Galvanised coils prior to removing wrappers

Plate 8. Edges of windings affected by white rust

Plate 9. Galvanised coils awaiting shipment in apparent good order and condition

Plate 10. White rust beneath wrappers

Plate 11. Hot rolled steel sheeting under production at Sidmar Works, Ghent, Belgium

(a)

(b)

Plate 12. Hot rolled coils stored in the open awaiting shipment. The aspect is normal for such goods—blue patches are the remnants of mill scale

Plate 13. Hot rolled coils (unwrapped) in prime condition (imported for the manufacture of pipes)

Plate 14. Damages to hot rolled coil through contact with seawater

(a) Aspect of hot rolled coil prior to decoiling

(b) Appearance of plate surface before entering pickling bath

(c) Note etch marks on surface of plating after pickling

Plate 15. Coil fitted with tailor-made end cover

Plate 16. Coil end edges fitted with protection rings

Plate 17. Plate awaiting shipment (note rusty condition)

Plate 18. Hot rolled steel coils—effects of variation of temperature

Plate 19. Steel coils affected by cargo sweat: overstowing packages unaffected

Plate 20. Disintegration of coils and tangling of wires

(a)

(b)

Plate 21. Q.R. disintegrating bundles. No claim for rust condition

BS	British Standards Institute
JIS	Japanese Industrial Standards
DIN	Deutsches Institut fur Normung
SAE	Society of Automotive Engineers (American)
AISI	American Iron & Steel Institute
API	American Petroleum Institute.

There is also the SIS (Swedish), AFNOR (French), ISO (International Standards Organisation) and the Euronorms, etc. The subjects covered by hundreds of volumes are tremendously extensive and the sections covering steel products are considerably involved. The organisations are scientific and technical, formed for the development of standards on characteristics and performance of materials, products, systems and services, also the promotion of related knowledge. We take, for example, three of the volumes of the DIN regulations which deal with steel:

Handbook No. 28—Iron and Steel Dimensional Standards.

This handbook deals with recommended measurements that concern permitted tolerances with regard to lengths, thicknesses and various gauges of steel products and allowable deviations from specifications. It is in fact a guide in the production of steel.

Handbook No. 4—Iron and Steel Quality Standards 1
Handbook No. 2—Iron and Steel Quality Standards 2

These books are used in quality control and cover the actual chemistry of the steel. They also cover to some extent the taking of samples for testing to determine the mechanical properties of the material. With the DIN number and steel grade particulars of any material the specified quantities of alloying elements can be determined.

A purchaser ordering steel would have stated in the contract of purchase, among other things such as price and quantity, etc., the chemical composition of the metal, mechanical characteristics, quality, which would conform with one of the standards, and the permissible tolerance in weight. A surveyor is usually appointed by the purchaser to check the shipment and determine that it conforms with the order contract; the surveyor would in due course issue a quality certificate. It is important to note that, although they might appear to be interrelated, which cannot be denied to some degree, where marine claims are concerned it is considered that rust damage is separate from quality. In other words quality is unaffected by rust. This argument is tied up with the definition of "Prime material" and whether goods are at time of shipment, or upon reception, in "Apparent good order and condition" (see "Prime material").

STEEL

The basis of steel is iron (Fe)—a very soft metal that is more or less unusable in its pure, or nearly pure, form. By alloying iron with carbon, so forming an iron carbide (Fe_3C) to make steel, a metal is produced which has sufficient hardness, strength and plasticity for it to be formed into numerous shapes which lend themselves to a tremendous number of uses.

In the composition of plain steel there are small quantities of basic elements which are not completely eliminated, such as silicon, sulphur and manganese. In the steel-making process the amount of carbon introduced will depend on whether it is intended to make low, medium or high carbon steel. Low carbon steel contains 0.02–0.25% carbon, and is referred to as mild steel. This type of steel has moderate strength and high ductility. Medium carbon steel contains 0.25–0.7% carbon and is used where greater constant wear and shock strength is involved, such as in machine parts. High carbon steel contains more than 0.7% carbon and is used for the manufacture of tools, springs, etc., where hardness and strength of the metal is all important, but ductility is less of a necessity. Plain carbon steel is highly responsive to heat treatment, which renders the material suitable for a diverse number of uses.

Alloy steel is a plain carbon steel into which has been blended controlled amounts of other base metals such as chromium, manganese, nickel, molybdenum, titanium and tungsten. By the blending together of these metals the resulting alloy possesses hardness, high resistance to wear, resistance to attack by acids and general imperviousness to corrosion, also great strength and resistance to shock.

STEEL HATCH CLOSING APPLIANCES

There is no doubt that the advent of steel hatch covers, used for the closing of openings in ships' decks, has made possible revolutionary changes and advancements in the shipping industry. Partly due to this type of hatch closing appliance, larger ships with, relatively speaking, mammoth deck openings have been made possible, so contributing to cheaper transport of goods by sea, a faster turnround of vessels in port and reduced deck crews on board ship. Where steel cargoes are concerned, the maximum area of deck opening is desirable in order to manoeuvre long lengths of steel quickly and easily into the furthest corners of the ship's hold; also to facilitate the handling of all types of steel cargoes which are usually composed of very heavy units.

MacGregor single-pull steel hatch covers

There are today numerous steel hatch cover patents in use. However, the most common is the MacGregor single-pull type. As the principle involved in achieving watertightness is more or less identical in every patent, it is the MacGregor type of cover which is used here as an example.

These hatches, broadly speaking, within the limits of the basic design, consist of a number of rectangular steel covers which when lowered into position close the openings in a vessel's weather deck, and through which openings cargo is worked into the holds of the ship. Around the peripheries of each cover or panel there is fitted a rubber joint, the purpose of which is to effect a watertight seal. The individual panels are supported on eccentric wheels which permit the panels to be raised and lowered within the confines of the hatch coaming. The panels are joined together by means of chains and, when opening the hatch, the panels are rolled to the ends of the coamings where they tip vertically and fold up in one position. In both opening and closing, the covers are moved by means of a wire, operated either directly, or indirectly, from a winch drum (see Fig. 50).

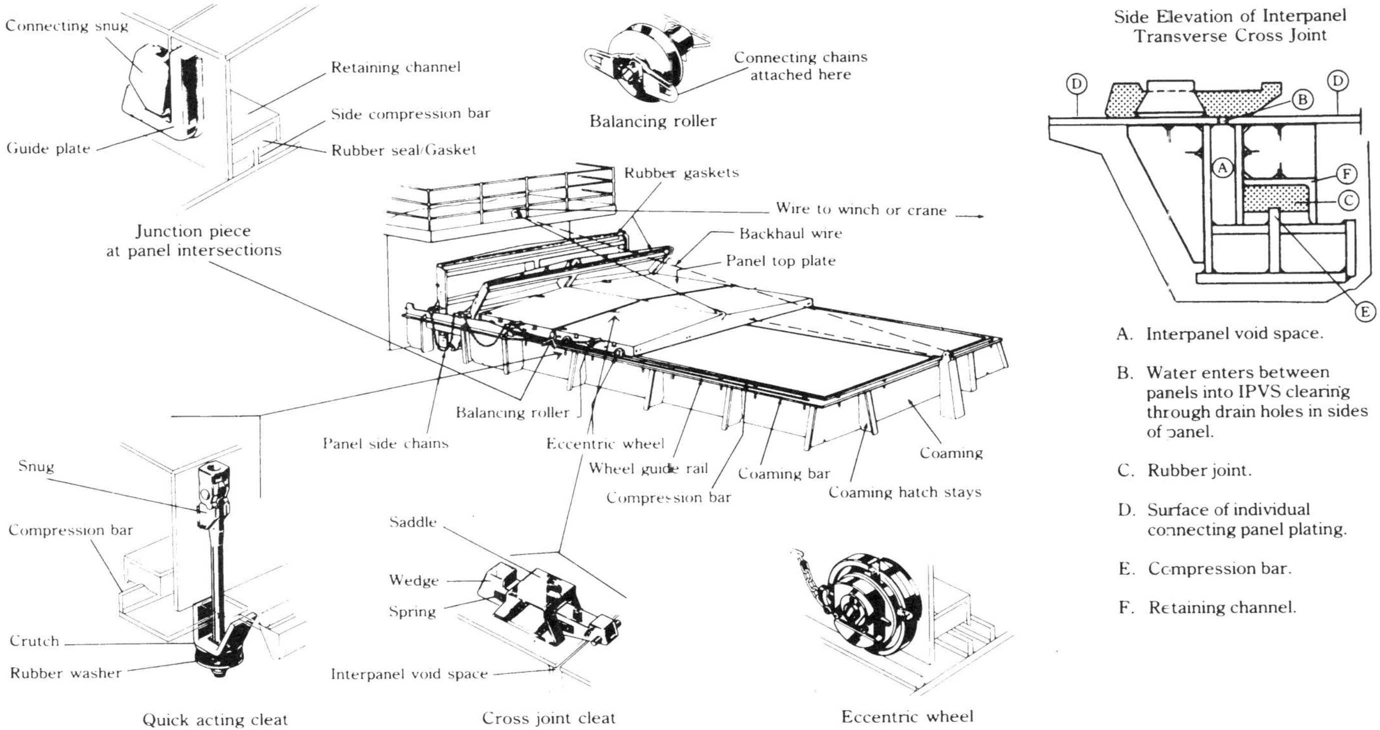

Fig. 50. Single pull hatch covers (MacGregor type).

Leakage of hatch covers—the carrier's defence

One of the worst things that can happen to steel products during a sea voyage is to sustain wetness through contact with seawater, and where mild steel is concerned the corrosive effects through contact with seawater may commence immediately and rapidly advance. In view of this, it can be understood that every possible effort should be made to ensure that watertightness of the steel hatch covers is maintained throughout the voyage.

Seawater contact with steel products will inevitably result in claims being entered against the carriers. In such circumstances, if the owners'/carriers' interests are to be successfully defended, it will be encumbent upon the master of the vessel to prove that maximum precautions were taken at the commencement of the voyage to ensure that the hatches were in a seaworthy condition. Such provisions enter within the scope of the carrier's obligation under the Carriage of Goods by Sea Act to exercise due diligence, before and at the beginning of the voyage, to ensure that the vessel is seaworthy.

The fact that damage claims in respect of steel cargoes can be successfully defended, in certain circumstances, when leakage of steel hatches is responsible for such damage, is proven by a hearing which took place in the United States Court of Appeals in 1970. The vessel concerned was the *Sabine Howald*, which loaded a cargo of steel products in Antwerp for discharge at United States East Coast ports. While crossing the Atlantic Ocean, the vessel encountered extremely heavy weather and suffered various structural damage. The trial court found that as a result of shipping heavy seas over decks and hatches, seawater in substantial quantities had penetrated the interpanel joints so causing rust damage to the steel cargo.

All arguments being concluded, the final decision of the court was that they were: "satisfied that the *Sabine Howald* was a seaworthy vessel when she left Antwerp on 5 December 1965; the ship being certified in the highest class with Germanischer Lloyd. Throughout the voyage, the ship was operated in a good and seamanlike manner. There was no negligence on the part of the carrier. The damage to the steel cargo was caused by violence of the wind, sea and particularly by the resulting cross-seas which, through wrenching and twisting the vessel, set up torsions within the hull which forced up the hatch covers and admitted seawater to the holds." The defendants—owners of the vessel—were not held liable for the damage. One important and principal point, without which the vessel would not have been successful, was that all the evidence showed that the MacGregor steel hatch closing appliances were in good condition at the commencement of the voyage.

Structural stresses in a seaway

What is referred to as racking stresses within a vessel's structure, are stresses brought about by the vessel rolling in a seaway, when the ship's structure has a tendency to distort transversally. Hogging stresses within the structure come into effect longitudinally when the vessel is supported by a high wave amidships, when both ends droop, and the structure tends to arch in the centre. In such circumstances the vessel is said to be hogged.

Sagging stresses operate when the vessel is supported by a high wave at each end, so that the centre of the hull of the ship tends to droop, likewise causing longitudinal stresses—in this condition the ship is said to be sagging. These stresses are active throughout the entire ship's structure, increasing and decreasing according to the force of the weather being experienced at sea. The effect of such stresses on the ship's hatchways is to cause the structure to develop a wavy motion consisting of temporary alternating distortions. The steel hatch panels are not part of the vessel's structure: they are a ship's fitting. Therefore, as they are only attached in places to the vessel's structural parts, when in a closed sea-going position, they do not move completely in unison with the hatchway when it temporarily distorts through being subjected to racking, hogging and sagging stresses. Furthermore, the panels are often pounded by heavy seas, causing further stress and heavy vibration. *In consequence of this, the rubber packing or jointing around the peripheries of the panels, which should effect a watertight seal, must be sufficiently resilient and in good condition in order that it can readily adjust to varying pressures which are imposed upon it.* At the same time, the resilience of the rubber packing should also be such as to fully compensate for any distortions within the area of the hatch coaming.

Steel cargoes make for an over-stressed and over-stable ship

A cargo of steel is, owing to its high density, an enormous concentration of weight in the bottom of a vessel. Usually, with a full cargo of steel products the vessel sails with full deadweight capacity, so leaving no margin which would permit taking ballast in the upper wing tanks, whereby the ship's range of stability might be reduced to within more reasonable and seamanlike proportions. On the contrary, a ship which proceeds to sea with a full load of steel products on board will be grossly over-stable, the period of roll being short, vicious and tremendously stressful to the ship's structure.

One of the aims in compiling a cargo plan is to place the weights, comprising the various parcels of cargo, in such positions in the vessel's holds as to achieve an even distribution of weight throughout the length of the ship. This often proves impossible with a steel cargo, if the voyage is to remain a viable proposition, when considering the types of cargo offered for shipment and the intended itinerary of the ship. In most instances, uneven weight distribution in a longitudinal direction is, in a manner of speaking, more the rule than the exception. Such an arrangement of weights will create localised concentration of stress in one particular area which during heavy weather periods may be magnified many times. Such stresses concentrated at one position across the vessel's structure can locally test the weathertightness of the hatches to the absolute limit.

High and low density cargoes compared

In the 1960s there was a lively two-way trade, across the Atlantic, in steel and grain between North America and northern European ports. In the Mexican Gulf vessels loaded grain for north European ports, and these cereal cargoes, in spite of extremely heavy weather experienced en route, were discharged and received without complaint. A full cargo of steel was loaded for the return voyage to the United States, and with similar heavy weather conditions being experienced the joints of the steel hatches leaked, seawater entered the cargo compartments and heavy claims in respect of the steel cargo followed. The larger the hatchway, the more susceptible the hatch panel joints will be to leakage. From what has been said, it is evident that if leakage of the hatches occurs during the course of the voyage it is necessary that their condition will be such as to withstand detailed scrutiny. Numerous things which seem unimportant, and are unimportant to a reasonably minded surveyor, can be built up in a report to give the most unfavourable impression to someone not directly and practically acquainted with the actual subject of ship's hatch closing appliances.

Hose, chalk and ultrasonic testing (watertight integrity)

Before commencing loading a cargo of steel products the steel hatches should be tested for watertightness. There are those who are of the opinion that these tests, performed when the vessel is in the static condition in port, serve no useful purpose, because after the test there is no guarantee that the hatches will remain watertight once the vessel puts to sea. This may be so, but the tests do prove that if the hatches are leaking from a test in port, then they will certainly leak once the vessel is at sea. Naturally, actual conditions which will be experienced at sea cannot be reproduced in port. Tests carried out in the loaded, sea-going, condition, when the ship's structure will have adjusted to the disposition of the various weights placed in the holds, etc, will produce the most reliable results. Testing the hatches just prior to commencement of the sea voyage ensures that all known measures have been taken, and that there has been shown no lack of due diligence, and that the vessel was seaworthy with regard to the watertight integrity of the hatch closing appliances.

The actual design of steel hatches by which watertightness is achieved is an ingenious invention. It should be realised that watertight integrity cannot be maintained only by painting, scraping and oiling. *Every component part of the hatches contributes to achieving watertightness and if this aspect is overlooked leakages will develop and never be properly arrested.*

Lack of appreciation by ships' crews in the proper method of manipulating the hatches is in many instances responsible for apparently incurable and constant leakage. In opening and closing the panels, continual banging and slamming must be avoided, as this can result in distortion, which once developed to some degree is usually not readily apparent, but nonetheless can contribute to the development of persistent leakage of the hatches.

From whatever cause leakage develops, something has to be done to prevent its continuance. At a point where leakage occurs the rubber gasket/seal or joint may appear to be heavily compressed, torn and damaged to some degree. *Often the damaged section is cut out and a new piece is fitted in its place. This is wrong. The entire joint should be renewed, as to part-renew creates a situation where at one position, where new rubber is fitted, there is a differing coefficient of compression to the adjacent old jointing. Part of the packing is therefore less resilient than the rest; this can lead to leakage when the ship is at sea.*

It may be discovered that the repair is unsuccessful and generally, in such circumstances, the order is given to tighten up all round. The quick release cleats (see Fig. 51(c)) are hardened down to the maximum; a practice which is not going to improve the situation. On the contrary, such action is a step towards impairing the good condition of the fittings and preventing them from properly performing their contribution to watertightness when the vessel is working in a seaway.

The purpose of the securing devices is to keep the hatches in place when the vessel is at sea; in the static condition they make no contribution to watertightness, e.g. when the vessel is in port. Recall that when the hatches are in the closed position it will be seen that the perifocal lower edges of the vertical side plating of the panels are resting on the coaming bar—this being referred to as steel-to-steel contact (see Fig. 51(a) and (c)). It is obvious that any amount of hardening up, or tightening of the quick release cleats cannot possibly change the situation by imparting more pressure to the transverse joint where the leakage problem is usually situated.

Additional tightening of the quick-release cleats, above and beyond what is required to hold the panels in place when the vessel is at sea, only contributes to preventing them from fulfilling their designed performance. By hardening up these cleats the rubber washers are crushed and in older vessels the crutch beneath the coaming bar may be distorted or even destroyed. When the vessel is working in a seaway there is a movement of the coamings which is not in complete unison with the hatches. Pliability of the quick-release cleat rubber washer affords compensation for this movement, so permitted the compensating effect of the transverse rubber gaskets to work unhindered. Without this function of the quick-release cleat watertight integrity of the hatches can become impaired. From this explanation it will be understood that when the vessel is at sea these fittings do indirectly contribute something to maintaining watertightness of the hatches.

Not having rendered the hatches watertight by hardening up the quick-release cleats, attention is turned to the cross-joint cleats/wedges (see Fig. 52) which bridge the transverse joints between the individual panels. If these are tightened more pressure can be applied to the transverse joints. In order to achieve this, in many instances, a piece of metal is welded to that part of the panel edge (the leading edge of the after panel) opposite to the actual wedge, which is then driven over the top of the applied metal by using a sledge hammer. Done on a regular basis this causes the wedge to become permanently bent and in due course pressure is released from the joint. Further to this, the first step has been taken towards destroying the purpose of the hatch design, preservation of which is necessary to ensure continuous watertightness of the joint.

With the hatches in the closed position, one part of the transverse rubber gasket retaining channel has steel-to-steel contact with a number of bearing plates (see Fig. 53). In such a position the composition, size and position of the rubber gasket is so arranged that a designed pressure is exerted to the upper edge of the compression bar against the rubber when steel-to-steel contact is made. Taking into consideration the steel-to-steel contact it is quite evident that any further downward pressure applied by manipulation of the wedge pressure at points along the surfaces of the panel (see Fig. 52) will not increase pressure on the joint unless

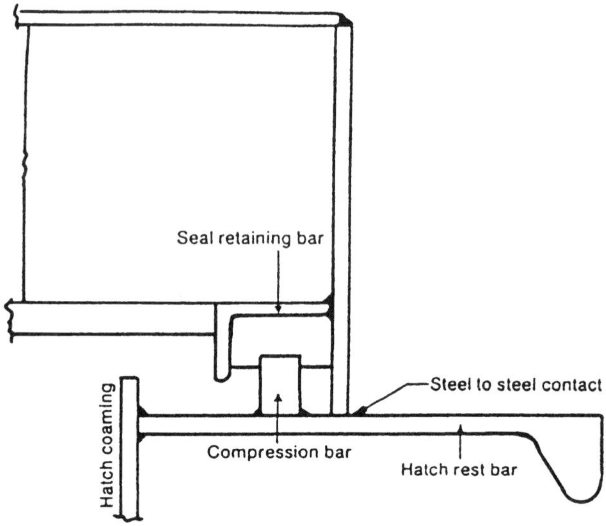
Seal retaining bar
Steel to steel contact
Hatch coaming
Compression bar
Hatch rest bar

(a)

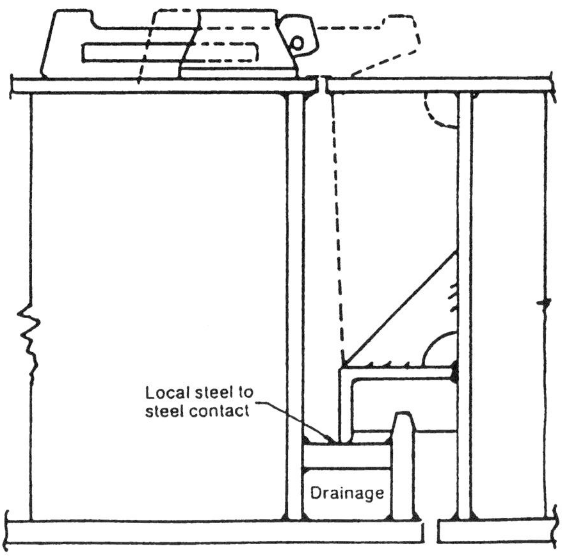
Local steel to
steel contact
Drainage

(b)

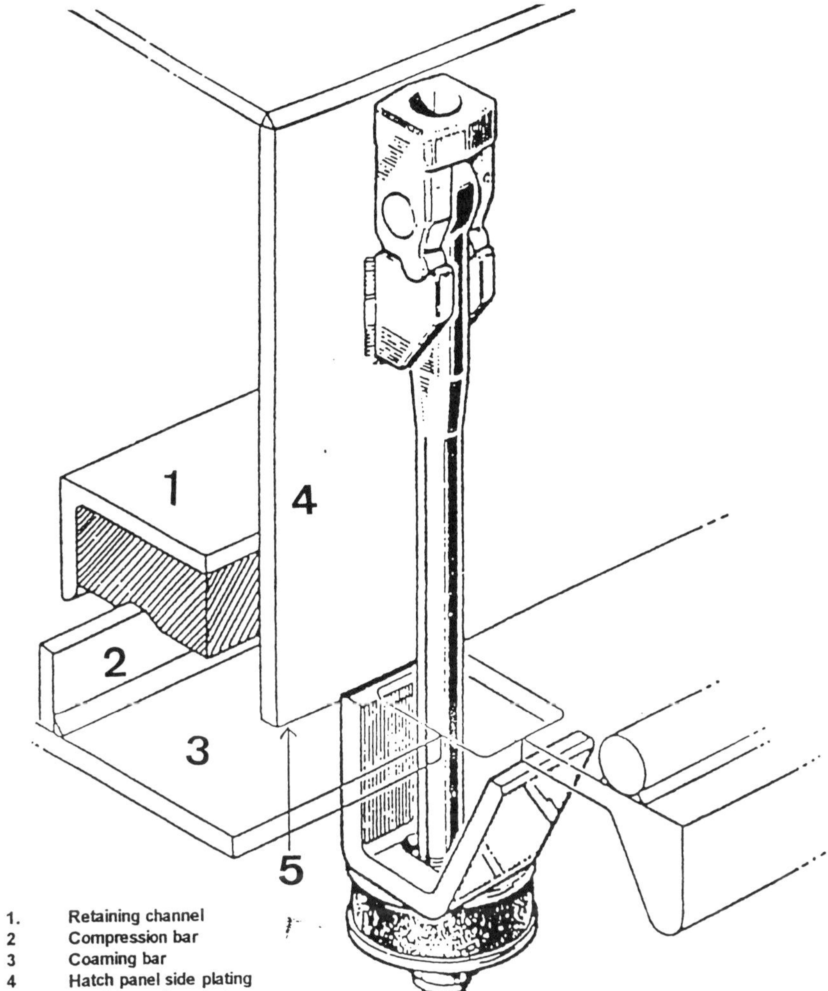

Fig. 51(a), (b) and (c). Design of steel watertight hatches.

Reproduced by kind permission of the MacGregor Group.

structural deformation takes place. When all is in a satisfactory condition as the designers intended, any movement of the hatch at sea will be compensated for by the resilience of the gasket, so maintaining a watertight seal.

Notwithstanding what has been said above, the undesirable measures taken to achieve watertightness can be effective while the vessel is in the static condition. They will not be so effective when the vessel is working in a seaway. Such corrective measures will upset the entire balance of the joint, crush the rubber gasket to some extent and at the same time in doing this the steel-to-steel bearing plates will be bent downwards or even broken—future leakage of a joint so manipulated, or

Fig. 52. View of cross-joint wedges.

Fig. 53. Steel-to-steel contact-bearing plates. Note: grooving on nearest plate.

treated, is ensured. Failure of ships' personnel to appreciate the intricacies involved in the design of ship's hatches and the method employed to achieve a watertight seal, leads to unintentional abuse of the fittings when leakage occurs—such leakage must for obvious reasons be arrested.

Lack of understanding of what damage is being done in achieving watertightness, in the short term, as related above, leads to continuing problems, with sporadic or on-going leakage and resulting heavy claims through moisture damage being sustained by the cargo.

The question of how the hatches should be tested for watertightness is of major importance. Depending upon what has gone on before and what amount of hammering is required to drive home the transverse cross wedges, *the very act of putting the securing devices in place, when testing in the static condition, may very well lead to a false impression as to the actual cargo-worthiness of the hatches.* If pressure applied by the cross-wedges is instrumental in achieving watertightness then the very principle of the design of the hatches is being violated, which through localised distortions applied in achieving watertightness in this way will certainly lead to leakage when the vessel is labouring in a seaway under severe weather conditions. The question arises as to what course of action should be taken if all is in apparent good condition, i.e. as far as the apparent condition of the component parts of the hatches is concerned. Items which might be overlooked are:

—Deformation of the steel-to-steel contact points in the area of the transverse joint (see Fig. 51(b)).
—Abnormal grooving of the steel-to-steel contact points (see Fig. 53).
—Unevenness of the coaming bar where steel-to-steel contact is made. Excessive corrosion can create such a condition (see Fig. 51(c)).
—Incorrect adjustment of the panel connecting chains which causes off-centre contact of the compression bar with the rubber seal.
—Slight deformation of the panel surface plating caused by the carriage of previous deck cargoes.

If general maintenance of the hatches is satisfactory in every respect, then something of a more sinister nature may be involved if leakages persist. Wire operated panels should be controlled when opening and closing; if not, the constant slamming and banging experienced can, given time, cause distortions which lead to a leakage problem. In any vessel, irrespective of age, where leakage is consistently experienced, in spite of all efforts towards rectification, the following of incorrect welding procedures at time of manufacture may be at the root of the problem. When all points mentioned have been checked and found in order and leakage still persists, hatch patentees should be called in to give advice.

Hose test

When this test is carried out a fire hose is used and coupled up to the deck service line on board the ship. According to one of the leading classification societies the hose should be held at a maximum distance of 1 m from the joint to be tested and the water pressure to be applied is stated as 20 kgf/cm^2. The rules also recommend the same test for watertight doors, watertight bulkheads, tunnels, flats and recesses etc. Unfortunately, when hose testing hatches, the stream of water cannot be played directly upon the actual joint, as it may well be when testing other watertight connections mentioned in the rules. In the closed position, the edges of the two adjacent panels form a transverse slit across the width of the hatch. The actual joint is situated some distance beneath this slit between the panels. Therefore, when the hose is directed towards the joint the pressure of the water is dissipated upon the surfaces of the panels.

In an effort to build up some pressure on the actual joint it is recommended that the drain holes of the joint under test be plugged with cotton waste, to port and to starboard, thereafter filling the interpanel void space above the joint with water. Opponents of this method of testing complain that such a test imposes hydraulic pressure upon the joint. This may be so to some small degree, but it will be nothing compared with the pressure applied when a large sea breaks across the hatches and buries them momentarily under water. The disadvantages with this type of test are that it is time consuming and two surveyors are necessary in witnessing and controlling the efficiency of the test. When cargo is on board, if the hatches leak the cargo may sustain damage. The results obtained are not accurate, because where the water is observed to penetrate the joint in the hold may not necessarily correspond with the actual defect causing the leakage (see Ultrasonic testing).

Chalk test

One of the disadvantages of a chalk test is that inconclusive results can eventuate from the test. With a chalk test the aim is to ensure that between the rubber joints and the compression bars sufficient contact and, more important, sufficient pressure exist. Chalk is applied to the compression bars after which the hatch covers are battened down. They are again opened and the individual joints inspected. When a clear regular chalk impression is observed on the joint it is assumed that sufficient pressure exists between the actual joint and the compression bar; however, when the chalk mark is irregular or less pronounced in some areas than it is in others doubt arises as to the watertight integrity of the appliances and endless inconclusive discussions can ensue. Further to this, chalk tests are very time-consuming in an age when time definitely waits for no man. This is a test which should be made together with other tests after repairs have taken place.

Ultrasonic testing

It is considered that hose tests and chalk tests are less reliable than ultrasonic tests but of course they are acceptable if ultrasonic testing equipment is not available.

Testing by ultrasonic means gives the exact location where defects are situated—not approximate areas of leakage but clearly defined precise points of leakage. Ultrasonic equipment is easy to use, easy to understand, does not require the assistance of crew members, is quick to operate and most importantly can be used on loaded/unloaded vessels without putting the cargo at risk. The conclusions reached with regard to ultrasonic hatch cover testing are:

—It can be used on loaded or partly loaded vessels without causing damage to cargoes.

—It can be used by one man.
—It reveals the exact location of any lack of tightness.
—It is less time consuming than either a hose or chalk test.
—It can be used when air temperatures are below zero.
—It is accurate.

The equipment used by one of the leading manufacturers in ultrasonic hatch-cover testing equipment consists of an ultrasonic transmitter and receiver. The ultrasonic transmitter transmits ultrasonic waves between 36.7 and 40.7 kHz. The ultrasonic receiver detects the ultrasonic frequencies and converts them to audible frequencies. The ultrasonic detector measures and shows the maximum value of an observed sound signal in dB. An analogue LED indicator helps the user with the calibration of the receiver in function of the ultrasonic source. The ultrasonic detector has an audio-exit by which the amplitudes of the received signals are shown by means of an oscilloscope.

In practice, the ultrasonic generator is placed inside the hold. The range of the generator is 60 metres and is equipped with 13 sound sensors. When all of these sensors are placed in operation, the density and the homogeneity of the sound waves produced in the hold increase by the crossing of the ultra sounds, as a result of which the density of the sound produced in the hold can be better controlled. The operator can now go on deck and check the hatch-cover tightness by using the detector. After testing, a visual inspection can be performed of the potential points of leakage and measures can be taken.

Whenever a test with ultrasonic equipment has to be performed, it has been found advisable, prior to testing, to explain the principles of the method in detail to the vessel's staff. The test is carried out in the presence of the ship's staff; the situation is then jointly determined and, depending on the results of the test, deficiencies ascertained, if any, are afterwards jointly established and agreed.

Two methods for testing the watertightness of the hatch-covers can be used. In the first method a reference minimum is used. The reference minimum is the lowest value achieved during preliminary testing around the panel's joints. Where the level of ultrasounds detected and measured is higher than 15 dB from the preliminary test's reference minimum value, this would indicate a position of leakage.

In the second method, the operator should primarily measure on the digital display the levels of ultra-sounds detected with an open hatch or at the open access hatch to the hold. He should then measure the ultrasounds detected on the hatch cover's joints when the hatch is closed. Where the level of ultra-sounds detected and measured is greater than half the value of the open measurement, this would indicate a position of leakage. The method to be employed evidences itself upon practical use of the equipment. Tests carried out have proved that sometimes the second method cannot be used. Usually, when testing, the hatches are closed, so open measurements can be registered only at the escape hatches to the holds. However, whenever the sound waves are hindered, for example by tweendecks which are open and the covers of which are standing in an upright position, the reception at the escapes is reduced. In such circumstances, when testing, the values received may not be reliable and this would, for example, manifest itself by higher values being received than reference values recorded at the escape hatches. Although the higher values recorded will signify that the hatches are not watertight at the position indicated, doubts will remain where the lower values are concerned.

Therefore, extra attention has to be given to the circumstances under which the tests are carried out. The second test method is an effective and reliable short-cut method of testing devised to save time, which should be used under ideal conditions, i.e. conditions evidenced from the results of testing as explained above. If the second test method proves to be unreliable, as determined by the experienced operator, then the more time-consuming first method must be used. Whichever method is employed, it is always less time-consuming and more accurate than any other known method of testing. The test is illustrated as follows: see Fig. 54(a), (b), (c).

What has been stated above with regard to the use of the equipment has stood the test of time and there has never been an instance where there was reason for doubt as to its reliability. However, as time has passed more interest has been generated in the system, which is now recognised by most of the classification societies. Many interested individuals have applied their minds to the use of the equipment and some have reached conclusions as a result of carrying out their own tests. It is for this reason that surveyors, when carrying out surveys on hatches for watertight integrity, should be aware of not only the views of the classification society concerned as to the use of the equipment but also whether a concerned vessel's P & I Association has formulated any idea; opinions do vary to some degree.

After testing, a Test Report with diagram should be made out, indicating the read-out of maximum and minimum recordings, and also showing areas of potential leakage. Only competent and certified operators using approved equipment should be allowed to use the ultrasonic equipment for testing hatches for watertight integrity. The equipment used must be accepted and certified by the vessel's classification society.

High standard of maintenance is necessary

Every component part of the structure and fittings of steel hatch panels contributes to their watertightness, which can only be achieved and maintained providing all is kept in good condition by affording a high standard of maintenance. If rubber packing is seen to be chafing or tearing, or if any defects of any nature develop from operating the panels, something is wrong with the assembly. In such circumstances, the panels need either adjusting, realigning or even some form of repair which will permit them to be operated smoothly. The cross joints between the panels are by far the most important and most susceptible to leakage; this applies especially to the ends of these joints near the sides of the panels which are inclined to suffer more heavily from wear and tear. In many instances, seawater leaks through these end areas and gathers in the hatch coaming inside the inboard drainage system to eventually run aft, bank up at the after end of the hatch, possibly because the drain holes are blocked, and spill over the coamings onto the cargo.

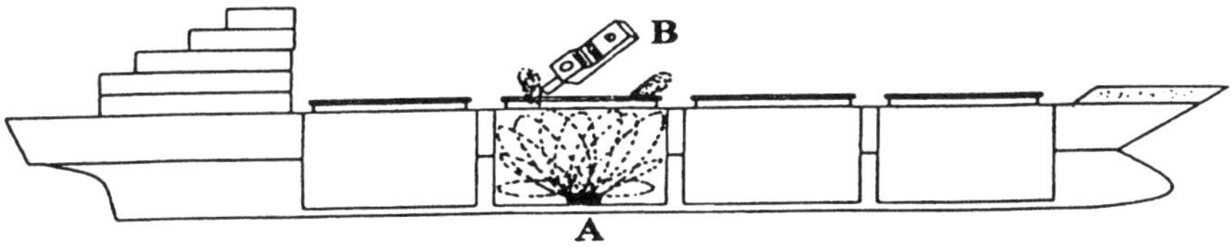

A = The Transmitter B = The Receiver

(a)

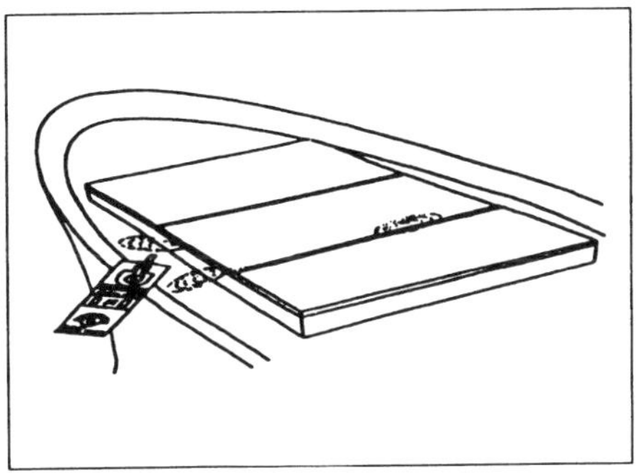

Testing of the hatchcover tightness is achieved by using the SDT 150 detector. Eventual leakages are found by detection of sound produced by the transmitter.

(b)

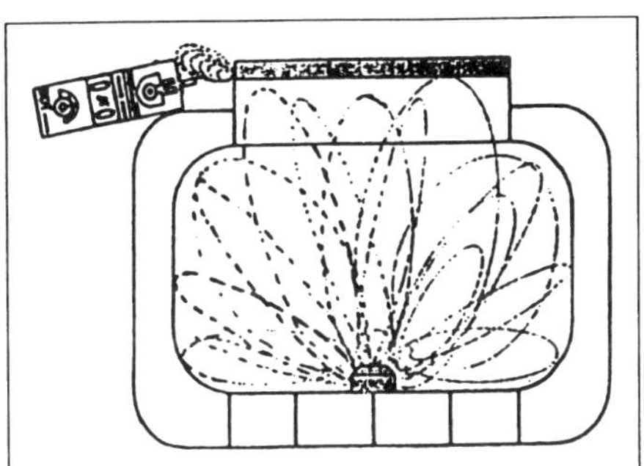

The ultrasonic generator SDT 13 placed in thc hold has a range of 60m and with 13 soundsensors it ensures a sufficient density of sound in the hold

(c)

Fig. 54(a), (b) and (c). Ultrasonic testing of hatch cover watertightness.

Repairs to rubber joints

With regard to the rubber jointings which are installed to effect a watertight seal, these must always be in good condition. They should present an even surface along their entire length and be free from any defects whatsoever. The imprint on the surface of the rubber seal must be even and should be directly in the centre—if it is not, then adjustment of the connecting chains is probably necessary. If any defect in the rubber does arise, damaged sections should not be cut out and replaced with new rubber packing. Such action results in uneven pressure on the joint, which promotes the possibility of leakage. In such circumstances the entire rubber seal should be replaced. The retaining channels, holding the rubber packing, must be kept free from corrosion or rust scale; likewise, the compression bar should be in perfect condition. The existence of physical defects, such as dents or bending damage to retaining channels and compression bars, should not be tolerated but be rectified immediately. This is because localised squeezing of the packing, within the retaining channel, or the development of an area where the retaining channel has been widened, results in varying degrees of pressure in the position of the damage as opposed to those areas unaffected by such damage—this leads to leakage in the joint and subsequent damage to cargo.

Compression of rubber joints

As an example, the size of the rubber forming the seal in the MacGregor single pull type of hatch cover may be 71 × 32 mm in cross section. Compression of the rubber seal, with the hatches in the closed position, should not exceed 25% of the thickness of the rubber— this is 8 mm.

Greater compression than this will result in the rubber seal not returning to its original form when pressure is released, so impairing the efficiency of the seal. With use, over a period of time, a permanent compression mark will develop. With rubber seals of the dimensions referred to, *it is considered that a permanent compression mark of a depth of 6 mm is a clear indication that the resilience of the rubber has become impaired and will no longer compensate for the movement of the hatchway when the vessel is working in a seaway.*

The rubber packing or seals, which form the watertight joint, should be kept constantly under surveillance. It should be ensured that they never come into contact with paint, grease or rust. Foreign matter, which might become attached to the joint surfaces, should be cleaned away. With regard to the maintenance of the covers the following should be considered.

Guide to maintenance

Quick acting cleats

—Keep clean and with threads well greased.
—Avoid painting over the threads or oil on the rubber.

Rubber seals, gaskets and packing

—Keep these clean and avoid paint coming into contact with them when painting the adjacent steel work. Clean away all foreign matter. Regularly check the depth of the permanent imprint caused by the compression bars.
—When renewing rubbers, use only the rubber specified by the hatch patentee, because the actual size, shape, compressibility, elasticity, hardness and construction of the material are all specifically so arranged to satisfy the hatch cover design, including size and weight of the covers. All rubber gasket replacements should be effected by professional repairers and not by the ship's crew, as there is a certain expertise involved in this type of work.

—*Never replace damaged sections of rubber with short lengths, always renew the entire length of the joint.*

—Keep the eccentric wheels clean, well greased and running smoothly. The trackways must be kept clean and free of rust.

—Retaining channels and compression bars should be completely free of rust or scale, which can impair the watertightness of the joint. Any dirt, grease or foreign matter should be cleaned away.

—Check that the compression bar imprint is always in the centre of the joint.

Cross wedges

—Check when hatches are in the closed position that the wedges are not slack and also that the spring clips are operating properly. Bent wedges should be replaced.

Connecting chains

—Check these chains for tension and if adjustment is necessary seek the assistance of hatch patentees.

Drainage system

—Keep all guttering and drainage holes free from rests of previous cargo, and clear so that they may function as drains.

Steel work

—The thickness of the top plating of the MacGregor single pull steel hatch panels is usually about 7–8 mm. This should be maintained by avoiding the development of penetrating rust spots, corrosion and rust scale

Records

—A consistent, conscientiously kept record of all maintenance work on the hatches should always be available. Such a record would be useful in the event of claims arising if the hatch joints leak.

Persistent leakage of rubber joints

Hatch cover joints which still leak, in spite of all measures taken to render them watertight, indicate that something of a more sinister nature might be involved, such as distortion of the panels, which can be caused by the following:

—Hatches manufactured in some countries under licence are distorted by incorrect welding procedures.

—Deformation of the panels can develop through insufficient control when opening, so resulting in banging and slamming of the panels against each other. This can be very much pronounced on opening hatches when the ship trims heavily by the stern.

—Dents of a serious nature in the surface of the panels may develop perhaps through contact with grabs, etc.
—Excessive overloading of the panels with cargo carried on deck can cause distortion of the panels.

Taping of cross joints

When watertightness of the joints is suspect it might be decided to apply one of the patent sealing tapes which consist of bitumastic type strips, having a width of about 75–150 mm. Such a decision might also be taken as an added precaution against leakage owing to the nature of the cargo being carried. For reasons already stated, a cargo of steel products might be considered to qualify for the application of such tape, sometimes referred to as "masking tape".

Preparation of panel surface

The remnants of previous tape applications should be completely removed.

—By means of a scraper and/or wire brush, all loose rust, scale and paint flecks should be removed.
—The surfaces in the area of the joint must be free of dust, dirt, oil or grease and rests of previous cargo.
—The surfaces to which the tape will be adhering must be completely dry before application.

Steel patent covers are provided with several fittings, e.g. wedges over the cross-joints. A tape can never seal perfectly tightly around these fittings. Seawater will always enter and will then be trapped inside until it finds its way into the hold.

After all of these points have been satisfied there comes the question of cutting the tapes to the required lengths, stripping off the cover of the adhesive, applying the tape correctly and treading it, by the action of the feet, onto the surfaces of the panels in the area of the joint.

Lack of time available to apply tapes

Even when properly applied, and given time to bond with the surface of the panel, the tape will not be effective in many instances because invariably seas wash away some of the tape, vibration of the panels causes the tape to loosen, and an efficient seal in the area of the cross wedges is very difficult to achieve. Preparation of the bonding area and application of the tape take time, and time is necessary if the tape is to be properly applied. In most instances the vessel's sailing time is set for an hour close to completion of loading the cargo.

Quite often hatches have to remain open in order to facilitate lashing and securing operations right up to the moment of departure. Furthermore, if panel surfaces are wet or damp, the tape cannot be effectively applied, and it may be necessary to use blow lamps to dry the areas of the panels involved. There is an ever present desire to reduce ship's crews, and with the limited number of personnel available to apply the tapes, preparation is inadequate and the entire operation is more often than not pushed through haphazardly, so that the tapes are not effectively put in place. Add to this gloomy picture the hours of darkness, and there is every indication that the purpose of applying the tapes is defeated before the actual sea passage commences.

Taping joints causes deterioration

It is also considered that, in the long term, these tapes have a detrimental effect upon the metalwork faces within the transverse interpanel sections. The space between the panels is sealed off by the tape and becomes an unventilated void, wherein there is an atmosphere created which promotes rust and corrosion of the metal. Experience seems to indicate that even if a high standard of maintenance is followed, the metal parts referred to will deteriorate much more rapidly than would be the case if tape was not used.

Tape applied to hatch panel joints should be carefully considered. If tape is used and the hatches leak with resulting damage to the cargo, claimants will contend that the tape was applied because at the outset of the voyage the watertightness of the joints was in doubt. In such circumstances, if exceptionally heavy weather was experienced the rights and immunities, which might have been exercised by the master under the Carriage of Goods by Sea Act, will be prejudiced.

In defence the use of sealing tape, it has been argued that tape was applied only for the express purpose of guarding against the eventuality of exceptionally heavy weather being encountered. The message implied is that in anticipation of abnormally heavy weather being experienced on voyage, and in the interests of the preservation of the cargo, which, being steel, places exceptional stresses upon the vessel's structure, the ship owner indulged in additional expenses over and above what might normally be expected of him.

On a voyage where no tape is applied and hatches have leaked owing to stress of heavy weather, the fact that marks on the hatch panels indicate that on previous voyages tape was used can be employed in future debate where liability is involved. The pros and cons of the entire subject lead to one conclusion. *There is no substitute for a high standard of maintenance, supported by records showing consistent attention to detail in an effort to preserve the operational functions of the hatches as was originally intended by the hatch patentees and manufacturers.*

Points for consideration

The International Load Line Convention. SOLAS regulations, and rules of the various classification societies all indicate that the maintenance of weathertightness of steel hatches shall be arranged in such a manner as to preclude penetration of water through openings in the freeboard deck under normal weather conditions or, as is stated by one society, in any sea conditions.

From the above, it would seem that any manifestation of lack of watertightness of the covers, such as might appear under a hose test in port, would mean that the vessel was out of class until such time that the rules of the classification society and international rules and regulations, had been satisfied. Unfortunately, this is not so in practice, and there are cases on record where within days of recertification of maintenance of class the steel hatch closing appliances have had to undergo major repairs in order that they be made watertight.

Fig. 55(a). The ship in question had been laid up. An attempt was made to re-class as is but without success.

Fig. 55(b). Beyond redemption.

Fig. 55(c). Deterioration of metalwork and rubber joints.

Fig. 55(d). Wastage of surface of hatch covers.

Fig. 55. Steel hatch closing appliances.

When the appliances are in good condition, hatch patentees guarantee watertightness with the hatches in a static condition. Owing to the stresses imposed upon a vessel's structure by various types of cargoes, and under changing loaded conditions, it is evident that integral movement of the component parts of the vessel's structure, in the area of the hatchways, cannot be compensated for by the resilience of the rubber seals to satisfy every situation. For this reason, there is no guarantee of weathertightness when the vessel is in a seaway, but weathertightness should be assured under normal weather conditions. Consequently, from the point of view of the hatch patentees, weathertightness does not preclude some leakage of the hatch joints. Notwithstanding this, the fact remains that if the hatches leak during the voyage and lack of care and maintenance of the hatches can be proved, it will be concluded that at the outset of the voyage the vessel was, at least, unseaworthy from a cargo carrying point of view.

With regard to the fittings on steel hatches which are installed for the purpose of securing the panels in place, e.g. quick acting cleats and cross joint cleats, the actual number fitted are the minimum required by the classification societies. At the commencement of a voyage, any of these fittings which are missing, or are inoperable, will affect the watertightness of the hatch closing appliances.

There is no doubt that the greater the dimensions of the hatchway as opposed to the length and beam of the ship, the greater the tendency towards leakage when the ship is working in a seaway. For instance, a coastal ship of 150 m in length was equipped with one single long hatchway of 50 m. The steel hatches were in excellent condition, but the cargo of steel beams and channels sustained heavy damage through contact with seawater when the hatches leaked during heavy weather. The master of the ship claimed that this was due to stress of heavy weather, but cargo interests contested this on the basis that the hatchway was too long, highly susceptible to leakage when the ship was carrying a high density cargo, and the ship was not suitable to transport a cargo of steel products.

Other types of joint

From time to time the type of MacGregor hatch referred to has some variations. For example, some are chain operated and that is usually the only difference. Others are hydraulically operated with half the panels folding at one end of the hatch and the other half at the opposite end. The intermediate panels are joined together with hinges and where they joint there is a different type of transverse joint to that shown on Fig. 51(a), which would tend to close with pressure applied from above by gravity and on-coming seas. The type of joint used between the intermediate panels of the hydraulically operated hatch arrangement in question, is fitted with a rubber seal placed vertically, which is penetrated with a compression bar exerting pressure horizontally as shown in Fig. 56. This type of joint, if not kept in top condition, is highly susceptible to leakage. Pressure is eased from the joints when the hatches are in the closed position if there is wear-down on the panels connecting the hinges. There are no cross-wedges fitted to these joints. The centre joint separating the panels may be a similar joint, or it may be the same joint as shown on Fig. 51(b), in which case the characteristic cross-wedges with this joint would be

Fig. 56. The type of joint often found between hinged panels which relies upon pressure applied horizontally for watertightness. This type of joint is highly susceptible to leakage if maintenance is neglected.

fitted. Whatever type of hatches are fitted and however operated, they are all of a design involving the system of steel to steel contact.

STEEL PRICES

The prices involved are FOB, CIF, C&F, SMV (see Appendix) and the export price. The first three mentioned can be found in the relevant commercial invoice associated with the sale of the goods. The SMV (sound market value) usually refers to that value upon the day of discharge at the port of destination of the goods. The SMV can usually be supplied by a steel trader in the port, or country, where the goods were landed at the end of the voyage. However, to obtain these values years after the incident, at the request of those in need of such information, is usually a formidable task, but the value can be estimated by a local surveyor, based on his experience, in the delivery port. If a local steel trader sold similar goods he may still have records available, or that value should be in the report of the surveyor who sold the goods around the time of the incident. To ask a surveyor in one country to estimate the SMV of goods sold in another country is something he cannot do with any degree of accuracy.

Situations can arise whereby, for example, a vessel for some reason has to discharge the cargo at a way port, such as for structural repairs to the vessel necessary to continue the voyage. The cargo may be eventually reloaded or transshipped. Whatever the case may be, the cargo must remain ashore for a period of time at the responsibility of the carrier. Therefore, he would normally insure the goods for the period in question and in such circumstance he would be searching for a value for the goods. The shipper could be contacted and asked to supply the invoice prices. Failing this, the export value could be used, as it would be extremely unusual for the entire cargo to be lost while it was stored ashore awaiting shipment. Any damages sustained by the goods, mechanical or otherwise, would not be too devastating and would in all probability be covered by the export prices taking the highest from the scale below.

The export prices can be obtained from certain periodicals (see Appendix). These prices do have hover values around the world, which fluctuate between more-or-less established highs and lows. The values, of course, depend upon supply and demand, the type of steel involved and the measurements of the material. Some of these prices could be expected to be as follows per metric ton.

CRSS in coils	US$ 350–430
HRSS in coils	US$280–310
HDG in coils	US$390–470
Re-bars	US$235–290
WRIC (HR)	US$240–290

The situation is that these prices can, from time-to-time, go temporarily much higher, and so volatile is the market that it is not beyond the realms of possibility for the prices to go temporarily lower. There are certain types of steel which are not shipped in large quantities, e.g. pre-stressing wire in coils, and associated export prices for these goods are often difficult to get hold of.

STOWAGE OF STEEL SHEETING IN COILS AND PACKAGES

A dangerous cargo

The loading operation of a cargo of steel coils is often surrounded by a psychology of mild fear. This probably stems from a number of reasons associated with, to a lesser extent, mythical stories of large coils flying into the air or through the ship's side during heavy weather. It is a fact that during heavy weather there is a tendency for coils in the top of the stow to lift under the influence of the pitching of the ship, and this will apply especially in the forward holds. If all of the strapping bands break, and a unit comes out of stow, usually in such circumstances the outer turns of plating will open up and the unit might be said to resemble a slack spring (see Fig. 57).

Such cargoes, if not loaded with care and attention, supported by a considerable measure of expertise, can become more than just troublesome during the voyage. In many circumstances, owing to slackness in stow, movement of cargo results in damage to the cargo and eventual claims being entered by receivers. Disasters with steel cargoes are not unknown. One such case involved a bulk carrier which for certain reasons was trimming heavily by the head in heavy weather. It can only be assumed that, at a certain moment, part or all of the steel cargo rushed forward, destroying the sub-divisional bulkheads so causing the ship to actually sail

Fig. 57. Hot rolled steel sheeting in coils—broken adrift during the voyage.

beneath the sea. The vessel disappeared in a matter of seconds, taking with her the entire crew.

Good stowage essential

The first requisite in the stowage of a steel coil cargo is that the stowage is extremely well performed, and arranged in such a manner that if inherent slackness exists when the vessel puts to sea this slackness will be taken up by the initial movement in the settling of the coil stowage, any further movement of the cargo being permanently arrested for the remainder of the voyage. In order to achieve this, an ideal stowage position must be found in the bottom of the ship where the effects of the vessel's motion in a seaway are least pronounced.

In the late 1950s and early 1960s, when seaborne trade in steel coils was beginning to gather momentum, there were many instances where masters of vessels, quite rightly, from a ship stability point of view, insisted on part cargoes of steel coils being stowed in the tween decks of their vessels. In many instances, serious problems developed and this practice was soon discontinued. Owing to the large radius of gyration experienced by a ship's tweendecks under the influence of rolling, coils occupying a high stowage position in a vessel cannot be secured in a manner which would be considered as affording maximum guarantee against shifting, with the exception of light steel coils adequately secured on wooden skids. In consequence of this, experience dictated that the best stowage position was in the lower holds where the coils could be stowed directly on to the tank-top.

Another important aspect which took time to develop was the orientation of the stowage. Stowages were attempted with the centre cores of the coils placed in an athwartship direction. On other occasions, the centre cores were all with their axis in a fore-aft position, and mixtures of the two types of stowage were not uncommon. None of these arrangements proved to be a complete success. Over a period of time, with perseverance and through trial and error, it became generally accepted that the safest and most reliable form of stowage was to arrange the stowage with the centre cores of the coils placed in a fore and aft direction with modifications, in certain circumstances, which involved inserting locking coils placed in a suitable manner in chosen positions.

Locking or key coils must be fitted

The insertion of locking coils in the stowage is of the utmost importance, and no vessel carrying coils should put to sea without this requisite being satisfied except in very exceptional circumstances. Situations have arisen when a single layer of steel coils has appeared to be an almost tailor-made fit when placed on a vessel's tank-top. In such circumstances, the coils seem to be a solid and tight fit transversely between the ship's sides. Nevertheless, slackness will exist and will become only too apparent once the vessel begins to work in a seaway. Single tier stowage of steel coils should be avoided if possible, and block stowage in two tiers or more should be the aim by shortening down the longitudinal spread of what would have been single tier stowage.

In those instances when a steel coil stowage will not cover the entire surface of the tank-top plating it should be endeavoured to place the coils in the after end of the compartment, a space being left between the forward face of the stowage and the forward bulkhead of the cargo hold or cargo stowed adjacently ahead of the coils. This is because the seagoing trim of a ship is usually by the stern, and the effect of this will manifest itself upon certain aspects connected with the preservation of the stowage, as the voyage progresses. Such an after stowage position should certainly be performed, if possible, in Nos. 1 and 2 holds where the effects of pounding and pitching are so pronounced. With part stowage of steel coils in the forward part of the hold, there will be a tendency for some of the top coils in the brow of the stow to move in an after direction, which does often occur.

Method of stowage

The method of stowage is to commence in the wings against the after bulkhead and work towards the centre of the hold from both sides. As each coil is placed with its centre core in a direction conforming to the longitudinal axis of the vessel, a wedge is placed beneath so as to prevent it from moving away from the adjacent coil. This is necessary to build initially as much tightness as possible into the stow, and at the same time it is a precaution against movement should the vessel develop a list whilst loading operations are in progress. Beneath the bottom line of coils, adjacent to the tank-top, it is usual to place two double lines of flat dunnage (approx. 6″ × 1″ wooden planks) and also in the wings against the lower wing tank plating or ship sides frames, as the case may be. This is done to hold the coils clear of the tank-top plating and permit drainage of any moisture which might accumulate. To a greater or lesser degree, some spreading of the weight of the stowage over the tank-top

Fig. 58. Single tier stowage of coils in the hold of a bulk carrier.

plating might be involved. However, with such thin timber being employed the effects are minimal, or more likely non-existent.

As the build up of the first line of coils progresses it may develop in such a way that two locking coils, or key coils, are built in, as indicated in Fig. 59(a). These key coils (locking coils) are not essential for multi-tier stowage; however, if it is done in this way, it will be found that some spacing will develop in top tier stowage which will facilitate securing. This is also how the stowage should be arranged when a single horizontal tier of coils cannot be avoided. This procedure is also to be adopted in the first horizontal tier of coils, when it is intended that the stowage will extend to two or three coils high. With stowages extending more than one tier high, the stowage is progressively built up to appear as shown in Fig. 59(a).

Compact, close fitting and neat looking stowages can be achieved when all of the coils are the same size. Unfortunately, units offered for shipment quite often vary in size and weight, and it is therefore necessary to ensure that the heaviest coils are stowed in the lower tiers if crushing damage is to be avoided.

Lashing and securing

It is often stated by those interests who would rather not have to invest in the materials necessary to the securing of the cargo that, if a steel coil cargo is properly stowed, it should jam itself off in the hold of the ship, so making securing more or less unnecessary. There is a measure of truth in this statement, in so far as, if a steel coil cargo is not properly stowed and it shifts when the vessel is rolling heavily, no amount of conventional securing will prevent it from breaking adrift completely.

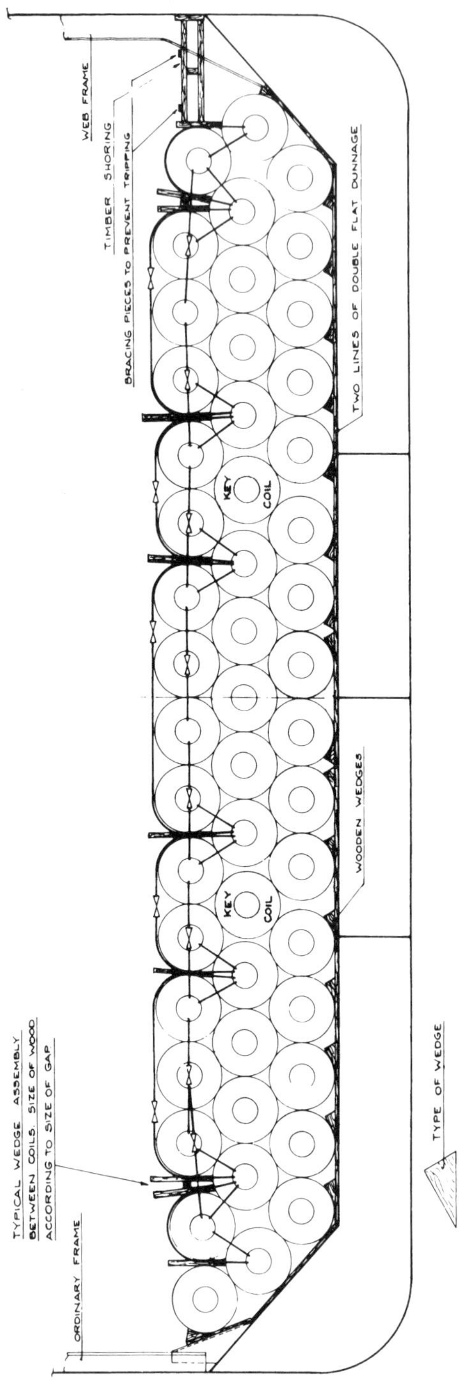

Fig. 59(a). Forward face of stow—cross-section.

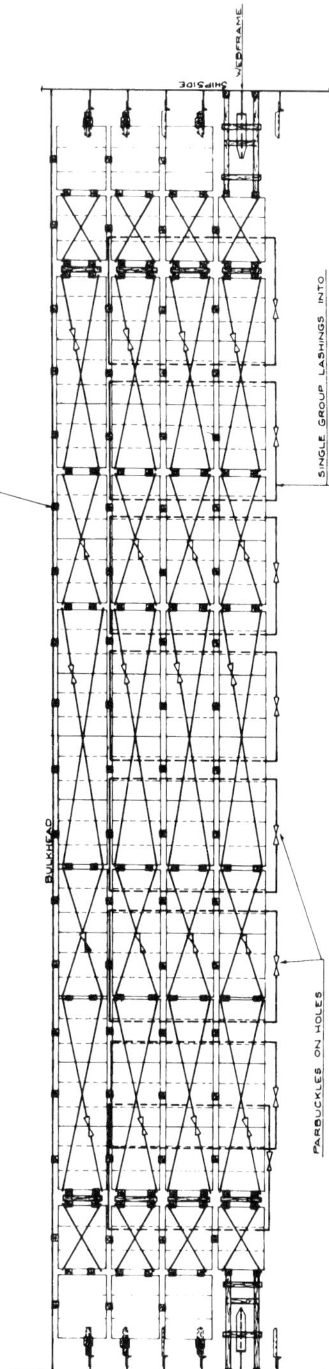

Fig. 59(b). Forward face of stow—plan view.

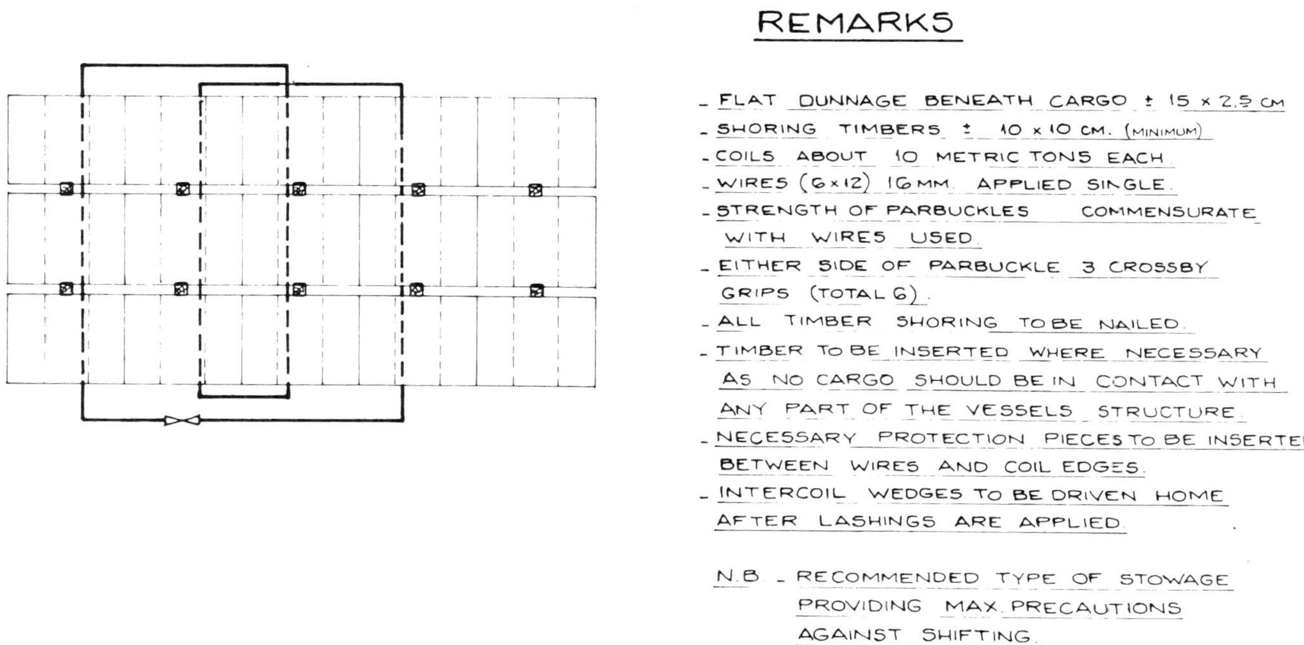

Fig. 59. Steel coil stowage.

In consideration of the securing necessary to be applied to a cargo of steel coils, it is prudent to keep in mind that, in a seaway, a combination of a number of movements and forces is involved. The ship moves vertically upwards and downwards when pitching into the sea, there is an angular movement alternately forwards and to aft, and there is also rolling—a tilting motion transversely to port and starboard. These combined forces transmit to the coils a slewing effect which can cause coils, in the uppermost tier, to turn in stow. This often results in breaking of the securing bands of the coils, heavy grinding and chafing with eventual partial unwinding of the turns of plating, with accompanying damage, and forthcoming claims being entered upon delivery of the cargo. Insufficient securing has on occasion been the cause of coils, positioned in the upper tier of the brow of the stow, rotating and eventually falling down onto the tank-top. Wrapped coils are particularly vulnerable, being loose in their wrappers. Therefore, some freedom to move inside the wrappers is afforded, so resulting in the metal wrapper being chafed and partially torn off with resulting damage to the plating as the coil rotates (see Fig. 60). Coils in upper stowage positions do have a tendency to lift when the ship is pitching, hence the reason for the cross lashings over the tops of the coils as shown in Fig. 59(b).

Single tier stowage and height of multi-tier stowage

Figure 54 shows a three tier stowage with wooden chocking placed as an obstacle to movement in a lateral direction. Figure 59(b) (view in plan of Fig. 59(a)) shows the securing arrangements of the top coils on the edge of the stowage where the steel coils are group-lashed into interconnecting blocks of nine coils per block. In these

Fig. 60. Cold rolled steel sheeting—results of coil turning in stow.

blocks, note the wedges inserted between individual coils to prevent them converging and causing the lashings to slacken off. These block lashings are unnecessary when the face of the stow terminates close to an end bulkhead. If single tier stowage is unavoidable then two locking coils as shown in Fig. 59(a) can be introduced into each line of coils stowed on the very wide tank-tops, e.g. bulk carrier of 40,000 m/tons deadweight or more. These coils may be secured as indicated with lashings as shown in Fig. 61. In smaller ships with less beam, depending on the diameter of the coils, one locking coil per row should be sufficient. If possible the locking coils should be kept in line in a fore and aft direction and fitted with wedges to avoid any possibility of convergence or movement in a fore and aft direction. It is to be noted that the lashing system shown in Fig. 61 for the locking coil in single tier stowage is equally adaptable to the top coils in multi-tier stowage. However, the suggested method for top tier coil securing in multi-tier stowage is shown in Fig. 65.

The question often arises as to how high the coils should be stowed. Normally, it will be found that in consideration of what pressure the tank-top of the ship can withstand, and the sizes and weights of coils involved, the maximum height of stowage will be three coils. As a rule of thumb, what might be referred to as the "empirical rule", the height of the tiers of cargo would be governed by the weight per coil and the tons per m^2 permissible load upon the tank top. The pressure exerted by the base coil in any vertical multi-tier of steel coils should not exceed three times the tank-top's permissible load per m^2. Therefore, if the tank-top permissible load is 15 tons per m^2, 15 m/ton coils could be loaded three high. The base coil would exert a pressure on the tank-top of 45 m/tons (normal in the industry). See also TANK-TOP STRENGTH IN RELATION TO STOWAGE OF STEEL CARGOES.

The total amount of cargo to be loaded into any compartment must never exceed the permissible load per square metre of tank-top surface multiplied by the area of the tank-top plus whatever load is permissible above the side tanks. The classification societies specify that the load should be evenly spread over the tank-tops in order to avoid spot overload of the tank-top plating (see under "Tank-top strength"). However, dunnage used does not achieve this objective.

Generally speaking, if the "empirical rule" in a vertical tier or tiers of coils is adhered to, experience has proved that the two lines of double dunnage beneath the coils has been sufficiently adequate to avoid any damage developing to the tank-top plating of the modern-day bulk carrier. In deciding whether or not thicker dunnage must be used, the age of the vessel and the condition of her plate work must be taken into consideration.

Where the carriage of coil cargoes is concerned, there is a tendency towards the unit weight of individual hot rolled steel coils to be heavier; 25-ton coils are not unusual. Apart from this, where not so long ago the largest ship carrying a mixed steel cargo, or a full cargo of coils, did not very often exceed 30,000 m/tons deadweight, ships of up to 40,000/70,000 m/tons deadweight are now not uncommon. As a result of this development of more weight per steel coil and ships of greater deadweight capacity, steel coils are being loaded three and in some instances four tiers high. In such circumstances the weight of the load in the area of the bearing surfaces of the bottom coils in the stow can be from 60 to 75 m/tons per individual unit. Even in the very large ships being used, the permissible tank-top load per square metre can be as low as 10 m/tons in some instances; more often it is

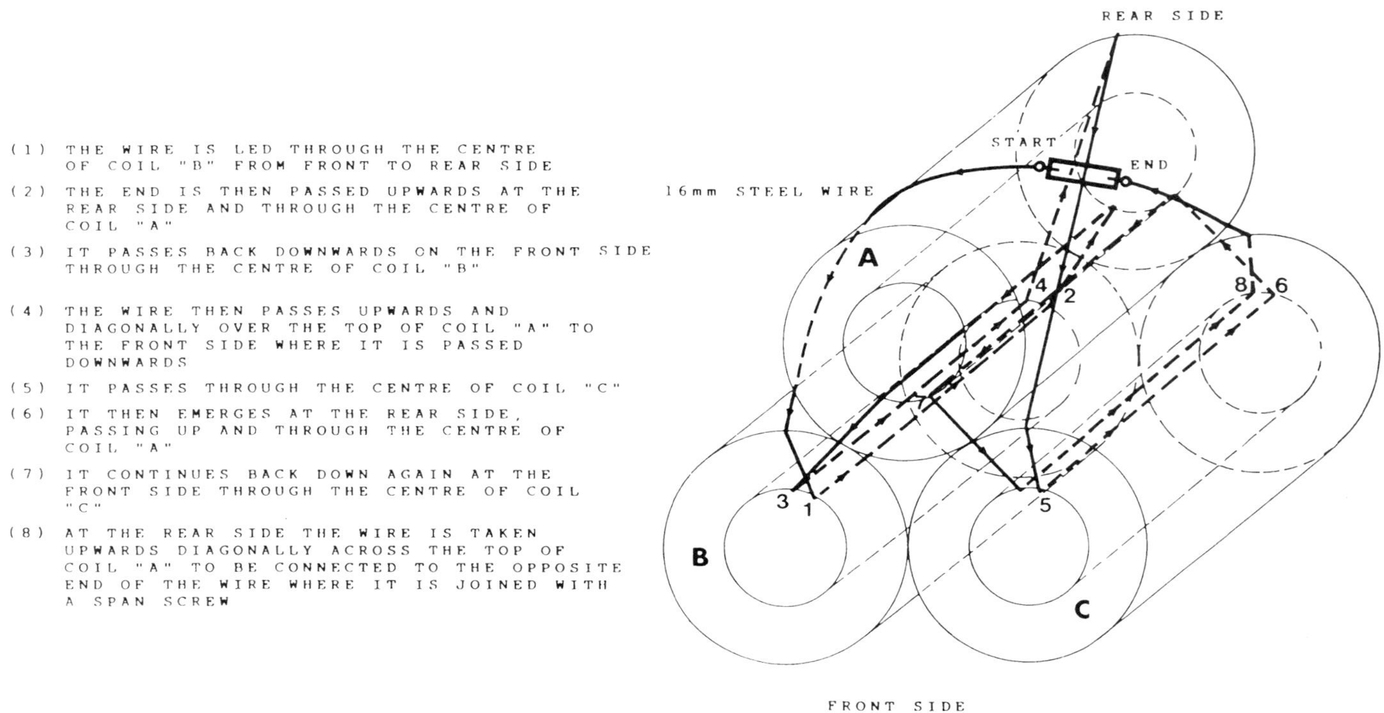

Fig. 61. Lashing system for locking coils in single tier stowage.

between ± 15 and 20 m/tons. Add to this scenario shippers who are reluctant to supply dunnage wood of strength and quality commensurate with these concentrated loads. In the circumstances it is understandable that concern is growing with regard to the stowage of these shipments of steel. For further information on this subject see under "Tank-top strength".

Securing with wire cables

With regard to the actual securing of the coils after they have been stowed in the hold of the ship, the conventional method of effecting this is by means of steel wire cables which are joined together and tightened by means of patent "Bulldog" or "Crosby" clips and turnbuckles.

Owing to its adequate strength, and ease of manipulation, 16 mm (6 × 12) wire cable is popular and normally used for this type of work. The wire is constructed from six strands of 12 wires per strand, each having a fibre core. The six strands are then wound around a central fibre and core, thus finally consisting of approximately 77% fibre and 23% wire, making up this cable. This large amount of fibre does afford a desirable degree of flexibility but at the expense of the overall strength of the wire. Further to this, it permits considerable elongation under tension when relatively low loads are applied. Under strain there is a reduction in diameter of the wire, which affects the holding power of the Bulldog or Crosby clips and overall effectiveness of the lashing.

It has been proved on test benches that there is a reduced strength of up to 40% in those instances where a wire lashing leads at an angle of 90 degrees around edges formed by items of cargo, and new wire will stretch up to 20% before breaking. In tests, which were carried out with new 16 mm wire which has a breaking strain of approximately 8,500 kg, the wire stretched 25% before breaking. In this test, the Crosby clips had been correctly fitted and hardened up to the wire. However, after the wire had stretched the clips failed and the wire pulled out under a load of 5,500 kg.

Because there is so much stretch in new wire, when wires are tightened before the ship departs they slacken off and, therefore, further tightenings will be necessary to take up this slack during the ocean voyage. This may have to be carried out more than once, as the tightness of the lashings is of extreme importance as a precaution against movement of the cargo. As already explained, owing to the tension on the wire after tightening, there is a reduction in the diameter so that the nuts on the grips must also be retightened. The surface of a coil cargo lashed with wire may comprise hundreds of span screws, clips and tightening nuts, attention to which, as suggested above, is inconceivable even with the ship's crew manning scales prior to the introduction of the bulk carrier. When joining the ends of the lashing, the turnbuckle should be extended to the maximum of thread before application of the wires in order to ensure that after tightening not more than one third of the thread is used; this will allow for further tightening. When the wire is turned back through the eye of the turnbuckle, it should be secured with three clips. Unfortunately, it is normal practice to use only two grips per fastening, which is incorrect according to the recommendations of the British Standards Institute (B.S. 462: 1983). Invariably, the grips are wrongly applied, so reducing the overall efficiency of the lashing.

Securing with flat metal strapping bands

In securing a steel coil cargo the wires are applied single, not double. Therefore, it is relatively simple to compare the wire method of securing steel coils with a new method which has been introduced, where wire rope is replaced with flat metal strapping bands. This new method has met with a considerable amount of success, and is increasingly being used in place of the conventional wires.

One well-known manufacturer, whose strapping bands are extensively used for securing steel coil cargoes, produces a band with cross sectional dimensions of 31.75 × 1.12 mm and guarantees a breaking strain—of a single band—of 4,200 kg with a maximum elongation of 9.8%. These strapping bands are extremely easy to handle when compared with wires, and thread through the coils without difficulty. After a band has been put in place, the ends are brought together and tightened with a special pneumatically operated tool which renders the band bar tight with a tension equal to 2,000 kg. Two crimp seals are applied to secure the ends together and these are alleged to have a joint strength of 90% of the actual band. This produces a breaking strain for the entire lashing of 3,780 kg. It is important that the proper pneumatically operated tools are used to tighten the bands. It is to be emphasised that on no account should the strapping bands be tightened by means of hand operated tools. There have been instances where the bands have been tightened by hand operated tools. This has resulted in the tension of individual bands throughout the stow varying considerably and either causing or contributing to shifts of cargo during boisterous weather conditions. It is imperative to safe carriage that all strapping bands are very tight and all similarly tensioned.

By comparison with the conventional wire lashing system, there are advantages in the use of strapping bands. In many instances the actual lashings can be considerably shortened down, which is a very desirable feature. Short core lashings can be achieved with ease, and such lashings properly applied are extremely effective, and if necessary may be used double without difficulty. Securing is completed much faster and less labour is required. This together with the fact that each individual lashing with strapping bands is considerably cheaper than a wire lashing, means that the final overall cost of the strapping band system is less than when wire is used.

With regard to the disadvantages of using strapping bands in place of wires, the securing operation requires a compressor and special expensive pneumatically operated tools, and trained labour has to be employed. The original idea was that strapping bands were to be used to replace wire cables in securing steel coil cargoes. In the interests of cutting costs yet further, the use of steel bands has been commercially advertised as a new system which lends itself to a number of variations on the old securing methods, but enables these to be performed more efficiently. In consequence of these professed advantages, the so-called strapping band system has been used, in many instances, as perhaps a subterfuge to dispense with timber securing.

Experience has proved that whether the coils are secured with steel cables or metal strapping bands, it is necessary, in most instances, that they are complemented by timber used as checking and/or as shores. Vessels loaded with steel cargoes are usually over-stable and roll violently in a seaway. At the termination of a roll to one side, forces develop, due to inertia, which can result in a short sudden pull (jerk) on a lashing or lashings. This might result in the lashings being overloaded. The timber securing absorbs the extra forces created at the termination of the roll, thereby assisting in the preservation of the effectiveness of the lashing material. Strapping bands are tensioned up to a load equivalent to 2,000 kg. The metal straps referred to have an elongation factor of 9.8% and must, therefore, under the load indicated when tightened and in place, be fully stretched and have very little, if any, room for further elongation. Any additional strain applied, over and above the breaking strain, could not be temporarily absorbed by the steel band. In view of this, it is considered that the insertion of timber securing is even more important when strapping bands are used with steel coil cargoes.

There is, on the other hand, an argument against the use of timber when the securing has been performed with strapping bands. it is claimed that for the strapping band system to be fully effective, tensioning of the strapping bands must be uniform throughout the stow in order to create an overall efficient securing system, all parts of which are complementary to the other. With all bands equally tensioned to 2,000 kg it is fairly evident that belting in timber between the coils will certainly upset the tensioning of the bands throughout the stow.

From the above, it is clear that there are two schools of thought regarding whether timber should or should not be used when securing steel coil cargoes with strapping bands. Indeed there are many people who condemn the system completely as being inadequate. Invariably, such opinions are advanced by those who have no experience of the strapping band system, have not studied it sufficiently or are just resistant to progress or any form of change. Admittedly, the system of securing with wires has a more dominating impact in appearance than steel strapping. This need not mean that wire securing is more efficient than strapping bands.

The use of metal strapping bands for securing steel coils on board ship seems to have first appeared at the steel production mills that export their products direct from their own premises. The practice has now spread to some of the world's larger ports, and in view of the fact that it is now firmly established, and because of its appeal from a financial point of view, it is logical to assume that the system and practice are here to stay. Since its inception, there have been thousands of successful voyages made with steel coils on board secured with metal strapping bands. Since all the controversy over whether or not timber chocking should be used with steel strapping bands, it has now become firmly established that no timber whatsoever should be used for securing. Using timber, as explained earlier, upsets the stability and efficiency of the band system and can lead to movement of cargo when the vessel is working in a seaway.

What is stated above covers all the past arguments and considerations regarding the use of wire securing, strapping bands and timber for chocking, bridging, strutting and tomming. The final conclusions are that the best method for securing steel coils is by means of steel strapping bands, without any securing timbers being used. Even when securing with wire cables wooden chocking, etc., can be dispensed with. However, with both methods if there is no timber chocking, etc., the securing wire must be looped through the under stowing coil and crossed over the top of the uppermost coil as indicated in Figs. 61, 64 & 65. The diagram shown in Fig 65B lends itself to a lashing involving the five coils, whereby coil 2 remains unlashed, as the same sytem of lashing crosses from coils 1 and 4 to coils 5 and 6. There is no possibility of coil 2 shifting.

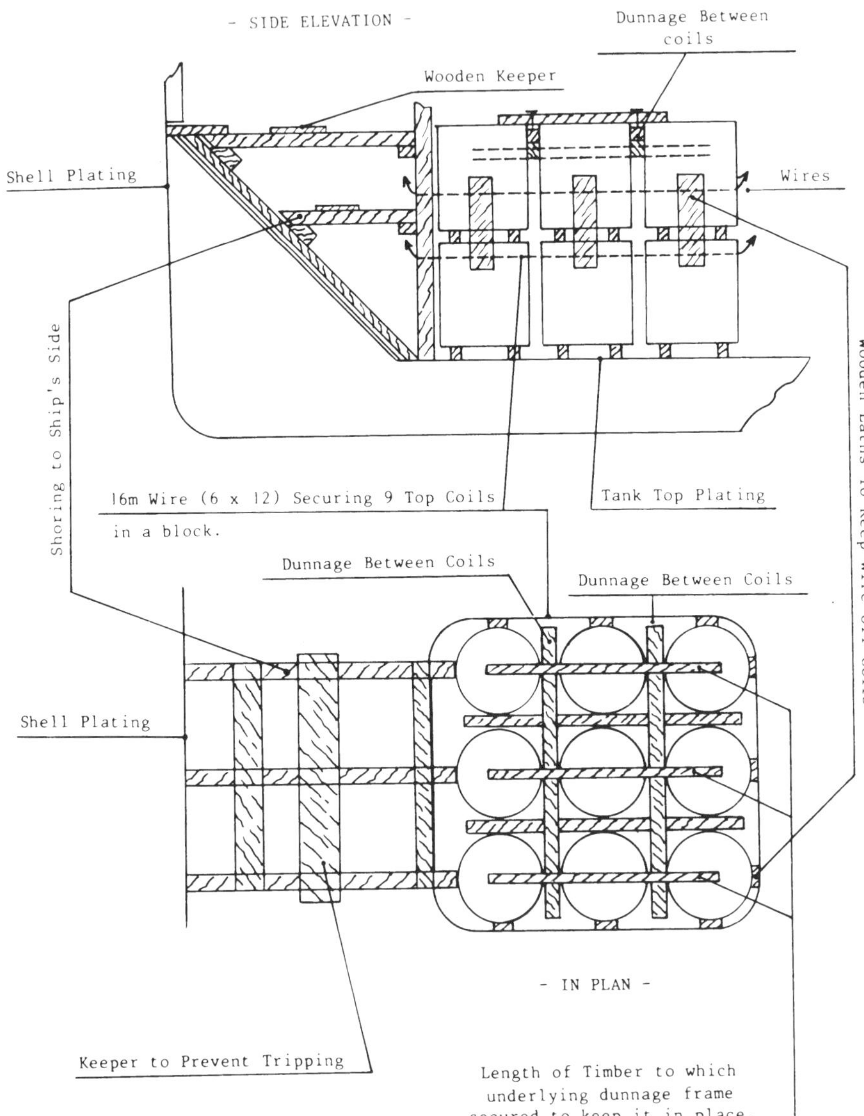

Fig. 62. Stowage and securing of coils placed vertically on skids.

Vertical stowage of coils (eye to the sky)

Although this can be said to be a secure method of stowing conventional-type steel coils on board ship against movement during the voyage, it has proven to be, in general, not very satisfactory. Special gear is required to manipulate the units with complete success, and in all probability the same type of gear is not available at the discharge port. It is a method of stowage which has to a great extent dropped out of use. From time to time enquiries are still made with regard to this type of stowage.

Notwithstanding what has been said above, specialised types of steel in coil are often transported stowed in the vertical position on wooden pallets. Care must be taken with such material, as imprints of securing wires around the packing can cause damage to the contents. Belting into place of wooden wedges and chocking can also be very damaging. The methods of securing shown in Fig. 62 suggest what might be done to avoid delivering damaged cargo. For the type of cargo referred to, see Fig. 63.

Packages

As with steel coils, and for reasons already stated, packages and bundles of steel plate should be provided with stowage in the after end of the cargo compartment. Multi-tier stowage is preferable to single tier stowage as this cuts down on the amount of securing necessary. Single tier stowage is often resorted to in tween decks, which is not the ideal stowage position. Stowage direct on to the tank-top is more ideal when multi-tier stowage will be inevitable.

In the lower hold, stowage would be commenced against the after bulkhead by laying two double lines of 6″ × 1″ dunnage athwartships. Similar dunnage would be

Fig. 63. Electrical steel sheeting—stowed vertically on wooden skids.

placed against the lower wing tank plating. If the vessel has to proceed to sea with the goods unsecured with other cargo, then an appropriate number of securing wires, placed in a fore and aft direction, should be laid upon the tank-top plating. The ends of these wires can eventually be passed back over the top of the cargo to secure the goods into one block.

As the stow develops, the horizontal tiers of cargo should be kept level, and flat dunnage should be placed between each horizontal layer of packages or bundles. Also, the units must normally be winged out to fit in with the slope of the wing tanks. Upon completion of the stow a wooden fencing, composed of flat boards criss-crossed vertically and horizontally, must be erected across the face of the stow. The fore and aft securing wires can then be taken up outside of the fencing and secured to their opposite ends to be tightened by means of par-buckles. Wedging and shoring should be driven into any gaps left between the top tier of packages.

All of the securing, as referred to above, is recommended on the assumption that none of the stowages mentioned are overstowed with other cargo during the course of any sea passage. When steel cargoes, or individual parcels of steel, are overstowed with suitable securing cargo, an on-the-spot assessment of the situation might indicate that part, if not all, of the securing of the steel cargo can be dispensed with.

Points for consideration in the stowage of steel coils

Coils of steel in the lowest tier should be stowed upon two athwartship lines of double flat dunnage. Where the breadth of the coil is considerably in excess of about 1¼ m, three lines of dunnage should be laid. Only hardwood dunnage should be used, with dimensions of between 6″ and 8″ wide and approximately 1″ thick.

It is essential that the largest and heaviest coils occupy the lower stowage position, with the lighter coils on top in the stow to prevent coils becoming partly crushed or distorted. With the smallest coils occupying an upper stowage position, there is more spacing between the top coils and this arrangement facilitates securing.

It is also essential that each coil sits in the cantline of the two coils beneath. There should be no mixing of coil sizes as this leads to a jumbled type of stowage which can result in, or be contributory to, a breakdown of the stowage in certain circumstances, and also broken strapping bands are often associated with such a stowage arrangement.

The key coils or locking coils should be positioned with from ¼ to ⅓ of their diameter below the level of the top of the line of coils which they are supposed to secure, thereby effectively providing sufficient pressure.

Adjacent athwartship rows of coils should not be stowed too close together, as this can hamper the discharging operation and also result in the prong of the forklift truck extending through the coil to contact the adjacent coil, and so cause damage. On the other hand, the rows should not be too far apart to prevent effective wedging. A space between these athwartship rows of 6″–9″ clearance is recommended.

In wedging the lowest tier of coils, the correct type of wedge should be used as shown in Fig. 59(a). Right-angled wedges are not recommended as these tend to trip.

Of the various methods used to secure steel coils the most effective method of lashing top tier coils in a multi-tier stowage is to secure through the core of the top coil to the cores of the two coils forming the cantline in which the top coil is positioned; this would amount to two short lashings. Owing to the shortness of these lashings, suitable tightness when using steel wire is difficult to achieve; strapping bands lend themselves admirably to this method of securing. Nonetheless, such a method of securing has proved not to be a viable proposition as the lashings have to be applied as the stow is built up, so necessitating workmen standing by continuously for this purpose: their activities may also interfere with the continuous and uninterrupted flow of the actual stowage operation. Notwithstanding this, it is undoubtedly the very best and most safe method of securing steel coils. See page 212 under **Securing with flat metal strapping bands**.

Contrary to what has been said on pages 212 and 213, with the recommended method of securing as explained below, the elimination of securing timbers from stowages is desirable. It will be recalled that the paramount requisite to ensuring that coil cargoes do not move or shift is efficient and proper stowage, whereby the individual units involved exert force laterally and vertically downwards so contributing to arresting movement in all directions, with the exception of upwards and to some degree in a fore and aft direction. An alternative and highly recommended method of securing top tier coils, with wire or strapping bands, is shown in Figs. 64 and 65. The wires or strapping bands used are applied in much the same way as those shown in Fig. 56 except that in the proposed method the cores of coils "A" and "C" only receive a single pass of the wire or strap to the right. In Fig. 65 the drawing is expanded in an effort to show the arrangement more clearly. Fig. 64 is a picture of a coil cargo similarly lashed; note the absence of securing timbers.

Figure 65A/B represents five coils in an athwartships multi-tier line of stowage in which the top tier must be secured. In A, for the purpose of ease of explanation

Fig. 64. Lashing applied as indicated in Fig. 61.

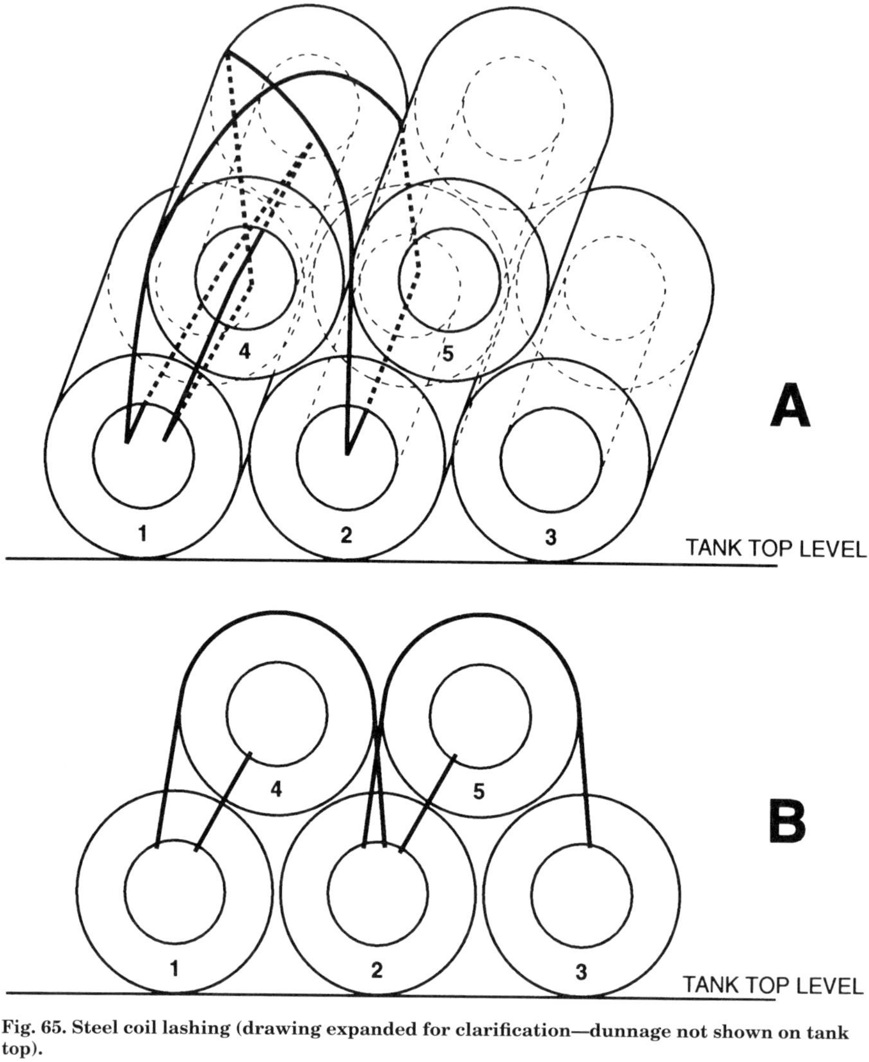

Fig. 65. Steel coil lashing (drawing expanded for clarification—dunnage not shown on tank top).

and understanding, the length of lashing wire is halved through coil (4), whereafter the two ends are taken down each side and passed through the core of (1). The two ends are then brought up in front of, and crossed over the top of, (4), then brought down each side to meet in the centre of coil (2). Naturally the wires or bands would be manipulated in such a manner as to prevent the ends meeting inside the core piece of coil (2). The same method of securing would be applied to coils 2, 3 and 5 as shown in B, and so forth along the line of coils.

More experience has been gained with the metal strapping band system of securing. If the stowage and securing is performed as indicated above, timber chocking can be dispensed with for athwartships securing. No timber should be used with strapping bands as the bands are uniformly tightened and tensioned throughout the stow. By forcing timbers in between the coils the tensioning is upset and the efficiency of the overall effect reduced. Further to this, since 1988 many voyages, with strapping band lashings and no timber securing, have been studied and monitored—the results have been satisfactory, so that the dependability of the arrangement has been confirmed without doubt.

Returning to the method of securing as indicated in Fig. 65, this arrangement of the lashings is highly recommended as the coils are held down by the core lashings and at the same time are prevented from moving in a fore and aft direction. Further holding down force is applied by passing the lashing over the top of the coil being secured (see Fig. 64).

Stowages to which fancy names have been attached, e.g. for slab cargoes "California Stowage", "the Oxelsund Stowage", and for coil cargoes the "Pyramid Stowage", are invariably devised to cut corners and save money at the expense of the principles of good stowage. The so-called "Pyramid Stowage" is achieved by omitting one coil either side in each transverse row in multi-tier stowage, so that in cross-section the stowage takes on a pyramid shape. It is safer to abide by the time-honoured method of filling to the best possible degree all space available transversely by completing all layers/tiers of cargo.

Loading, stowing and securing

From the point of view of the ship master and in the interest of claims prevention, the following are some cardinal points to be kept in mind when carrying steel cargoes.

(a) Cargo compartments should be scrupulously clean and free of rests of previous cargo. This applies essentially to the overhead members of the component parts of the ship's structure.

(b) If the cargo holds are to be washed out and salt water is used, a final thorough wash should be made with chlorine-free fresh water evaporated on board the ship. Fact to be entered in the deck log book.

(c) Before commencement of loading it must be ascertained that there is adequate suction on the bilge lines. The date of the test and the result should be entered in the deck log book.

(d) A reasonable inspection of the cargo holds must be carried out before loading commences. The name of the vessel's officer who inspected the holds and the date must be entered in the deck log book.

(e) Never go to sea with the uppermost horizontal tier of a steel cargo incomplete, e.g. billets, beams, etc. If the top tier cannot be completed it should not be loaded, since securing with wires to the ship's sides and tomming with timber cannot be considered maximum precautions against shifting with this type of cargo and in this particular situation. Naturally, in certain circumstances this will not apply to steel coil cargoes, e.g. pyramid stowage of steel coils. During the build up of the stow it should at all times be kept level by judicious use of dunnage wood.

(f) Any dunnage used must be dimensionally and structurally adequate. Dunnage with a moisture content of more than 14% on weight should not be

used in the same cargo hold as steel products. The drier the dunnage the better.
(g) Steel products should never be permitted to rest against the ship's structural parts. Dunnage should always be used to prevent this occurring.
(h) Underdeck steel cargoes should never be secured to the component parts of the vessel's structure with the exception of wire rod cargoes in certain circumstances, e.g. half hatch stowage: face of the stow, athwartships securing wires must be bowsed in to tighten them against the cargo.

STRESS AND STRAIN

Associated with tensile strength. Strain is the measure of distortion in a material due to an applied force which produces stress—may be tensile, compressive or shear in nature—in the material and is the force exerted per unit area.

STRESS OF HEAVY WEATHER

Leaking hatches, with subsequent damage to steel cargoes, have often been attributed to stress of heavy weather, force majeure, perils of the sea—meaning an Act of God which is something beyond human control. Such pleadings by the ship in the case of the *Sabine Howald* (see Leaking of hatch covers—the carrier's defence, page 173) was successful. However, in the more recent case of the S/S *Eurounity*, now *Seamusic II*, argued in the United States Court of Appeals on 13 December 1993 and decided on 8 April 1994, it was stated, in respect of a voyage with steel products from Antwerp to Charleston, Jacksonville, Savannah and Houston, that:

> "During the transportation of the cargo, the vessel encountered a severe storm on 4 January 1989. This storm was classified as an 'ultra bomb' (extra-tropical cyclone) because its central pressure plummeted sixty millibars in twenty-four hours. The storm resulted in Beaufort Scale winds of between force 10 and 11, waves between 10 and 11.5 metres in height and chaotic cross seas. During the storm, the vessel was hove to (ship's bow pointed into wind without forward motion) and its weather deck was awash. There is no dispute that seawater entered the cargo holds through the vessel's cargo hatches during the storm. The vessel's owner, Licetus, presented evidence and expert testimony to support its claim that the severe weather placed such torsional stress on the hull and hatches that seawater entry was inevitable."

In the judgment a definition of peril of the sea is given as: "When conditions are of an extraordinary nature or arise from irresistible forces or overwhelming power, and which cannot be guarded against by the ordinary exertions of human skill and prudence."

Later the judgment states:

> "We cannot agree, however, that the weather conditions created by this storm constituted a peril of the sea. The vessel's bridge log book, which the district

court relied upon, recorded Beaufort Scale winds that did not exceed a level of 10–11 on 4 January 1989. Expert testimony at trial indicates that there were significant wave heights of between 10 and 11.5 metres. We find nothing of an extraordinary nature, nor do we find irresistible force of overwhelming power in these conditions. Indeed, the testimony of the meteorological expert witnesses for both sides revealed that, for the most part, the weather conditions experienced by the vessel were not unusual in the North Atlantic in the Wintertime."

Deliberating the intensity of the weather and the final judgment, in such cases it seems reasonable to assume that irrespective of what cargo is carried by the ship, neither this, nor the size of the ship, will be taken into consideration and an owner's plea for exoneration from liability based upon peril of the seas, with weather forces below force 12, must fail. Further to this, we must accordingly conclude that a ship of 3,000 tons deadweight can be no more seriously affected by these severe weather conditions than a vessel of 30,000 tons deadweight. Such views have been a contentious point of discussion between owners and cargo interests for many years and will undoubtedly continue to be so into the foreseeable future.

It cannot be disputed that the hull of a ship loaded with a high density homogeneous cargo, e.g. steel slabs, billets, steel coils, etc., will be subjected to much more severe constructional racking, hogging, sagging and sheer stresses at any time, in a seaway, than a vessel loaded with a low density homogeneous cargo such as bulk grain. The ideal stowage situation is to position the cargo in the various holds in such a manner as to distribute the weight evenly along the length of the vessel. With a cargo consisting of a variety of steel products—a heterogeneous cargo—comprising relatively high and low density material, e.g. LD wire rods, mesh, steel fencing, HD steel coils, slabs, billets, etc., the weight of the latter mentioned being concentrated at a low level in the ship's holds, imposes tremendous shearing forces upon the hull of the ship. The combined effects of all the stressful forces will manifest themselves in those areas furthest from the vessel's neutral axis, which will result in extraordinary movement of the hatchways and on account of this the hatches will be more prone to leakage than would be the case if the vessel had been loaded with a low density cargo.

There is no doubt that under heavy weather conditions the hull of a ship loaded with grain, or any other low density cargo, will suffer less from the adverse forces imposed upon its structure than would a ship loaded with steel. The stressful differences between these loaded conditions can be proved by calculation. Such circumstances suggest that sights should be set lower on the Beaufort Scale when deciding what constitutes exceptional weather conditions for vessels loaded with steel products.

STRETCHER STRAINS

A furrowed roughening of the surface which may occur with low-carbon sheet or strip steel, due to uneven yielding in the early stages of cold deformation after annealing or, in a lesser degree, after normalising, or hot rolling. See also LUDER'S LINES.

STRUCTURAL STEEL

The Deutsches Institut für Normung, whose rules are often referred to as the DIN Standards, and which are the German standards for steel classification, describes structural steel, for general structural purposes, as:

> "Steels for general structural purposes are unalloyed steels which are characterised essentially by their tensile strength and yield point at ambient temperature and are used for example, in building above and below ground, bridge building, hydraulic engineering, tank and bunker construction and in automotive and mechanical engineering."

As far as ships' cargoes are concerned, generally the above definition would cover—as per the DIN norms:

> "Steel sections (including channels, angles and beams), steel bars, wire rods, flat products (strip, plate, wide flats) seamless and welded, square and rectangular hollow sections, forgings and semi-finished products."

It can be stated that structural steel covers all types of steel sections, whether large or small, which are used in structural work, as the name implies. All profiles mentioned under Merchant iron would also be involved. In order to avoid going into too much detail, and for practical purposes where ships' cargoes are concerned, structural steel embraces those items of steel cargo which are in long lengths, usually 20 feet (6 m), 40 feet (12 m) and 60 feet (18 m), shipped as unwrapped single units, e.g. heavy beams or in the form of unwrapped bundles, where weight per piece demands that they be bundled. Within the scope of the term "structural steel", must also be included hot rolled plates whether shipped in unprotected single units or unwrapped bundles. Quite often tubes are used in the construction of, for instance, shed roofs, and so in such circumstances they must also be classed as structural steel.

Stowage

It will naturally be appreciated that heavy structural pieces should not overstow lighter material in order to avoid crushing damage. Therefore, long steel would normally be expected to occupy the lower stowage position within a ship's hold.

The stowage of long structural steel is commenced by placing dunnage on the vessel's tank-top or tweendeck plating, to hold the goods clear of the ship's metalwork. This assists, to a certain extent, in spreading the weight, protects the ship's structural parts and guards against the goods being contacted by any moisture which might collect beneath the cargo. The dunnage used with structural material has to be fairly robust as there is always a degree of compression involved, which causes weak dunnage to break.

The purpose of using dunnage throughout the stow is twofold. The first consideration is to bind the mass of steel into a solid block within the cargo space. For this to be achieved long lengths of stout timber are necessary, possibly 3″ × 3″ pieces in cross-section, and these timbers are positioned athwartships, in each longitudinally placed layer of steel, about 3 m apart. The other purpose of using dunnage is to facilitate slinging of the material at the port of discharge. In order to avoid warping or bending of the long lengths of steel the dunnage must be kept vertically directly in line throughout the stow (see Fig. 66(a)).

Necessary to a compact and level stow is the arrangement with dunnage in the area of the slope of the lower wing tank plating in the holds of some bulk carriers. Incorrect dunnaging or failure to insert dunnage can upset the stow and cause cargo profiles to become twisted and bent. The results of some of these types of haphazard stowages are shown in Fig. 67. As the stow is built up progressively against the tank side, gaps develop extending the entire length of the stow. These gaps must be filled in with timber, in the way of the athwartships dunnage, in order to permit safe extension of the stow out to the tank side (see Fig. 66(b)).

Long steel, whether bundled or not, can be secured against movement by block lashing the uppermost tiers of cargo. Depending upon the type of steel forming the upper layer, wedges can be driven between any gaps, as found practical or necessary, after the block lashings have been tightened. This will ensure that there is no pile up of the steel through rolling of the ship or movement in a fore and aft direction. On the other hand, if the block of cargo occupies a low stowage position sufficiently distanced from the ends of the ship, e.g. amidships stowage, the wire lashings might be dispensed with.

Athwartships stowage

It is customary to use flat pieces of timber to hold the steel clear of ship side frames and tank sides, and in no instance is it admissible to permit the goods to be in contact with any component parts of the vessel's structure—this is particularly so with regard to the ship's shell plating. Long lengths of steel should be stowed in a fore and aft direction. Situations do arise when in order to prevent part of a cargo being shut out from shipment, some long steel does have to be stowed athwartships. Should this "unseamanlike" type of stowage be unavoidable it is recommended that some lengths of steel are placed fore and aft in the wings, adjacent to the ship's side frames, before the layer of athwartships steel is loaded (see Fig. 66(c)). In the case of one vessel, the end of an "I" beam placed athwartships was vibrating against the shell plating and eventually caused a hole to develop which permitted seawater to enter the cargo compartment.

Further to the above, the ends of steel placed athwartships must be held well clear of the slope of lower wing tank plating, and it should be remembered that dunnage compresses. If the wing tank plating obstructs the level of settling of the ends of the steel, the steel can become bent over its length.

Top stowage securing will naturally not be necessary if the block of steel is overstowed, and remains overstowed throughout any sea passage while it is on board. The overstowing goods must consist of a suitable cargo, both in type and amount, which will ensure that there is no possibility of the steel moving when the vessel is in a seaway.

The ideal method of stowing long steel is in a for and aft direction in the vessel's hold. The lifts are loaded in layers and the dunnage per layer is continuous in an athwartships direction.

However, the ideal method, as explained above, is rarely, if ever, performed anymore. The system now is to stack the steel, still in a fore and aft direction, first in the wings by use of fork lift trucks, so leaving the centre of the hold free for these

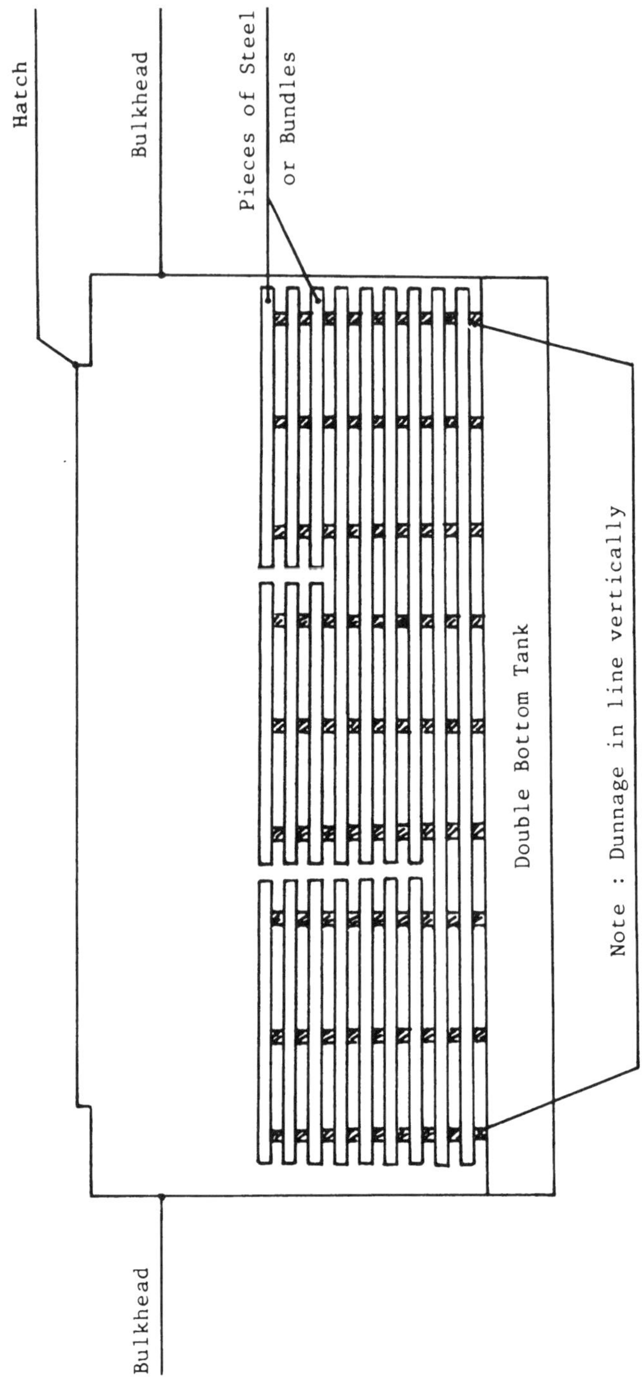

Fig. 66(a). Side elevation through ship's hold.

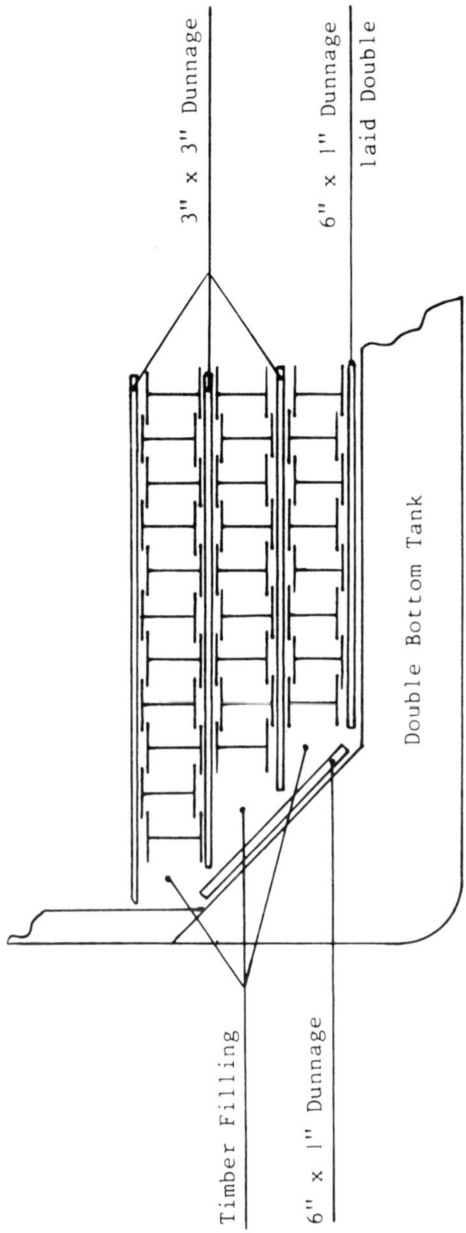

Fig. 66(b). Cross-section through area of bilge.

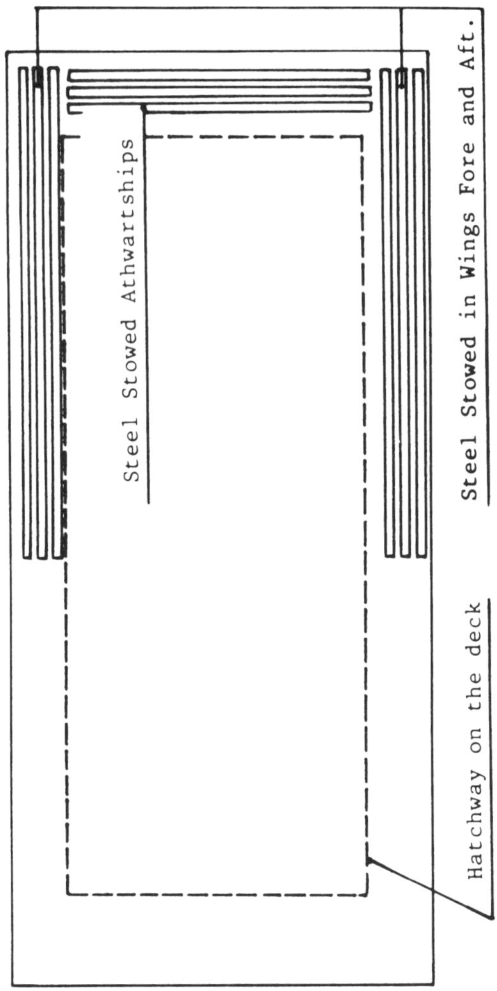

Fig. 66(c). Ship's hold—plan.

Fig. 66. Stowage of structural steel.

vehicles to work. Most of the bulk carriers employed in the trade are fitted with lower wing tanks, and manoeuvring the lifts out over the slope of these tanks, to reach the ship's sides, is not easy, as the stowage is built up vertically against the lower end of the slope plate. Consequently, it is not unusual for a space to be left between termination of the stow and the ship's sides. The system is not perfect with regard to keeping the dunnage continuous, as upon the stow being built up, and out, towards the square of the hatch, upon the stow clearing the edges of the hatch coamings the fork lifts are removed,and the centre of the hold is filled in by just lowering the lifts into place.

The danger associated with this method of stowage has manifested itself on numerous occasions. When the centre hatch lifts are lowered into place there is a

Fig. 67(a).

Fig. 67(b).

Fig. 67. Examples of bad stowage and incorrect dunnaging.

break in the continuity of the athwartships dunnage where cargo commences to be loaded without the use of fork lift trucks. Consequently, the wing stowages are not connected, by dunnage, to the centre stowage. In the case of one voyage, the vessel encountered heavy weather and the wing stowages fell over against the vessel's sides, and the centre stowages also shifted. This concerned a cargo of flanged beams, a large proportion of which became bent and twisted. The beams became entangled and had to be literally removed from the stow by force. The claim for damage was in excess of US$1,000,000. Special attention should be given to the stowage of such cargoes whose preservation should take precedence over a quicker turnround of the vessel.

SULPHUR

Apart from sulphur occurring naturally in iron ore which is used in the manufacture of iron and steel, much sulphur finds its way into steel through the coke used in the blast furnace. It is an undesirable impurity which is reduced to a minimum by the introduction of a small quantity of manganese, which chemically combines with the sulphur so preventing the sulphur from blending with the iron to form iron sulphide. Sulphur in steel causes brittleness at high temperatures, which affects the hot working properties of the material.

SURFACE PREPARATION BEFORE COATING

Generally speaking, most of the more serious claims which arise with regard to steel products carried in seaborne trades, concern rust development which has impaired, or is alleged to have impaired, the surface condition of the material. There are some situations where claims for a rust condition are unwarranted because, before the steel is used for whatever purpose it was intended for, it has to undergo surface preparation before a protective coating is applied. In other instances during the normal progression of further processing the steel would be subjected to acid bath pickling, which removes rust, scale and other undesirable matter. This being so, situations do arise where the surface of the steel has become pitted or damaged to such an extent that additional surface preparation is necessary, over and above what would normally be required, in order to restore the goods to a condition which is acceptable for what they were originally intended.

If steel surfaces are to be protected in such a manner as to provide indefinite service life for their intended uses, it is necessary that all mill scale is removed, as well as rust, grease and any other extraneous matters, before protective coatings are applied. If this pre-coating preparation is not efficiently carried out, as well as being complete, the coating will eventually fail. The consequences of painting over rust and mill scale, and improper preparation of steel surfaces before applying protective coatings, were very much in evidence in the rebuilding of Britain's Merchant Fleet after the last war. In those days the steel was weathered in order to remove the mill scale. Owing to excessive demand for ship's plates, in many instances their weathering was incomplete and paint was applied over mill scale. The metalwork of the ships was in an extremely advanced state of deterioration long before it should have been.

Hand cleaning of steel is a very unsatisfactory method of preparation of steel surfaces, and is always less likely to be successful than the mechanical means generally in use these days. The rust itself is not really a problem, which is evident from the fact that iron oxide is used in the pigments of paints. Pure hydrated iron oxide is expressed by the chemical formula $Fe_2O_3H_2O$, which in reality rarely exists. The rust deposits which normally develop upon steel surfaces are composed of other elements which are part of the chemistry of the steel, as well as atmospheric pollutants. These develop into salts which are of a corrosive nature.

Steel surfaces are not as smooth as they might appear, and at the anode of each electric cell, necessary to the development of rust—see under Corrosion and oxidation of steel surfaces—a pit in the surface of the metal exists. These surface irregularities harbour the corrosive salts referred to, and it is their complete removal in surface preparation that proves difficult. Mill scale must be removed because it is not part of the steel and will, given time, be undermined by the development of rust on the surface of the metal and become detached. The most widely used methods of descaling and cleaning steel surfaces are acid bath pickling and blast cleaning.

Acid bath pickling involves submerging the steel for a period of time in a liquid solution of sulphuric or hydrochloric acid. The advantages and disadvantages of using either of the acids referred to revolves around temperature control, acid consumption, build-up of iron in the bath, quality of finish and acid disposal or recovery. Sulphuric acid is relatively cheap and practically no fumes are emitted up to a temperature of 60°C. When heated it is quick in action and can be easily mechanically agitated. Its disadvantages are that it has to be heated to obtain best results, and the activity of the bath slows down considerably with the build up of ferrous salts. The finish of the metal lacks the brightness of that produced by hydrochloric acid.

Hydrochloric acid is much more efficient than sulphuric acid. The objections to its use in the past were its considerable evaporation during pickling and its highly corrosive vapours, which prevented agitation of the bath. These difficulties have been to a great extent overcome—however, the cost of hydrochloric acid still remains much higher than that of sulphuric acid. On the plus side hydrochloric acid does not require heating and very satisfactory results are obtained at temperatures of 40°C. This temperature is normally generated during the reaction of the bath in the process of pickling. The surface finish is much brighter than that produced by sulphuric acid.

Acid bath pickling is usually used where fine finished surfaces are required before applying protective coatings. Hot rolled steel is so treated before cold rolling. The pre-coating condition of hot rolled steel surfaces is customarily achieved by what is referred to as blast cleaning. This is an abrasive method of cleaning steel surfaces which amounts to bombarding the surface of the metal with metallic shot, sand and grit etc., depending on the surfaces to be treated. Not only is the purpose to clean the surface but also to roughen it and create a suitable bond with the coating to be applied.

It is appropriate to mention here the situation which arises when steel surfaces are attacked by aggressive mediums such as salt water. If one pass through the acid pickling bath indicates that there is no etch mark development on plating being treated, the material should be considered sound. On the other hand, if etch marks are visible after one pass through the acid bath, or if the surface of the plating was pitted, then further treatment will be necessary. In the case of cold rolled steel sheeting the steel cannot be restored to its original condition as pickling and re-rolling will result in losses of material and the production of another type of material. In such circumstances, the damaged parts will be unrepairable and probably have to be reduced to the value of scrap.

Where hot rolled steel sheeting is concerned the prospects of restoring the goods to their original good condition are much better by acid bath pickling. In any case, if too many passes through the acid bath are envisaged, or if such a course of action proves not to be a viable proposition, there are at least other options such as declassing the goods to the value of second choice material.

The situation with hot rolled structural material is again better than with goods destined to be cold rolled. A fine surface is not necessary. All that is required is to remove the damaging salts, mill scale and rust. This can be achieved by normal blast cleaning of the material. One of the advantages of blast cleaning is that it can be performed more or less anywhere. After cleaning it is usual to apply a blast primer to the material.

SURVEYOR'S SURVEY REPORTS

Although it is not intended to tell surveyors how to write their reports, many survey reports relating to cargo damage claims contain far too little information necessary to the efficient and successful negotiation of these claims. Surveyors should always be conscious of the fact that the liability for large amounts of money may depend upon what they have put into their reports. In many instances, owing to casualness in reporting, the claims negotiator, claims settler, has little to manoeuvre with and is often misguided both in his legal and practical assessment of the case owing to poor quality survey reports. For instance, the following may appear as a description of the goods and the damage: "goods wrapped in metal envelopes which were rusty"; this is not a detailed description. It is hoped that what is said here will, apart from other things, be useful as a checklist.

The commodity report

Where goods are landed from a ship and are found to be in a damaged condition, i.e. in some serious particular respect, not conforming to the apparent condition as shown in the relevant bill of lading, the receiver of the goods, upon determining that they are damaged in a manner which will detract from the sound value as per the relevant commercial invoice, will initiate a survey by claiming against his insurance who will put the carrier on notice; or if not insured he will put the carrier on notice direct by holding him responsible for the damage and notifying him of when and where a survey on the goods is to be held. The cargo interest's surveyor will attend the survey and in the absence of a surveyor appointed by the carrier assess the damage unilaterally. If the carrier appoints a surveyor, the survey then becomes a joint survey: the purpose of a joint survey is to reach an agreement, if possible; not to reach a disagreement. The commodity report is a survey record of all information relative to the assessment of loss sustained by goods; the main function of it will be to assist in the eventual adjustment of any entered third party claims. The purpose of any commodity survey is to determine and to report upon the following:

(1) Name of applicant, his function, e.g. representing owners, underwriters, town or city where he resides.
(2) Date and place where the survey was held.
(3) Purpose of attending the survey, e.g. to ascertain the extent of the damage alleged sustained by a consignment of bundles of steel beams.
(4) Name of the carrying vessel/details of transhipment if any. Date of arrival of the vessel and date of discharge of the goods.
(5) Bill of lading details concerning the goods to be surveyed.
(6) The names of all parties present at the survey, who they represented and in what capacity.
(7) The name of the receiver.
(8) Where the goods were stored and how they were presented for survey, e.g. the goods were stored in an open-sided warehouse and presented for survey without any disturbance to wrappers or securing. *Identification*: It is necessary that the surveyor makes clear the fact that he identified the goods before the survey commenced. He must state how he did this, e.g. by checking marks and numbers as stencilled on the metal packing and/or metal identification tags attached to the metal strapping etc.
(9) *Description of the goods*. The actual goods should be fully described, stating the type of packing and the method of securing. Any relevant measurements or weight per unit to be reported upon.
(10) *Aspect of the damage*. A full and descriptive report of all relevant aspects of the defects and/or damages to the packing which gave rise to the survey being called. It is not sufficient to state that the covers were rusty; the configuration of the rust and severity of such must be fully described. During removal of the packing it must be reported upon how the defects/damages to the packing conform or correlate to/are adjacent to/the defects/damages sustained by the actual goods. A full description of the damage, if any, to the contents must be included in the report.

    Whether the cause of the damage is mentioned in the commodity report or is referred to in a separate report, the aspect of the damage should be described and worded in such a manner as to tie in with the reported cause of the damage. This would especially apply if the reported cause of the damage was derived directly only from what was observed at the time when the commodity survey took place.
(11) *Nature of the damage*. If the damage has been sustained through contact with some foreign substance or liquid it is incumbent upon cargo interest's surveyor to determine the nature of the offending substance or liquid, if possible; e.g. if rust development through contact with water is involved, as is usually the case, the cargo surveyor is obliged to prove that it was either fresh water or salt water. If the presence of chlorides is indicated the surveyor should prove whether it was seawater or some other chloride-bearing liquid/substance.

(12) *Extent of the damage*. The survey must provide a detailed explanation as to what transpired in arriving at an agreed depreciation, the normal end uses of the material and how the receiver proposes to dispose of the damaged goods. He needs also to relate the reason, or reasons, why he granted an allowance for the purpose of justifying his agreement. The old cliche "after long and laborious discussions an allowance of 75% was granted and considered by all present as being fair and reasonable" is no longer acceptable.

(13) *The Cause of damage report*

(a) Name of applicant and who he represents.
(b) Name of the ship.
(c) Gross tonnage.
(d) Type of ship.
(e) Port of registry.
(f) Classification society.
(g) Date built.
(h) Place built.
(i) Details of safety construction certificate.
(j) Details of last load line survey.
(k) Details of last special survey.
(l) Number of cargo holds.
(m) Number of deck hatches per hold.
(n) Hatch closing system.
(o) Last two cargoes carried.
(p) Name and city of the ship owner.
(q) Name and city of manager/operator.
(r) Name and city of charterer.
(s) Name of charterer's local agents.
(t) Name of owner's local agents.
(u) Name of the master.
(v) Name of the chief officer.
(w) Name of the super cargo.
(x) Name of stevedoring company.
(y) Name of owner's superintendent if in attendance.

Transport damage to steel products can result from many causes, most of which are well known and recurring; for example, if upon arrival water is found in the hold of the ship there are only so many possibilities whereby this water could have penetrated in at the bottom of the cargo hold. If these known possibilities are properly and thoroughly investigated the cause of the damage must be found. Having discovered the source of ingress of the water, certain facts surrounding the incident need always to be included in the report in order that it be a document of the highest factual value when used in conjunction with ensuing claims.

Leaking hatches

Having performed a full and complete survey of the hatch closing appliances and conducted tests for watertight integrity, the surveyor should include full details in his report (see section on Steel hatch closing appliances). This survey would only be necessary if the surveyor was of the opinion that leakage of the hatches was involved in the development of any damage. Where such a survey forms part of a pre-shipment survey and the hatches are found to be defective, the surveyor is expected to recommend repairs which would restore the hatches to a seaworthy condition.

Leaking tank lids

The condition of the tank lid packing, type of packing, and also completeness of securing studs and their condition must be reported upon. It must also be stated as to how leakage was discovered, confirmation must be obtained as to how it was caused, and it must be stated whether in the process of discovery the tank was subjected to an hydraulic pressure test witnessed by the surveyor. In such circumstances, as is found possible, remarks covering the condition of all tank-top lids in the cargo holds have to be included in the report.

Leakage from pipes

Leakage from cargo hold piping through defective connections and actual defects to the metal work developing are a regular occurrence. When flanged and bolted connections are involved it is usual for the flange packing to fail, so permitting ballast water or seawater to enter the cargo compartment. The condition of the bolts and packing, the type of packing, the exact circumstances whereby the packing failed, also when, where and by whom the packing was installed is necessary information. When pipes develop holes, the observed general condition of the pipe will be determined, also whether the hole came into existence through the effects of corrosion or erosion. Details of the position of any defects resulting in leakage are of paramount importance and the exact position of any defect must be stated as well as, in the surveyor's opinion, whether the position of the defect was accessible or inaccessible. If it is evident that the defect existed at the commencement of the voyage it is essential that this be made known together with an opinion as to the possibilities of the defect being observable by a normal, reasonable, visual inspection of the cargo space prior to the commencement of loading.

Bilge well overflow

It is necessary to report upon the condition of the bilge pipe sounding caps on deck and how they were fitted including the condition of the threads. Were these caps hose pressure tested and what results were obtained from the test? This test would be isolated to the bilge well suspected of overflow; nonetheless, the condition in general of all bilge sounding caps is to be included. Where leakage from the bilge lines are suspected of being the cause of the overflow, it is necessary to know if

(1) A ship's piping plan was consulted. If so was it determined whether or not by original construction the bilge and ballast systems were completely isolated from each other?
(2) The number and position of all non-return valves, also the type involved, situated between the bilge pumps and the bilge well in the specific cargo compartment.

(3) Were all non-return valves properly seated and working correctly?
(4) Was the bilge line between the engine room and the cargo hold tested under pressure?

Structural defects

Defects such as cracks in the weather deck plating, cracks in ballast tank boundary plating and in the ship's shell plating are not unknown as sources of water entering ships' cargo holds. Recalling what has been said with regard to pipes above, here again the accessibility of the area of the defect is of great importance, also how and when the defect came into existence. Naturally any defects caused by the action of stevedores' grabs or shifting cargo will be reasonably straightforward to deal with, but defects in the structure brought about through lack of maintenance or structural stress or both pose more of a problem, in which the following reported information would be required:

(1) have such defects developed before, if so what was their location and what steps were taken to rectify the situation?
(2) In any previous incidents, was the classification society involved, what were their recommendations and were these followed up to the satisfaction of the class surveyor?
(3) Depending upon the circumstances, which may involve various voyage histories of the vessel, also the current voyage, stress calculations may have to be included in the report.
(4) As deemed appropriate and relative to the incidence, the surveyor will find it necessary to include copies of cargo plans, also details of weight distribution in general for more voyages than the current voyage.
(5) Taking into consideration the four points above mentioned, also the weather conditions encountered on the voyage, and/or comparison voyages, the surveyor will report conclusions reached with regard to the cause of the damage.

Stowage report

Surveyors are often instructed to carry out a loading stowage survey, in which case they will be obliged to advise the master in connection with this. Where steel cargoes are concerned, in most instances a "New York Produce Exchange" time charter will be involved, into which will probably be incorporated an Inter-Club agreement. In such circumstances, under this agreement, charterers will *ultimately* be responsible for stowage of the cargo. The liability of the master will only be involved, under such an agreement, if during the course of loading he interferes with stowage operations to such an extent that a charterer's intended method of stowage is changed, whereby, as a result of this, damage is sustained by the cargo. Naturally, if in the opinion of the master, at any time, safety of the ship or crew might be adversely affected by the manner in which the cargo is being stowed, he must intervene and insist upon some change in the method of stowage which will rectify the situation to his satisfaction—that will be the extent of his responsibility.

If an Inter-Club agreement is not incorporated in the charterparty the wording of clause No. 8 of the C/P will probably be "Charterers are to load, stow and trim the cargo at their expense under the supervision of the Captain". Here again the master's responsibility would be confined to safety of the vessel and crew as above mentioned. However, if in clause No. 8 as referred to above it is stated "under the supervision and *responsibility* of the Captain" the addition of these words significantly affects the balance of responsibility for the actual loading and discharging of the cargo and for any cargo claims that might arise. Since the master's responsibilities have been increased he will now have to take a much wider and more active role for the entire operation of loading, stowing and discharging the cargo. The situation will now be that there is a joint operation between the charterers and the shipowner, the charterers being responsible for loading and stowing the cargo, and the master, on behalf of the shipowner, taking responsibility in seeing that the charterers carry out their duties in such a manner as to ensure that the cargo does not become damaged.

Since the second edition of this book was published the Inter-Club agreement has been re-negotiated, however, that part of the agreement concerning stowage is unaltered; what is stated above is unaffected and therefore still applicable. When a stowage surveyor is in a situation whereby he is to advise the master on stowage, it is necessary that he understands how the Inter-Club agreement can affect the advice he is providing. It is recommended that the above is kept in mind when attending such a survey and it would be considered prudent of the surveyor to take the matter up with his principals and obtain guidance in the intricacies of the matter.

Where chartered ships are concerned the charterers nearly always appoint a super-cargo to follow loading operations. Prior to commencement of loading, a pre-loading cargo plan will be submitted by the charterers to the master for his signature of approval. In accordance with what has been said above, the surveyor must mention in his report what suggestions were advanced by the master in relation to stowage and why such suggestions were made, e.g. to rectify a foreseen unfavourable end loading trim or incompatibility of various commodities in stowage, etc. In any case, it must be clearly stated as to whether the master insisted upon change or merely made a suggestion for charterers to consider. Broadly speaking, if intervention of the master is not required, providing the safety of vessel and/or crew is not involved, the surveyor in his report should make it clear that all ultimate stowage arrangements were left to the discretion of the charterers whose activities in supervising the stowage were followed by the surveyor and the vessel's staff. Any discussions between surveyor/master on one side and charterers on the other, relating to changes in stowage, should be reported in such a manner as to make it quite clear that the master did not insist on any changes that were made, with regard to the actual preservation of the cargo, but merely suggested them and such suggestions were eventually followed or went unheeded by charterers. If there is no Inter-Club agreement in the charterparty, or some other charterparty or form of charter involved, then in all probability the approach to the stowage situation would be such that full supervision and control of loading and stowage would be required of the master.

It has been found that stowage reports should contain the following information:

—Name of person/organisation commissioning survey.
—Name of the ship.

—Gross tons register.
—Home port of the ship.
—Reasons for attending and details of appointment, e.g. following stowage operations and advising the master.
—Place where the ship was berthed.
—Date proceeded on board.
—Name of the master.
—Give details of stevedore damages and whether these were pursued against the stevedores. Refer to any correspondence and attach copies to the report.
—Give details of any discussions or disputes revolving around the loading, stowage, dunnaging or securing of the cargo.
—Report on weather conditions during loading. During periods of rain was work stopped and hatch openings closed—give details.
—Date and time of completion of loading.
—Date and time of completion of securing.
—Finally, was the master approached and asked to sign a document to the effect that the loading, stowage, lashing, and securing was carried out to his satisfaction (or something similar to this); did he sign? Give details.
—Date and time of sailing.
—Attachments to the report, if possible; copy of cargo plan, copy of statement of facts, copy of any letters relative to the loading of the cargo sent or received by the master.

Sweat damage

The following details, apart from the usual introductory remarks, as indicated on page 232, with regard to the vessel's ventilation system are to be entered in any report dealing with import cargoes damaged through contact with sweat water.

—The type of ventilation system involved and the position of the ventilators. Height of the ventilators, type of ventilator, whether fitted with fans (suction or reversible), fan capacities. Diameter of vents and state whether or not the intakes are trunked to the bottom of the hold or whether they terminate at deck head level. Condition of method employed in sealing off the vents, e.g. flaps, screw-down dampers, etc. Copy of ventilation plan.
—Copies of ventilation records for the voyage showing wet and dry bulb thermometer recordings for cargo holds and for the atmosphere. Time and date when observations were taken. When ventilation was operative. If thermo-couples used to obtain cargo temperatures, etc., give details.
—Were any special measures taken in sealing off the cargo holds or ventilators when moving from a cold to a relatively warmer climate? Give details.
—How were temperatures taken (a) by whirling psychrometer, (b) thermometers mounted in a portable louvred screen box, (c) thermometers mounted in a fixed screen box usually positioned one to port and one to starboard on the navigation bridge (see Fig. 75 on p. 261). Report on the situation as indicated and on the condition of the instruments.
—A detailed description to be provided of the pattern and type of rust affecting the cargo, preferably as sighted in stow on board the ship. Where the affected cargo was situated in the hold of the ship.

Fig. 68(a). Relative positions of coil and dunnage on tank-top.

Fig. 68(b). Detail of contact area between coil and dunnage.

Fig. 68. Steel coil stowed on dunnage placed on uneven tank-top.

Fig. 69(a). Coil in position on tank-top.

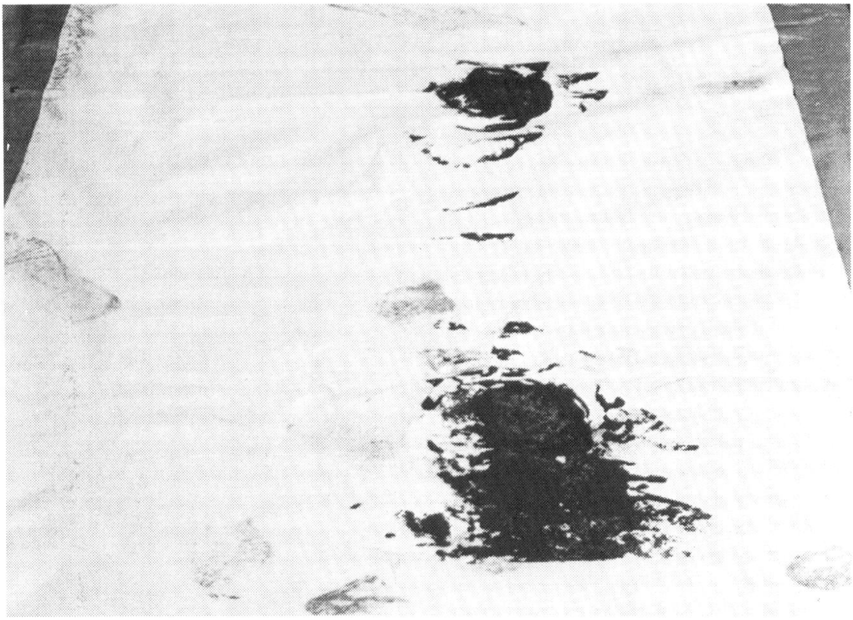

Fig. 69(b). Detail of contact area between coil and tank-top.

Fig. 69. Steel coil stowed directly onto uneven tank-top.

The position of the double-bottom internals, i.e. longitudinal beams, floors, can clearly be seen beneath the tank-top plating. As the plating is set down between internals, the shiny appearance of the ridges formed by the internal members suggest that the forklift trucks are mainly in contact with these areas and not the tank-top plating.

Fig. 70. Position of double-bottom internals clearly seen beneath tank-top plating.

—Provide reasons as to why ship or cargo sweat, or both, were considered to be involved.

Hatch condition survey report

There have been many complaints regarding the standard of reporting, in that surveyors provide insufficient detail as to what they actually observed or surveyed. Some survey reports cover and name only defective parts, so leaving the reader of the report to assume that all else was in good condition and he may wonder whether all component parts were actually surveyed. Others might report upon the actual parts surveyed but fail to report comprehensively: such remarks as "acceptable", "OK" or "good condition" are not always regarded as being sufficiently descriptive. In the hope of avoiding criticism and bearing in mind that not everyone's whims and fancies can be catered for, the following method of remarking is suggested. Such a list, or form, it is hoped, will provide guidelines which will ensure that nothing is missed.

Part involved	Good condition remarks	Adverse condition remarks
Panels, Hinges, Hydraulics	Panels well painted up, free from rust, rust scale or indentations. Affected by rusty patches and scattered dents but general condition not considered to affect the strength of the appliances. Hinges in apparent good order and condition. Panels not misaligned. Hydraulic system functioning correctly. No leakage of oil from joint connections or hoses.	Panels rusty (or partly rusty) in appearance. Affected by rust scaling. Corrosion (light) (heavy). Affected by scattered dents. General condition poor and considered to affect the strength of the appliances. Hinges rusty and adjacent areas rust streaked. Wear down of hinge spindles and/or bushes apparent with resulting misalignment of panels. Hydraulic system malfunctioning. Oil leaking from hydraulic system.
Rubber Seals (Gaskets)	Rubber jointing pliable, not unduly compressed. Not torn, chafed or defective in any respect.	Locally chafed, torn and/or distorted (squeezed or pinched in the retaining channels). Not firmly affixed in the retaining channels. Rubber jointing hanging out of retaining channels in places. Rubbers perished, hardened and lacking necessary pliability. Rubbers affected by heavy and unusually deep compression marks.
Compression Bars	Free of rust scale or any deformation.	Locally bent and/or torn in places. Waved along length. Locally waved in places. Distorted. Affected by heavy (light) corrosion (points of corrosion).
Gasket Retaining Channels	Free of rust scale or any deformation	Locally bent and/or torn in places. Waved along length. Locally waved in places. Affected by heavy (light) corrosion (points of corrosion). Distorted (locally) or along length. Affected by heavy (light) corrosion. Metal parts seriously wasting away through corrosion.
(a) Transverse Gutterings and (b) Hatch Rims	(a) Free of rust scale, serious. (b) Free of corrosion or any physical damage.	Channel bars locally bent and/or waved. Metal work generally corroded. Locally wasted away through corrosion. Hatch rims locally wasted and reduced in height through corrosion. Deeply scored.

Part involved	Good condition remarks	Adverse condition remarks
Steel to Steel Contact Points	No distortion, bending or excessive grooving.	Sections bent. Heavily grooved, distorted and/or partly broken.
Crutches	Free of scale or corrosion.	Heavily rusted or corroded. Partly crushed. Completely crushed. Distorted.
Snugs	Free of rust scale or any defects.	Corroded—diminished in size. Holding properties diminished. N.B. There are cases on record where quick release cleats have come loose during heavy weather. This may be attributable to the deteriorated condition of the snugs.
Quick Acting Cleats, Rods, Nuts and Washers	Free of rust, scale or any defects. Rubber washers pliable. Corroded but corrosion not serious—general condition acceptable.	Corroded—cross sections area of rod diminished. Strength of rod impaired. Rod bent. Washer hardened and perished. Washer crushed. Tightening nut rusted and inoperable. N.B. It must be kept in mind that any reduction in cross sectional area of the bar forming the rod reduces its strength. Rods have been known to break.
Cross Wedges (Cross Joint Cleats)	None missing—all in place when surveyed. (Sometimes, for various reasons, all or some wedges are removed when the ship is in port.) Wedges not bent, side springs complete and in good working order.	(Number) cross wedges bent. (Number) cross wedges with inoperable springs. (Number) cross wedges with side springs missing. All cross wedges in place at time of survey. Wedges tightened over welded pads on adjacent panel edges.
Inboard Drainage System	Coaming drain channels and all gutterings free of loose scale, foreign matters and/or obstructions. Drain holes, unobstructed, in good order and condition also in good working order. Drain valves free and working correctly.	Coaming drain channels partly (completely) obstructed with rests of previous cargo. Coaming drain channels corroded (lightly) (heavily). Drain holes blocked with rests of previous cargo (with rust scales). Drain valves jammed (or partly jammed), not in good working order. Drain pipe fitted with a piece of canvas hose—an acceptable arrangement.
Wheel Tracks (Guide rails)	Free of loose scale, clean and unobstructed.	Worn, grooved, uneven, waved. Corroded (heavily) (lightly) and weakened.

Part involved	Good condition remarks	Adverse condition remarks
Wheels	Well greased and free running.	Seized up. Spindle bent, wheel not working.
Connecting Chains	Free of scale and correctly adjusted.	Affected by rust scale and requiring adjustment.

Where hatch surveys are concerned it is becoming increasingly more common practice for escape hatches and hold access doors to be included in the survey. The following are suggested as suitable remarks:

Note: For access doors and escape hatch lids, the same remarks, with slight adjustment, are applicable as mentioned above for gasket channels and rubber gaskets:

Escape Hatches	Clear bolts and wing nuts working freely: wing nut thread well greased up. All cleats in place and complete with wing nuts.	Cleat bolt section seized up on operating spindle (all or give number). (Number) of cleats wing nuts missing. (Number) of wing nuts seized up. All (or give number) cleats rusty, corroded and completely inoperable. (Number) cleats bent and inoperable.
Access Doors	All securing lugs complete, working freely and fully operable.	Out of (give number) lugs, all (or give number) seized up and completely inoperable. Out of (give number) P: (give number) lugs missing.

Conclusions

If the results of a hatch survey are reported as indicated above, it is reassuring to the applicant that all parts relative to the efficient working of the hatches were given attention. Further to this, such remarks are of value in tracing back the history of the appliances at a later date. It has often been stated that to report that something is in good condition is less desirable than to state why it was in good condition. Therefore, remarks such as "reasonably good condition", "good condition", "acceptable condition", "normal working condition", "good working condition" and such like, should never be used because they are too vague in meaning.

General remarks on survey reporting

—Where damage is caused, or suspected of being caused, through contact with moisture, survey reports should always confirm the results of silver nitrate tests for the presence of chlorides.

—An adequate amount of photography must accompany the report and dated exposures are recommended.

—When the master of a ship is interrogated in relation to the cause of the damage full details of the voyage should be recorded. On every occasion details of weather conditions encountered must be obtained; where serious damages are involved copies of relevant log book pages need to be taken.

—As many, if not most, steel cargoes are subject of a pre-shipment survey, it may be necessary to obtain a copy of this report as its contents could have some influence upon conclusions eventually to be reported in either the commodity or cause of damage report.

—When considered appropriate and relative to the incident, the contents, or part thereof, of the last ship condition survey report can contribute to a more detailed and complete cause of damage report.

—Commodity reports are best concluded by a recapitulation of the loss, i.e. depreciations granted, extra costs involved, final results of any sales, details of which if occupying a prominent position on the first page of the report can readily be observed without having to be searched for.

—The surveyor should always finish the cause of damage report by making a statement of his conclusions explaining how the damage was caused; supported by any other qualifying remarks pertinent to the cause such as, intensity of the weather experienced, storm damage to the vessel's structure which will support the intensity of the weather conditions claimed and inaccessibility of structural defects involved etc.

—Steel hatch survey reports must be concluded by giving recommendations, not suggestions, for repairs. Naturally, if these recommendations are carried out and the surveyor is instructed to follow the repairs, he would be obliged to ensure that the repairs were satisfactorily completed, also that the final watertight tests were performed to his satisfaction: these end results to be included in the report.

SWARF

This consists of scrap, ferrous metal borings, shavings and cuttings, drillings, filings and turnings. It is the waste material of machine shop operations from lathes, drill presses and other machine tools. The goods are a mixture of many types of carbon steels and may include organic materials besides being contaminated by cutting oils and affected by rust.

In preparation for shipment the material is crushed. Owing to its high surface to mass ratio it becomes an extremely dangerous cargo. Oxidation of the large exposed surface areas releases heat, and there have been incidents where complete conflagration on board ship has been involved. Swarf is often stockpiled in open areas where it is totally exposed to the elements, and experience has shown that this type of cargo loaded in wet weather areas can be the most troublesome.

The cargo might be heating prior to shipment; however, the serious heat build up has a greater propensity to develop within the confines of a ship's hold after the goods have been loaded. The oxidation process with accompanying susceptibility to spontaneous heating, possibly developing into spontaneous combustion, is greatly intensified if the cargo sustains wetness through contact with rain or, worse, snow, and contact with seawater can be the worst situation of all.

Before and during loading the temperature of the goods should be continuously controlled. If the temperature of the cargo exceeds 55°C it should not be taken on board the ship. If after the goods have been taken on board the surface temperature rises above 90°C, cargo operations should cease until the temperature falls below

STEEL CARGO PRE-SHIPMENT SURVEY REPORT

CHECKLIST FOR STEEL CARGO SURVEY REPORT

Survey
1.1 Name of surveyor(s)
1.2 Date of report
1.3 Name of person/organisation commissioning survey
1.4 Name of organisation survey commissioned for
1.5 Summary of surveyor's terms of reference

Vessel
2.1 Name
2.2 Type
2.3 Gross tonnage
2.4 Port of registry
2.5 Classification society
2.6 Date built
2.7 Place built
2.8 Details of safety construction certificate
2.9 Details of classification certificate
2.10 Details of last load line survey
2.11 Details of last special survey
2.12 Number of cargo holds
2.13 Number of deck hatches per hold
2.14 Hatch closing system
2.15 Number of deck ventilators per hold
2.16 Ventilation system
2.17 Type of cargo carried on last voyage
2.18 Type of cargo carried on voyage before last voyage

Companies/Personnel
3.1 Name and city of owner
3.2 Name and city of manager/operator
3.3 Name and city of cherterer
3.4 Name of charterer's agent at port of loading
3.5 Name of owner's agents/P&I correspondents at port of loading
3.6 Name of master
3.7 Name of chief officer
3.8 Name of supercargo
3.9 Name of stevedoring company
3.10 Name of tallying company
3.11 Names and business of others in attendance

Activity Report
4.1 Location of loading berth
4.2 Time and date vessel arrived
4.3 Time and date surveyor proceeded on board
4.4 Time and date loading commenced
4.5 Time and date loading completed
4.6 Time and date vessel departed
4.7 Brief description of steel cargo hold cleansing
4.8 Brief description of steel cargo loaded
4.9 Total weight of steel cargo loaded
4.10 Weight of steel cargo loaded per discharge port
4.11 Brief description of other cargo stowed in steel cargo holds and where stowed in relation to steel
4.12 Full details of disputes arising during loading concerning bills of lading (attach all relevant correspondence)
4.13 Full details of stevedore damages occurring during loading (attach all relevant correspondence)
4.14 Full details of weather conditions during loading and of stoppages and protective action taken during wet weather†
4.15 Full details of silver nitrate tests carried out and of any disputes arising
4.16 Full deatils of surveyor's involvement with clausing bills of lading/mate's receipts/exception lists*
4.17 State if surveyor present when bill of lading/mate's receipts/exception lists* signed

Cargo Report
5.1 Port of destination
5.2 Hold number(s)
5.3 Bill of lading/mate's receipt/exception list* number(s)
5.4 Description of items as shown in bill of lading/mate's receipt/exception list*
5.5 Number of items
5.6 Surface condition and mechanical damage clauses superimposed on bill of lading/mate's receipt/exception list*
5.7 Brief description of storage immediately prior to loading
5.8 List of attached photographs illustrating surface condition and mechanical damage clauses, and any loading problems encountered

†Official weather reports ordered and retained by the surveyor. *Delete as appropriate.

Extract from *Steel Preshipment Surveys* by Arthur Sparks, MNI, and North of England P&I Association, reproduced with permission of Anchorage Press.

65°C, and shows a steady downward trend. The cargo should be compacted in the ship at frequent intervals during loading. Further to this the stow must be kept level.

Preparation of the holds before loading will include removal of all combustible material, such as sweat boards, wooden pipe guards, etc. The cargo spaces must be dry, as must the bilges. Every possible effort should be made to ensure that the hatch closing appliances are in good condition. Where steel hatches are concerned the interpanel joints should be sealed with masking tape, properly applied, as an additional precaution against leakage.

If during the sea passage the surface temperature should rise to 80°C the possibility of a fire will exist. The ship should make for the nearest port equipped to handle the situation. The ship should not sail on an international voyage if the temperature is above 65°C. The temperature must be stable below 65°C or observed to be falling below this temperature for at least eight hours. If heating develops, its upward or downward temperature trend, observed over eight to twelve hours, is an indicator as to whether it can remain on board the ship.

Should fire-fighting become necessary, it is advocated that water should not be used at sea, because, in certain circumstances, there is a tendency for hydrogen to evolve, which in the presence of oxygen can produce an explosive mixture. Copious water can be applied for fire-fighting when the ship is in port, due consideration being given to the vessel's stability. In the event of heating developing at sea to such as extent that there is overheating with outbreak of fire being imminent, it is recommended that inert gas be discharged into the cargo compartments concerned.

TANK-TOP STRENGTH IN RELATION TO STOWAGE OF STEEL CARGOES

During the 1980s the largest type of bulk carrier vessel usually employed for the carriage of steel cargo was around 30,000 m/tons deadweight capacity. It is now not unusual for vessels of up to 40,000/70,000 m/tons to be used in the steel trade. In previous years the carriage of any hot rolled steel coils in excess of 17 m/tons were the exception rather than the rule. Today, these unit tonnages have increased to 25 m/tons and in some instances 30 m/tons. It is anticipated that there will be further increase in deadweight tonnage of vessels used in the trade and also heavier cargo units. When comparing the two sizes of bulk carrier referred to above, it will often be found that there is no great difference in the permissible load limits per square metre of tank-top area, which are usually in some instances as low as 15 m/tons per m^2 and as high as 25 m/tons per m^2. These tonnages per cargo space will vary in vessels which have some of their cargo holds strengthened.

The tank-tops in the cargo holds of ships are designed, using Classification Society Rules, to have a maximum permissible loading in tons per square metre of surface area of tank-top and this loading is calculated for cargo evenly distributed. Bulk carriers were designed to carry evenly spread homogeneous bulk cargoes; they are used for carrying steel because they are capable of lifting large tonnages and are equipped with hatchways and unobstructed cargo spaces. Their main disadvantage is the limited strength of their tank-tops, which when carrying some steel cargo are subjected to severe concentrations of weight which greatly exceeds the permissible load limit per m^2. This is rarely discussed when a ship is loading; however, should it become a subject of discussion between charterers' representatives and the ship master it can develop and has often developed into a serious situation. The cost of extra dunnaging of the cargo ostensibly to spread the weight may be unacceptable to charterers. Cargo may have to be left behind with the vessel not completing her deadweight, thus resulting in dead freight claims and eventual legal proceedings or hearings in arbitration.

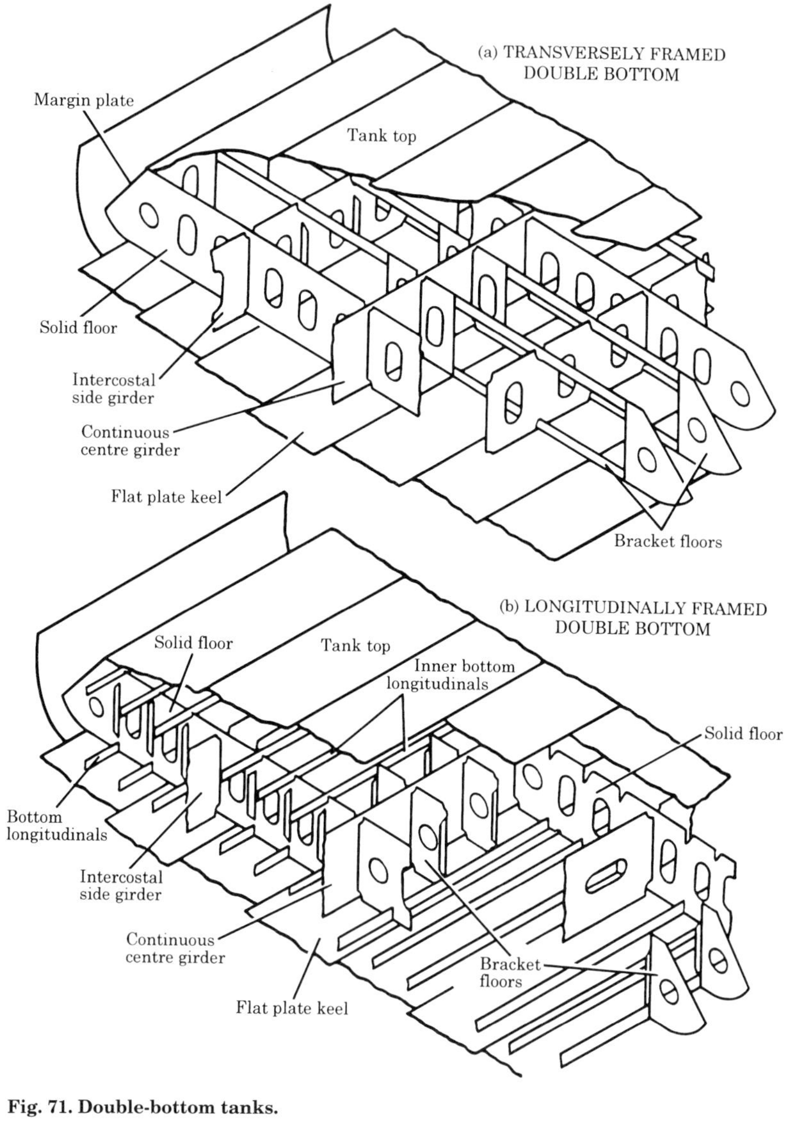

Fig. 71. Double-bottom tanks.

The area of a bulk carrier, or any other seagoing vessel, situated between the outer bottom shell plating and the inner bottom (tank-top) in the cargo hold is referred to as the double bottom. The complexity of its structure can be seen from Fig. 71. The various compartments within the double bottom system are used for the storage of fuel oil bunkers and water ballast (see Fig. 72 showing a

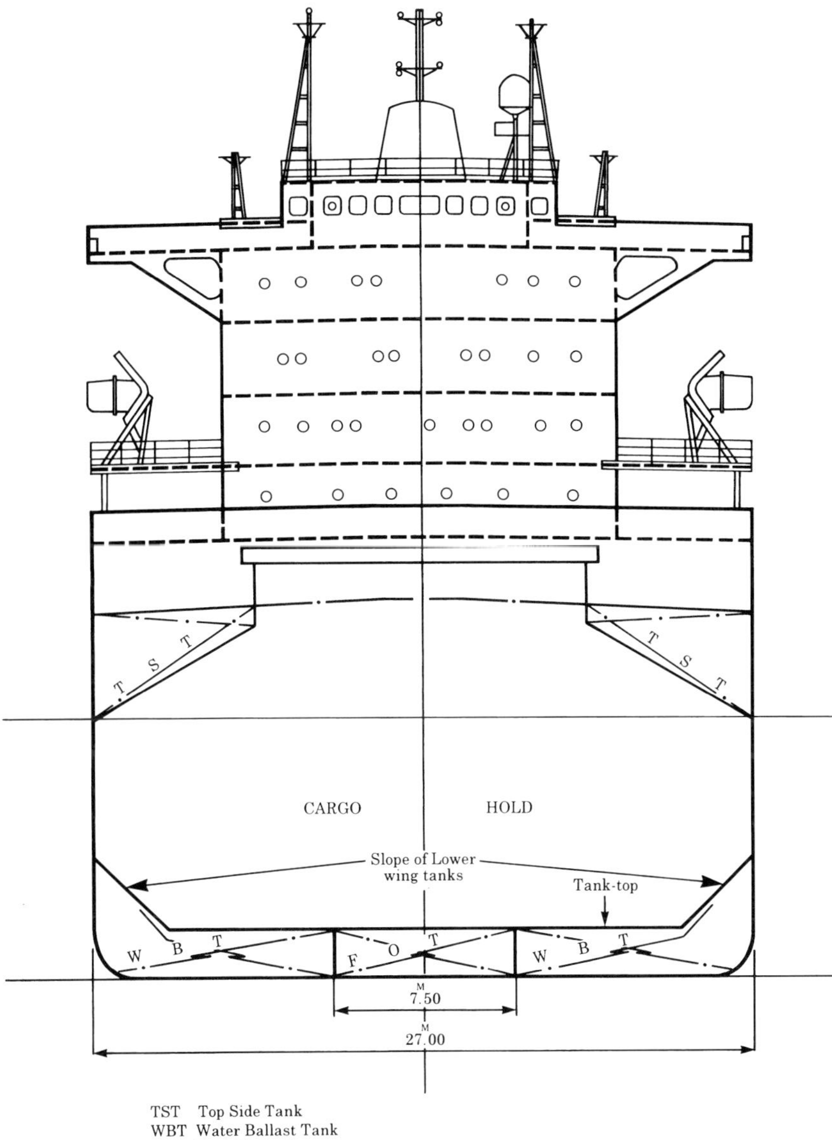

Fig. 72. Tanks.

cross-sectional view of a typical bulk carrier). Double bottoms may be constructed with longitudinal or transverse framing—a part of the structure which is a prominent feature of discussions relating to the distribution of weight of steel cargoes upon the tank-top plating. Where the ship exceeds 120 metres in length it is usual for longitudinal framing to be used and all further reference to double bottoms for the purpose of this article will be in relation to double bottoms so constructed. The scantlings of the component parts forming the double bottom is determined when the ship is built and approved by the classification society involved. This also applies to the thickness of the tank-top plating and the permissible load per m^2.

If the ship/bulk carrier loads, as she was designed to do, a homogeneous bulk cargo of, for example, iron, ore, the maximum amount of cargo permitted to be loaded in any cargo hold would be determined by multiplying the surface area of the vessel's tank-top by the permissible load per m^2. The load given by this calculation should never be exceeded whatever the cargo. Once the bulk ore has been loaded it can be trimmed and the weight evenly spread over the tank-top to exert a pressure per m^2 within the tank-top limitations. With steel the situation changes dramatically owing to the weight and nature of the cargo. A steel coil is a good example, as owing to its cylindrical form there is usually a great concentration of weight, which is referred to as "spot overload"; the following calculation should make this clear:

(*Involvement of dunnage ignored*)
Coil weight: 17m/tons; diameter: 160 cm; width: 186 cm
Coils stowed 3 tiers high = 3 × 17 = 51 m/tons.
Tank top limits: 21.94 m/tons per m^2.

Width of bearing surface of base coil = 15°/360 × 3.142 × 160 cm = 21 cm
Area of tank top covered by coil contact = 21 cm × 186 cm = 3906 cm^2
Permissible load on area covered = 2.194 kgs × 3906 cm^2= 8.569.8 kgs

	= 8.570 m/tons
Load imposed upon the tank top by the 3 coils	= 51.00 m/tons
Spot Overload	= 32.431 m/tons

There is a spot overload of 32.431 m/tons covering the bearing surface of the base coil. It is also necessary to determine whether or not per m^2 of the tank-top itself is overloaded; this can be arrived at as follows:

Amount of kgs over one m^2 is	51,000 kg/186 cm × 100 = 27,419 m/tons
Area covered is	21 cm × 100 cm = 2100 cm^2
Permissible load for the area covered	2100 cm^2 × 2.194 kgs = 4.607 m/tons
Weight on one M2 of tank top plating	= 22,812 m/tons

See also **Single tier stowage and height of multi-tier stowage** on p. 208.

From the above it will be seen that using the "empirical rule" creates an enormous spot overload upon the tank-top. It is a rule which when followed does not in practice appear to have any serious adverse effects, if any at all, upon the vessel's structure, although it is a well-established fact that unequal spreading of weight in high concentrations can in the long term contribute to deterioration of a vessel's structure.

Steel cargoes have to be dunnaged for the purpose of holding the cargo clear of the tank-top plating in order to avoid cargo damage should moisture, from whatever source, find its way beneath the cargo. Furthermore, dunnage is placed between layers of cargo to bind the stow together and also to facilitate slinging for discharge. As far as dunnage placed on the tank-top is concerned, the weight of the cargo will be spread through the dunnage across the area of the tank-top which it occupies, providing it is of adequate strength and sufficiently rigid to absorb the pressure applied to it. Otherwise, the weight of the cargo will be concentrated over the area where the item of cargo contacts the dunnage.

No tank-top on board any vessel is completely level, since in the first instance tensions set up by welding the numerous seams joining the plates and attaching them to the component parts of the vessel's inner bottom cause distortion. Within a short period of time after a vessel enters the bulk cargo trades the tank-top plating becomes slightly or moderately set down between internals; the severity of this depends on the types of cargoes carried and how they are loaded and discharged. In the case of vessels which have been consistently operating in the bulk trades over many years it is not uncommon that the tank-tops are battered and heavily corrugated in appearance. The fact remains that the surfaces of ships' tank-tops are not level and may be described as affected by scattered bumps all over; many give the appearance of being corrugated. On board one very well maintained bulk carrier an experiment was performed by choosing what appeared to be a level area of tank-top. A sample steel coil of 16.565 m/tons was selected, which, after painting the anticipated contact area, was lowered onto the surface of the inner bottom plating. The points/area of contact were so small that it was obvious that the coil rested only upon the few most prominent, highest points/areas of the plating (see Figs. 69(a) and (b)). A similar experiment was carried out ashore using dunnage beneath the coil, placed as is normal in the industry; the area of contact on the dunnage was by measurement 21 cm along the plank, i.e. the bearing surface coincided with the width of the plank × 24 cm (see Figs. 68(a) and (b)).

From what is stated above, it is evident that dunnage used to spread the weight of cargo, unless unusually strong and thick, will deflect owing to the tank-top being set down between double-bottom internals. The deflections will vary considerably from one area of tank-top to another and from vessel to vessel. There must be no delays in loading a ship and it is therefore impractical to require depth of all tank-top indents for the purpose of including these measurements in any form of dunnage calculations. In the circumstances, the best that can be done is to use adequate-sized dunnage in an attempt to spread the weight and ignore the deflections, unless of exceptional severity, in any calculations which have to be made with regard to the dimensions and strength of dunnage.

In the various discussions which take place between ship masters, carriers and shippers, when strength of the tank-tops and the unit weight of cargo are an issue, the charterers invariably refer to the vast number of similar cargoes carried on similar voyages in similar-type ships without any difficulties being experienced. It is a fact that with dunnage normally used in the industry with steel cargoes, the carrying vessel's tank-tops are severely locally overloaded and such vessels may be considered technically unseaworthy. To indicate how serious the situation can become, the master of a 45,000 m/ton bulk carrier had to refuse to load 25,000 m/tons of steel coils offered for shipment. In view of the tank-top' permissible load per m^2 the master refused to load the steel coils more than one tier high.

For the business to be a viable proposition ships must be loaded to their full deadweight capacity, and taking all the facts into consideration it is reasonable to assume that nearly every vessel carrying steel coil cargo has tank-tops which are what is referred to as spot-overloaded. Many of these overloads are severe and it is surprising that there have been no noteworthy incidents involving severely distorted structures. It does prove that the double-bottom structures of seagoing vessels are incredibly strong. The reason why the plating of the tank-top does not suffer more than might be expected may be partly attributable to the fact that in a vessel of about 45,000 m/tons deadweight the distance between the longitudinal beams situated beneath the tank-top plating (see Fig. 71) does not exceed 800 mm, which means that most of the concentration of weight, when inadequate dunnage is used, is centred directly over a beam or close alongside it. Further to this, if the item of cargo is directly over an unsupported section of plating, one end will be above one or more parts of the major double-bottom internals. With regard to the unsupported areas of tank-top plating these are always dented to some degree (see Fig. 70). The following has to be considered:

—From practical experience it is known that a steel plate clamped on all sides is stiffer than a beam of the same thickness clamped on two sides.
—Deformations in the tank-top panelling, i.e. set down between internals, up to at least the thickness of the plating, have no prejudicial effect on the structural strength of the plate. In fact, these deformations result in additional strength imparted to the plate, which is mainly due to the presence of membrane stress which plays an increasing role as deflections increase above 20% of the plate thickness.

From Fig. 70 there is very little evidence which would suggest that there is a great deal of contact between cargo and the areas of unsupported tank-top panelling. The fork lift, travelling on four wheels, appears to be working mainly across the ridges formed by the inner bottom internals.

In general, where tank-top strength is concerned, it is steel coils which give rise to difficulties and sometimes confrontations. Other steel cargoes, perhaps with the exception of steel slabs, do not create problems in relation to tank-tops.

Steel coil cargoes

There is no doubt that dunnage must be used with coil cargoes; it is customary and necessary with regard to the preservation of the cargo if some free moisture collects on the tank-top, e.g. sweat water. If the coil is placed directly onto the tank-top plating the weight is concentrated over a very small area, as already discussed. When the coil is supported on dunnage the weight is even more concentrated. There is a tremendous spot overload, which in most instances, with heavy coils, does exist with only one tier of coils; to a greater or lesser extent, in most instances, this is unavoidable if this type of cargo is to be carried by any merchant vessel. In spite of this, from many years' experience and what has taken place for a long time in the real world of carrying steel coil cargoes, if the carrying vessel is well maintained and tank-tops are in good condition, loading heavy steel coils to three and, depending on circumstances, sometimes four tiers high, has no serious adverse effect, if any, upon the vessel's structural parts. It is, however, necessary to keep in mind that the aggregate permissible load per cargo hold must never be exceeded.

Notwithstanding what has been said above, there are occasions when conflict of interest will arise between the ship master and charterers when the issue of tank-top strength comes to be discussed. It is anticipated that this will be a growing problem in the future in view of the tendency to increase individual coil weights and the size of bulk carriers employed in the industry. The calculations of one prominent classification society indicate that the average-sized bulk carrier can only load heavy coils one tier high, no key coil to be inserted, and the society stipulated that if more than one tier is loaded the vessel will be withdrawn from class. Many vessels entered in this society which are loading full cargoes of steel coils would in all probability be similarly restricted if the class were consulted with regard to tank-top strength and the loading of steel coil cargoes.

Disregarding the intervention of classification societies, the question arises as to what can be done when these disputes arise. In many such confrontations between the ship master and charterers the dispute is resolved by increasing the size and amount of dunnage used—invariably by a technically unsupported mutual agreement. Dunnage costs money, and it is understandable that, in the light of previous experience with steel coil cargoes, charterers/shippers are reluctant to be forced into providing large amounts of costly, good quality wooden dunnage which they consider to be an unnecessary waste of money. On the other hand, the ship master does have a valid, convincingly powerful argument backed up by his classification society, who will usually go no further than confirming that the weight of the cargo loaded should be evenly spread and the pressure exerted upon the tank-top from the loaded cargo should not exceed the permissible load per square metre of tank-top surface. The situation with regard to this advice is changing and classification societies are being increasingly confronted with the problem. Many vessels' loading manuals make some reference to the loading of steel coils. However, in view of the spot overload associated with these coils, what is proposed in the loading manual with regard to coil weight and permissible height of the stowage does not enable the vessel to complete the required maximum permissible cargo deadweight. As a result of this, it is now often the case that owners fail to reveal these restrictions to charterers when hire of the vessel is being negotiated. In such circumstances claims for dead freight ensue.

The use of inadequate strength dunnage beneath the base coils in the stow does not in any way assist in spreading the weight of the coils, e.g. 1″ × 6″ planks. Such dunnaging only concentrates the weight into small areas where the coil rests upon the dunnage, however, some dunnage is necessary in order to protect the coils from any moisture accumulating upon the tank-top during the voyage. If 17.5 m/ton coils are stacked three-high—base coil against tank-top 17 × 3 = 51 m/tons—what is above stated regarding the dunnaging of one 17.5 m/ton coil means that enormous baulks of timber would be required in supporting 51 m/tons pressure that would be exerted through the base coil. Such an arrangement would be unacceptable to charterers or shippers, because the cost of the timber would be exorbitantly high; also from long experience it is known that these cargoes can be carried without resorting to such draconian measures.

Usually, there is a 15 cm gap between adjacent athwartships stowed rows of coils. The width of each coil will be usually 1 m or 1.25 m in length/breadth, some are wider, and in most instances part of each coil is above and/or close to, either an inner bottom floor, longitudinal intercostals and/or longitudinal strength beams. The worst scenario would be for a coil to be resting on an internally unsupported section of tank-top panelling, but such a situation is much more the exception than the rule. Even so, the official permissible tons per m^2 limit still applies, which leads one to believe that, in fact, unofficially much higher limits exist when directly overstowing double bottom internal members, so relieving the pressure from the panelling.

When conflicts of opinion arise between charterers and owners, the vessel's owner may call in a class surveyor to decide the issue. He will make a calculation often based on the following:

Thickness of inner bottom plates for the ships of longitudinal framing system is to be not less than the value obtained from the following formula:

$$\sqrt{kQ\{(1.65\beta - 2.3)\alpha - 6\beta + 12.2\}} + 1.5\ (mm)$$

where,

k: Coefficient, to be used 1.65 for mild steel in general.

Q: Mass of steel coils loaded per panel of inner bottom plating, in general, as obtained from the following formula:

$$\frac{Wn_1n_2}{1000n_3}\ \text{(ton)}$$

In cases where steel coils are lined up in one tier with a key coil, Q is to be 1.4 times the value obtained from the formula.

- *W*: Mass on one steel coil (kg)
- n_1: Number of tiers of steel coils.
- n_2: Number of load points per panel of inner bottom plates, as given according to the value of n_3 and a/l_1
- n_3: Number of dunnages supporting one steel coil

α: Aspect ratio of panel of inner bottom platings. When a exceeds 3.0, a is to be taken as 3.0.

β: As obtained from the following formula:

c/a

where

- *a*: Spacing of floors (mm)
- *c*: Distance between load points per panel of inner bottom plating in ship.

The calculation will produce a thickness for the tank-top plating in accordance with the proposed coil height stowage. If the answer for two-tier stowage is a tank top thickness of 22 mm, but the tank top thickness is only 1.8mm, the class will rule that only one-tier stowage of coils is permitted. From this it is understood, and expected, that the one-tier stowage is strictly within the limits of the classification society rules. The rules dictate that any weight placed upon the tank-top must be evenly spread in order to keep within the limits of the permissible load per m^2 of tank-top plating. The coils are evenly distributed across the area of the tank-top that they occupy, but the weight of each individual coil cannot be evenly spread owing to its cylindrical shape. Suitable dunnaging only redistributes the weight and concentrates it somewhere else, to adjacent inner bottom internal members.

A practical approach to the loading and stowage of steel coils, based on many years experience, is to use the "empirical rule method", whereby, the pressure exerted by the base coil on the tank-top, in either single-or multi-tier stowage, should not exceed three times the tank-top's official permissible load per m^2. In such circumstances, if a vessel's tank-top permissible load is 15 m/tons per m^2, irrespective of the dunnage used, the spot overload should not exceed 45 m/tons. It is also agreed that such a pressure exerted, contrary to the class rules, cannot be justified by any mathematical calculations. Nonetheless, such a situation must arise, and prevail, if any vessel loading steel coils is to complete her designed deadweight carrying capacity. It is hoped that sooner or later the classification societies can come to terms with this.

Finally, the charterers/shippers may have to supply dunnage on the tank-top to spread the weight of the cargo. It will have to be indicated that the dunnage used was of adequate strength and suitably employed to achieve the necessary spreading of the weight as required by class. This may be achieved by disregarding the existence of the tank-top plating. It would then be necessary to imagine that the pieces of dunnage timber, equally spaced, are placed directly across and spanning the ± 800 mm spaced longitudinal beams, whereafter, with two or three pieces of dunnage wood in place, the coil would be supported on the centre of the span of timber after calculating that the aggregate strength of the dunnage system in place would support the load. Thereafter, the steel tank-top plating will be reintroduced and if it is not unduly locally set down between the double-bottom internals its contributing support may be taken into consideration.

The calculation to determine the size of the timber to use involves the static bending value of the timber, and to be super accurate it will therefore be necessary to know the type of timber used. It is perhaps more practical to take an average static bending value for hardwood or softwood, whichever the case may be. The calculation is based on the simple beam theory. Where a steel coil of 17.5 m/tons is concerned the calculation gives for timber of 10 cm × 20 cm spanning longitudinal beams, spaced 800 mm apart, a permissible load of 9.153 m/tons per piece. Therefore, two pieces of dunnage of these dimensions would be sufficient for a 17 ton coil. Naturally, the dunnage used would have to be increased, in amount or dimensionally, as the rows of coils went into the second or third tier. If the surface of the tank-top plating between internals is not too uneven it will meet the deflection of the dunnage and the support it offers can be calculated; in consequence of this the size/quantity of the dunnage may be reduced accordingly. The principle behind the above idea, i.e. bearing timber across rails, is not new. This subject grew out of the practical problem of a railroad track when the railroads were being built. It is to be noted that two planks placed or nailed together do not have the same strength as a solid piece of timber of the same size. Adequate sized dunnage divides the weight of the coil principally in two directions, between the two fore and aft beams. If inadequate strength dunnage is used the weight is principally spread in four directions, through the panelling, between the longitudinal beams and the transverse floors, a scenario that does not arise very often, as more support is on offer depending on the positioning of each coil in relation to the position of the internal members of the inner bottom.

Steel slabs

Steel slabs may be loaded and handled in single pieces or grouped into unit lifts depending on weight and method of stowage. Where tank-top strength is concerned, the problem is the same for both in that dunnage must be laid beneath the slabs but only for the sole purpose of facilitating handling, i.e. passing slings beneath or inserting the prongs of forklift trucks. There would be less of a problem if no dunnage was used at all and the bottom slab was directly in contact with the tank-top plating. This will be understood if we consider a mild steel slab of the following dimensions: 1.25 m × 0.25 m × 6 m. Such a slab would weigh 14.7 m/tons having a surface area of 7.5 m^2. If the permissible load of the tank-top was 10 m/tons per m^2 then the aggregate weight of the slabs to be loaded over an area of 7.5 m^2 would be 75 m/tons. This figure divided by the unit weight per slab would mean that slabs could be loaded to five high.

Using the dimensions of the slab above-mentioned it would be customary to insert three pieces of dunnage beneath the lower slab in the stow. The largest dunnage which might be used may be 3″ × 3″ timber; however, apparently for the purpose of economy less thick timber is more commonly used and usually timber of inferior quality. Placing timber below the lowest slab concentrates the weight of the stow at the points where the dunnage contacts the tank-top. The pressure must radiate out to some extent into the inner bottom structure but to what extent is unknown. The dunnage should be placed directly above the component members of the inner bottom structure, not upon unsupported areas of tank-top steel panelling/plating.

Structural steel

Mainly comprising 40-foot and 60-pieces of steel in single lifts or in 20-foot bundles. The usual procedure is to lay 3″ × 3″ pieces of timber in athwartship lines upon the tank-top spaced about 3 metres apart in a fore and aft direction. Entire cargo holds are often filled with this material. As the dunnage spans the longitudinal beams beneath the tank-top and is invariably placed directly upon or adjacent to the inner bottom floors which are spaced about 2.7 metres apart there is usually no problem with this stowage procedure In such deep stowages the dunnage usually finally becomes crushed.

Weight over the wing tanks

When we talk of permissible load per m^2 on the tank-top with regard to calculating the amount of cargo which can be loaded, the figure obtained from this calculation is the total quantity that can be loaded on board those ships where the tank-top extends out to the ship's sides. This is not the situation on board the average bulk carrier, since most of these vessels are fitted with lower wing tanks with sloping sides (see Fig. 72). More cargo can be loaded above this tank side plating and the quantity must be added to the total figure obtained by multiplying the tank-top area by the permissible load per square metre. One of the well known classification societies calculates the wing tank load as shown in Fig. 73 for a mid-ship section of a cargo hold 35 metres in length with a frame spacing of 0.755 metres; the calculation follows, based on measurements taken from the general arrangement plan:

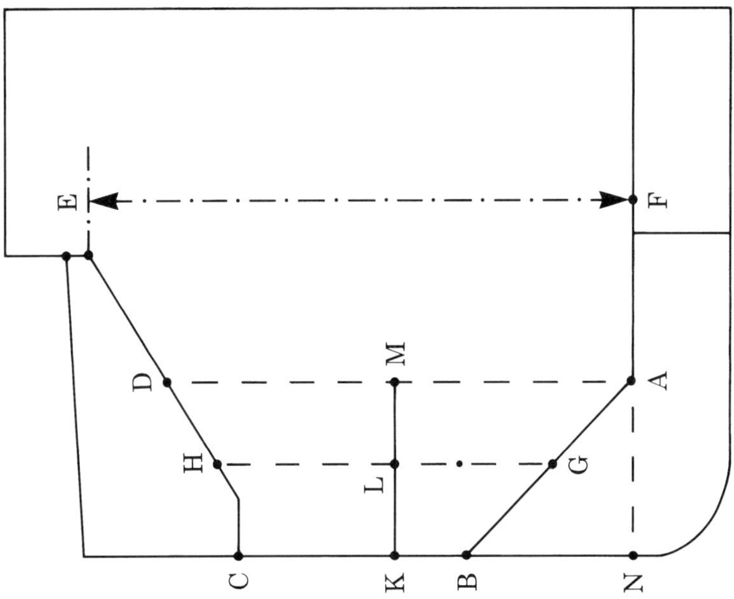

Fig. 73. Mid-ship section of cargo hold

Vertical dimensions
B–C = 4.00 m
A–D = 8.90 m
G–H = 6.20 m
E–F = 9.90 m

Horizontal dimensions
K–L = 1.70 m
L–M = 1.70 m
Frame spacing = 0.755 m

Tank-top Strength = 9.08 t/m

Loading Point Factor = 9.08/[E–F]
= 9.08/9.90
= 0.917

Bulk Cargo Hold Length = 35 × 0.755
= 26.42 m

Loading Point 'A' t/m = 8.90 × 0.917 = 8.16 t/m
Loading Point 'G' t/m = 6.20 × 0.917 = 5.69 t/m
Loading Point 'B' t/m = 4.00 × 0.917 = 3.67 t/m

Total Maximum Weight on Sloping Plating =
[G–H] × Loading Point Factor × [K–M] × Hold Length =
6.20 × 0.917 × 3.40 × 26.42 = 510.71 m/tns on each side.

Total Maximum Weight on Horizontal Tank-top = 14.40 × 26.42 × 9.08
= 3,454 m/tns

Total cargo that can be loaded in the hold = 3,454 + (510.7 × 2)
= 4,475 m/tns

With a deadweight cargo such as steel coils, slabs, packages and bundles of plates etc., the height of the stow is usually not more than about 3–4 metres. These cargoes when winged out over the hopper tanks by no means overload these areas.

With regard to the question of weight concentration on tank-tops where steel cargoes are concerned, it is doubtful whether any of the terrifying formulas produced which combine the strength of the tank-top with the dunnage are reliable. A committee composed of naval architects and others produced a report which stated in its summary that the report looked at load stresses on the tank-tops of a ship's cargo hold, and investigated whether the stresses caused by heavy localised loads could be reduced by the use of wooden planks to spread the load over a greater area. The theory of plates in the small deflection elastic range was used to determine maximum stresses caused by loads, varying from point loads at the plate centre to a load evenly distributed over the whole plate area. A model using the theory of beams on an elastic foundation was then developed to assess the effectiveness of a wooden beam in distributing point loads and thereby reducing stresses. The results showed that stresses are reduced by load distribution and that within limits wooden beams can be used to distribute point loads. A general procedure was formulated from the findings of the investigation. The procedure could be used to calculate the beam dimensions necessary to ensure that the maximum allowed stress on a particular tanktop is not exceeded. However, the exercise proved to be inconclusive, because towards the end of the report it was stated that it was considered that further theoretical and experimental work was necessary in order to verify certain methods used. The committee agreed that their report did not solve the complexities of the problem.

This article pinpoints the problems; it does not purport to solve them in a manner which will be acceptable to all of the many interests who participate in the operations of exporting steel cargoes. Its aim is to provide the surveyor, shipmaster and others involved in the loading of ships with a practical basis upon which to approach the subject when discussions arise which, without some solid basis for negotiation, can easily develop into confrontation. In most instances, if classification societies are approached for advice, their advice will be to the effect that the aggregate permissible load upon the tank-top must not be exceeded and the weight must be evenly spread in order not to exceed the permissible tank-top load per square metre. Everyone directly involved in the real world of loading ships' cargoes knows that, in order to keep the wheels of industry turning, disputes over stowage problems are usually resolved by way of compromise, unless the root of the problem is so glaringly serious that any compromise is out of the question.

TAP HOLE

This is a hole situated at the base of the blast furnace which is plugged with fire clay. A pointed bar is used to pierce the hole in order to permit the molten iron to flow out. The iron is then formed into pig iron or transferred in its molten state direct to the converter vessel for making steel.

TEEMING LADLE

The molten metal is poured from the converter directly into a refractory lined ladle, which is referred to as a "teeming ladle". Thereafter, the ladle is transferred to an area where it is tilted, and the molten metal is deposited in ingot moulds. In the "continuous casting" method of steel making, the teeming ladle is taken to the top of a tower-type construction where the molten metal gravitates down through a cooling chamber and then through rollers which form the metal into bars and slabs. With this method, the ingot stage is by-passed and savings are made, as no pre-heating is required if the slabs or bars are directly to be processed further.

TIN PLATE

Thin sheet steel coated with a fine layer of tin is used to make many millions of tin boxes for the canning industry, especially where food preservation is concerned. Some of the qualities of tin are that it is corrosion resistant, non-toxic, has good solderability, will not react chemically when in contact with the acids in food, has good electrical conductivity and imparts a bright finish to the product. Tinned steel sheets are used in electronics, textile, electrical and food processing industries and also the chemical and automobile industry.

Tin plate is very thin cold rolled steel sheeting which, after being prepared with acids and scrubbed and cleaned, is covered with a thin layer of tin. The tin coating is applied by passing the steel sheet through a bath of molten tin (hot dip method). However, in recent times this method has been replaced by depositing the tin on the surface of the steel through the electrolytic tinning process, whereby the thickness of the coating can be more accurately controlled. The thickness of the steel plate used can be as narrow as 0.18 mm and the tin coating can be gauged at 0.000038 mm.

These goods are shipped in metal boxes and also in the form of coils. The metal boxes are usually smooth sided and sizes vary but a typical box may weigh 1,200 kg and contain anything up to 1,300 sheets. The boxes are fitted with bearers beneath to facilitate handling. The coils are usually secured to a wooden pallet with the core of the coil vertical, all being secured in place with metal strapping bands. Tin plate is very expensive material which should always be kept dry and handled with great care (see Fig. 74(a)).

VENTILATION OF STEEL CARGOES

In the Shipping Industry the term ventilation is used to indicate the steps taken to prevent damage to ship's cargoes through condensation moisture within the cargo holds. The issue of ventilation has over time slowly but surely becoming increasingly important, when liability for delivery of moisture and rust-damaged

Fig. 74(a). Tin plate in coil form.

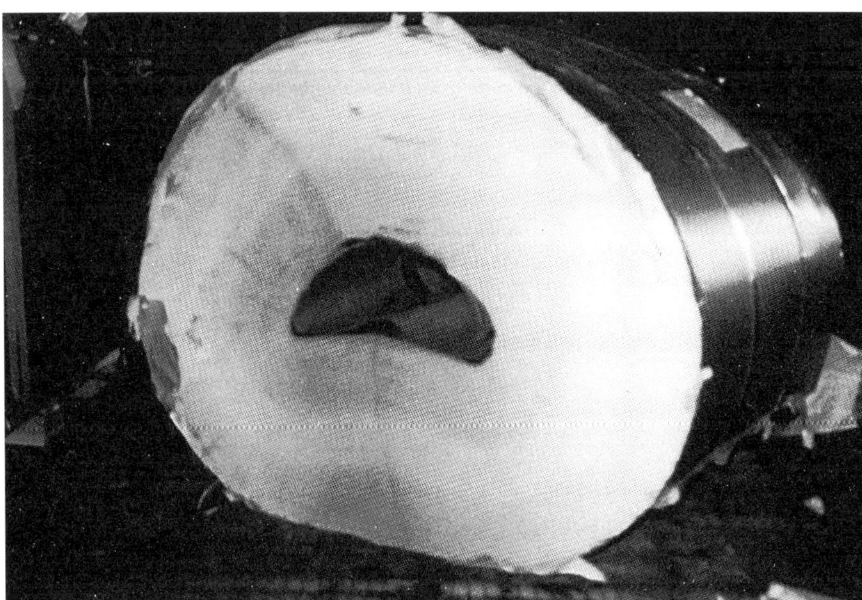

Fig. 74(b). One coil was dropped.

Fig. 74. Tin plate.

steel products is being decided, or for that matter nearly any moisture damage cargo delivered. Whether the cargo is either non-hygroscopic, of an inert nature such as steel products, or hygroscopic, such as items of cargo of vegetable origins, i.e. timber, cotton, paper in rolls etc., the maritime industry, including the various P & I Clubs, agree that the decision as to when, or when not, to ventilate cargo should be decided by comparing the dew point temperature of the outside atmosphere with that within the confines of the cargo hold, also the surface temperature of the cargo will in most instances have to be involved, especially with steel cargoes.

Although the scientific instruments necessary to determine the temperature values required are nearly always part of navigation bridge equipment, it is not unusual for the maintenance of these instruments to be badly neglected. Further to this, in many instances vessels' staff and owners have shown insufficient interest in the subject of cargo ventilation on specific voyages, until heavy claims are entered against them when, of course, their interest has come too late. Where steel cargoes are concerned, the failure in maintaining the necessary ventilation procedures is often associated with bulk carrier vessels being transferred from the bulk cargo trades to break bulk cargo carrying. The Master of the vessel and probably all of the vessel's staff, have never before had any experience of carrying a break bulk cargo and therefore have no experience of the intricacies of stowage, break bulk commodities and their associated ventilation requirements. This being so, any excuse for failing to "carefully load, stow, carry, care for and discharge" the cargo is unacceptable in law and also to shippers/receivers of the cargo.

When contemplating carrying any cargo consisting of steel products the question of cargo ventilation is of paramount importance to all concerned in the shipping venture, and especially the ship master/shipowner. The various P & I Associations in their Loss Prevention publications support what is stated above. In claims negotiation cargo interests demand proof of all measures taken to care for and protect the cargo on voyage, with emphasis on production of "cargo ventilation records" (see Appendix, p. 317). The main bulk of steel transported in sea borne trades is carried in bulk carrier vessels, which are not constructed to carry break bulk cargoes. The ventilation arrangements provided are for ventilating bulk cargoes and are usually totally inadequate for the ventilation of steel products. Nonetheless, certain measures must be taken with ventilation to protect a steel cargo. In order to appreciate the necessary ventilation procedures an understanding of the following paragraphs in this article should be sufficient.

The hygrometer

This is an instrument consisting of two thermometers, mounted side by side in a louvred screen, designed to shield the thermometers from rain and radiation, but at the same time allowing free passage of air. One of the thermometers is a normal dry thermometer, the other has a piece of cotton attached to the bulb on the lower end and the free end of the cotton is immersed in a small cistern containing water. The two thermometers are referred to as the "wet and dry bulb thermometers". The dry bulb thermometer registers the correct air temperature. However, through the process of evaporation heat is lost from the wet bulb and this thermometer usually registers a lower reading than the dry bulb. Under conditions of complete saturation of the air surrounding the instrument (100% relative humidity) the thermometers will read the same. When the relative humidity is less than 100% the wet bulb will have a lower reading than the dry bulb thermometer. It is customary to refer to this as the depression of the wet bulb.

From readings of the wet and dry bulb temperatures, the relative humidity, vapour pressure and the dew point of the air can be obtained. This is achieved by entering in suitable hygrometric tables or by consulting a psychrometric chart (see Appendix, p. 318).

The thermometer

Air temperature is usually measured by means of a mercury thermometer. Owing to the fact that mercury freezes at −38.9°C the mercury thermometer is unsuitable for use in very cold climates. Thermometers should be kept clean, and, in particular, no deposit should be allowed to accumulate on the bulb. The presence of moisture on the bulb of a thermometer will cause it to behave as a wet bulb; the moisture should be wiped of and the bulb must be dry for at least 15 minutes before a reading is taken. Thermometers should be read as rapidly as possible to minimise the effects of the observer, and from radiation of surrounding objects. They should be read with the top of the mercury thread at the same level as the observer's eye to avoid errors of parallax.

The two thermometers are mounted side-by-side, and the completed assembly is referred to as the Hygrometer or Psychrometer (Non Aspirated). The wick of the wet-bulb must hang straight down into the water at least 7 cm must be exposed to the air. The dry-bulb should be positioned to windward of the wet-bulb for a period of 15 minutes before reading. Pure distilled water should be used in the reservoir of the wet-bulb, which must be renewed once a week, and if salt water comes into contact with the wick, or the reservoir, all must be renewed immediately if erroneous readings are to be avoided.

The Screening of Thermometers: The measurement of true air temperature can be difficult, owing to the fact that a thermometer freely exposed to the air is affected by radiation of heat to and from surrounding objects. The passage of radiant heat through the air leaves its temperature practically unaffected. Therefore, the thermometer is enclosed within a screen, which will permit the air to circulate freely past the bulb of the thermometer but at the same time shield the instrument from the radiation emitted by its surroundings. Usually the type of enclosure used to house the thermometer(s) is a wooden box painted white and fitted with louvred sides.

Two thermometers, one wet bulb and the other dry bulb, are mounted vertically side-by-side inside the box, and can be viewed by opening the louvred door in the front of the screen. It should not be a permanent fixture anywhere, as it must be transportable to any suitable part of the navigation bridge, or elsewhere, to obtain accurate readings of the thermometers independent of any local heat. The types of fixed screens housing the thermometers on board many vessels, see Fig. 75, are completely inadequate and would contribute to faulty readings. The proper marine screen has an unimpeded circulation of air through it.

The Aspirated Psychometer: From what is above stated it is clear that if reliable readings are to be obtained from an ordinary non-aspirated psychometer the

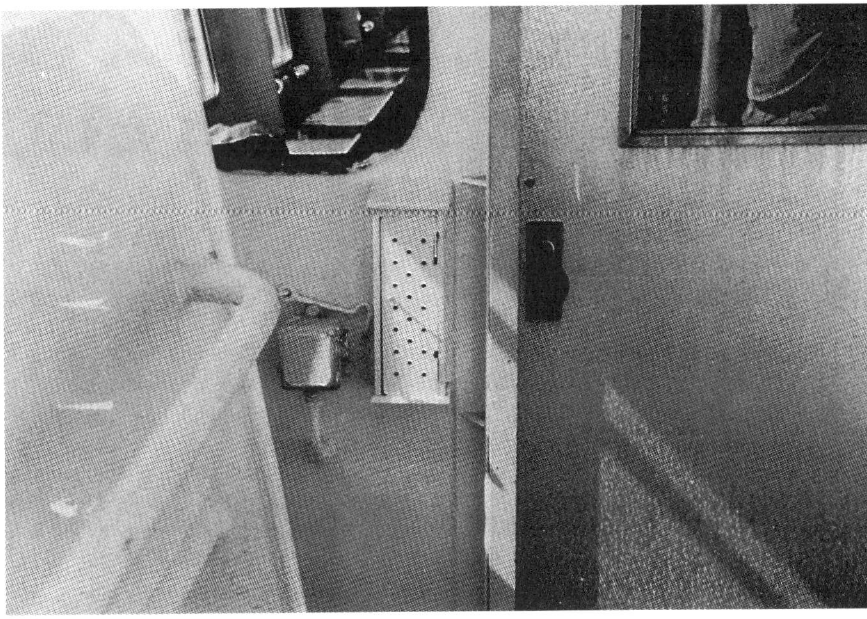

Fig. 75. Wet and dry bulb thermometer screen inadequately constructed and badly positioned for air flow.

wet-bulb thermometer must have been exposed and acted upon by a circulating air flow for at least 15 minutes before the readings are taken. In the cargo holds of a vessel which are not fitted with trunked, mechanically operated, air circulation, the readings obtained from the non aspirated psychrometer will be of no use whatsoever for determining the dew point temperature of the ambient air. What is required is an Aspirated Thermometer, which is principally the same as the non-aspirated thermometer except that it is fitted with a battery or clockwork operated fan, which circulates a flow of air across the wet-bulb. The fan should be run for about 7 minutes. Two minutes after commencing aspiration, and again after a further half-minute, the thermometers should be read, first the wet bulb and then the dry bulb. Corrections to the readings may have to be made in accordance with the operating instructions.

The Whirling Psychrometer, may be used for measuring wet and dry-bulb temperatures. It consists of two thermometers mounted in a box wood or plastic frame attached to a handle, with a spindle, by means of which the frame and thermometers may be whirled. The bulbs of the thermometers are subjected to a continuous current of air, the temperature of which is rapidly taken up by the dry-bulb thermometer. The instrument is exactly the same as the normal non-aspirated hygrometer/psychrometer without its screen, but suspended on a handle so arranged that it can rotate. The instrument should be whirled for not less than two minutes, after which the thermometers should be read as quickly as possible.

Dew Point Temperature (DPT): For cargo hold ventilation purposes, the recognized method employed in comparing the volume of air surrounding the vessel, also that within the cargo hold, is by the determination of the dew point temperature of each volume of air. The dew point temperature of any volume of air is the temperature at which the air is completely saturated, i.e. 100% relevant humidity. At this point any further reduction of temperature will result in super saturation of the air, and precipitation of its moisture will follow until the state of super saturation is reduced to 100% saturation of the air or less. The effect is somewhat analogous to the boiling kettle spouting vapour in the form of steam (100% RH). If a surface, relatively and sufficiently colder to the vapour such as the surface of a mirror, is placed in its path, the temperature of the vapour striking the surface will be reduced below its dew point temperature and beads of moisture will be seen to condense on the surface of the glass.

In appreciation of the above, it is evident, and also important to remember, that if two volumes of air of dissimilar DPT are mixed, the resulting dew point temperature will be a compromise of the previous two differing dew point temperature. The important point is that atmospheric ventilating air having a higher dew point temperature than the air within the cargo hold being ventilated will raise the dew point temperature of the air within the cargo compartment, in all probability above the temperature of the cargo.

Absolute humidity

The absolute humidity of a sample of air is the mass of water in a given volume of moist air. It is expressed in grams per cubic metre.

Relative Humidity (RH)

All air contains moisture in the form of water vapour. When any given volume of air contains the maximum amount of water which it can absorb at a particular temperature, it is said to be saturated and to have a relative humidity of 100%. If the temperature of this volume of air is lowered it will contract and become super-saturated and incapable of containing or holding its entire amount of moisture; some of the moisture will therefore be precipitated. If the temperature of the air is raised it will expand, become less dense and capable of holding more water vapour. Whereas in the first instance a reduction of temperature caused the air to be saturated so producing 100% "relative humidity", in the second instance, when the temperature was raised, expansion of the air reduced the "relative humidity". By definition, "relative humidity" is the ratio of the actual water vapour present in the air to the amount which the same volume of air would hold if it were saturated; this is usually expressed as a percentage. At the dew point the percentage is 100%.

Vapour Pressure (VP)

Water in the air, in the form of vapour, acts as another gas in the mixture of gases which compose the air. This water vapour exerts a pressure in all directions that is independent of all other gases in the mixture. This pressure is known as "vapour pressure of the air". The water vapour in the air exerts a pressure that varies directly with the amount of water vapour pressure and is commonly expressed in millibars of mercury per unit of volume of dry air.

Humidity

Water Vapour in the Atmosphere: Atmospheric air always contains, mixed with it, small quantities of water in the form of invisible vapour. In warm damp climates the proportion of water vapour is relatively large and is small in cold climates and in desert regions. The water vapour mixed with the air behaves like any other gas. In particular it contributes its own pressure to the total pressure of the atmosphere. The pressure of a sample of air containing water vapour is the sum of the partial pressure of the air and vapour. These are the same as would be exerted by the air and the water vapour occupying the whole volume separately.

Dry and Moist Air. Saturation: Air that contains no water vapour whatsoever is called dry air. The term "dry" is somewhat loosely applied, as it also covers air that has only a small proportion of moisture, but at the same time the proportion of moisture is so small that it can be considered dry. Air containing some water vapour is called "moist air" or "damp air", the term "damp" usually signifying that the air is nearly saturated. For a given temperature there is an upper limit to the proportion of water vapour that the air can contain and the limit increases with increasing temperature. For example, air at 0°C (32°F) is capable of holding water vapour in the proportion of 4.8 grams per cubic metre; at 10°C (50°F) the proportion is 9 grams per cubic metre. Air which contains the maximum possible amount of water vapour appropriate to its temperature is said to be saturated or "in the state of saturation". Saturated air and moist air are always lighter than dry air at the same pressure and temperature, and this is evident from the fact that otherwise the moisture could not remain in suspension in the air. If a water surface is available, evaporation constantly takes place from it provided that the air in contact with it is not already saturated. It is possible for very pure air to contain a larger proportion of water vapour than is necessary for saturation; such air is said to be "supersaturated". When such a condition takes place in less pure air the excess moisture is precipitated.

Fresh water damage claims against unwrapped hot-rolled semi-finished products are rare but not unknown. One such claim concerning hot-rolled steel coils and steel plates from Santos, Brazil, to Japan amounted to US$3,000,000. The cargo was kept under cover before shipment and was shipped in a factory blue condition, recent production. Heavy rain was experienced during loading. The Master of the vessel agreed not to interrupt loading operations and to sign a clean bill of lading in exchange for a letter of indemnity. The cargo was wet and rusty after being about one month on board the vessel, but was physically undamaged; there was no pitting or heavy scale formation. The case went to court and the carrier lost.

Cargo sweat

When the air ventilating a cargo in a ship's hold has a dew point temperature higher than that of the surface temperature of the cargo the ventilating air, in contact with the cargo, may be cooled to its saturation point. Further cooling will render the ambient air super saturated when it will commence to deposit moisture in the form of condensation upon the surface of the cargo. This condensed moisture is referred to as "Cargo Sweat". When the DPT of the air in the cargo hold exceeds the temperature of the cargo condensation may take place directly upon the surfaces of the cargo both externally and internally. Where wrapped steel cargo is concerned, the wrappers warm up ahead of the mass of the steel covered, therefore it is possible for the internal steel to sweat but not the wrapper covering the material. In such circumstances, it is not expected that the damage sustained through internal sweating of the material would be significant (e.g., cold to warm voyage).

Ship's sweat

Situations do arise, when atmospheric and sea temperatures are falling, whereby the component parts of the vessel's structure take up a temperature that is lower than the dew point temperature of the air inside the cargo hold. In such circumstances condensation takes place upon the cargo compartment overheads and sweat water may rain back upon the cargo. Sweat water forming on the ship's sides streams down onto the tank-top often causing the dunnage to become saturated, also to create a zone of high RH air, directly above the tank-top plating, capable of causing localised "cargo sweat" damage to the cargo. Ship sweat can be controlled by an efficient ventilation system, and such action may be particularly necessary in the case of wrapped steel (e.g., warm to cold voyage).

Hoar frost

A deposit of ice that forms on objects and is generally crystalline in appearance and produced by the direct sublimation of water vapour from the surrounding air. Hoar frost sometimes affects steel in store awaiting shipment, especially if the sheds are open sided or the goods are given outside storage.

Hygroscopic cargo

All goods of vegetable origin, such as cotton, timber, jute, raw coffee, to mention but a few, are in certain circumstances capable of adsorbing or desorbing moisture to and from the atmosphere. Such commodities are referred to as being hygroscopic.

Non-hygroscopic cargoes

In this category are included all inert cargoes such as steel, glass, the cans of canned goods, and all other goods basically manufactured from minerals. These materials do not either take up or give off moisture, they are inert and are of such a nature as to possess no inherent characteristics similar to those of hygroscopic cargoes as explained above. Often, but depending very much upon circumstances, hygroscopic and non-hygroscopic cargoes are in many instances incompatible when stowed together in the same cargo compartment on board a ship.

Cold steel cargo moving into a warmer climate

Condensation fresh water claims entered against wrapped material is a common occurrence, and is usually associated with voyages where the vessel loads in a relatively cooler climate than will be experienced during the voyage. Many such voyages, where these damages develop, are experienced during the northern winter period such as South America to the East and West Coasts of the USA and Northern Europe, from Japan, Korea to Europe via the Suez Canal, etc. Most of the damages are confined to cold rolled and galvanised steel coils—any wrapped cargo is particularly vulnerable—as these materials are most sensitive to changes of temperature and when contacted by moisture nearly always immediately sustain rusting damage.

In the first instance, fresh water moisture damage (real, or conveniently imagined) to hot-rolled semi-finished products is not ruled out; the same type of damage to wrapped material is a high probability; and both types of material must be provided with the same care and attention when being carried by sea. The various P & I Associations are strongly in favour of cargo being ventilated on the basis of efficiently kept ventilation records (see p. 317); some have confirmed these opinions in their "loss prevention" bulletins. When cargo interests enter sweat damage claims against the carrier, the main thrust of their arguments is care of the cargo while in the possession of the carrier, and properly kept ventilation records are a vital part of the carriers defence.

As has already been mentioned, "cargo sweat" damage to any cargo can develop and be most prolific when a vessel is voyaging from a cool to a relatively warmer climate. The amount of sweat developing directly upon the cargo will be controlled by the temperature differential between the temperature of the cargo and the "dew point temperature" of the ambient air contacting such cargo.

In the case of a vessel loading a cold rolled steel coil cargo at a temperature of 0°C (32°F). The hatches must be clean and dry, any dunnage used should likewise be dry, i.e. moisture content not more than 14%, and the wrappers of the cargo must also be seen to be dry in every respect. Once the cargo is loaded the cargo holds should be battened down and made as airtight as possible. In such circumstances, as the atmospheric temperature and sea temperature rise, by the vessel moving into warmer weather, the hold temperature will also rise. The temperature of the steel cargo will also rise, but much more slowly. If the hold is properly sealed off from the outside atmosphere and there are no sources of additional moisture within the cargo hold, the ambient air dew point temperature should remain static. Any change to the situation, such as appreciable leaks into the cargo compartments, will result in an increase in the dew point temperature and possible eventual damage to the cargo. If dew point temperatures outside and within the cargo hold remain constantly the same, the exercise is not working. If the measures taken are working, the depression of the wet bulb must continuously increase for an increase in hold temperature and static DPT. Under such conditions, however much the cargo hold temperature increases, the cargo is safe, and no condensation (cargo sweat) can develop.

Warm steel cargo moving into a colder climate

Perhaps later in the voyage atmospheric temperatures will begin to fall, and opening up the ventilation of the cargo holds might have to be considered. This could be worth considering if the vessel's holds were equipped with trunked forced ventilation. Unfortunately, these days a vessel so equipped is rare. Seeing that bulk carrier ventilation is incapable of dealing with a ship sweat situation, also in view of the small amount of the air within the cargo hold, any ship sweat that might form within the cargo compartment may be considered negligible.

If all of the above is accepted, when carrying a cargo of steel products cargo holds should be battened down and made as airtight as possible. They should remain battened down until the vessel arrives at the port of discharge. This procedure has always had the support of the various P & I Associations and is universally accepted as being the proper ventilation procedure in deciding when or when not to ventilate a cargo of steel products. However, from experience it is known that sweat water forming on the inside of the shell plating drifts down to the tank-top plating, saturates the dunnage and creates a microclimate zone above the tank-top that is damaging to the base items of the cargo stowage. This alone in mind suggests that a haphazard form of ventilating the cargo should not be followed.

Warm cargo transportation through and into areas of similar high temperature

In such circumstances there is probably only a small chance of either ship's or cargo sweat developing. In the month of July a steel cargo was loaded in Japan when the air temperature was 27°C, dew point temperature 24°C and the temperature of the cargo was recorded as being 22°C. The voyage was across the north Pacific Ocean, through the Panama Canal and onwards to northern Europe, where the ship arrived at the end of August. En route, the dew point temperature of the atmosphere, with which the cargo was ventilated, was nearly always below the temperature of the cargo. Only in one instance did the dew point temperature of the atmosphere rise above the surface temperature of the cargo when ventilation was suspended because of this. The entire cargo was delivered without complaint. This is an example of the value of keeping proper temperature records irrespective of personal experience of any particular trade. As a matter of fact, this cargo need not have been ventilated at all.

Conclusions on the issue of ventilation (cargo consisting of steel only)

It is evident that there is nothing to be gained by ventilating a cargo of steel products provided it is loaded in a dry condition and that the quantity of dunnage used to stow the goods is kept to a minimum. Further to this, the dunnage can be checked by a surveyor in order to establish that its moisture content is not more than 14%. In the event of claims for moisture damage, cargo interests will expect the vessel to produce efficiently kept ventilation records for the voyage in question. If these records cannot be made available, the vessel's defence will be seriously prejudiced. These records must be kept from start to finish of the voyage—commencement of loading to completion of discharge—in order that it may be proved that the vessel's staff were abreast of the situation at all times and followed an acceptable and reliable system of ventilation in caring for the cargo.

Steel stowed together with other cargoes

Other cargoes carried in combination with steel should not be problematic if such goods come within the range of inert cargoes. Cases and crates of machinery, crates of glass and many other commodities which are not of vegetable (non-hygroscopic) origin may be carried safely with steel and given the same consideration with regard to ventilation. However, it will be necessary that the wooded packing of manufactured goods has a low moisture content. Goods with wet packing, or packing with excessive moisture levels, should not be stowed in the same compartment as sensitive steel products.

Hygroscopic cargoes involve those goods which are of vegetable origin such as tobacco, timber, cereal products, jute, wood pulp, etc. Two of the characteristics of these goods, which are of great importance where ship's hold ventilation is concerned, are their ability to absorb and desorb moisture, thereby in due course equalising their moisture content with the atmosphere in which they reside. Each individual hygroscopic cargo has an optimum residual moisture level peculiar to that particular commodity, e.g. it would normally be expected that maize and wheat grains would have a moisture level of about 14%, raw cotton 9%, kiln dried timber 14–16%, air dried timber 22–24% or more, depending on country of origin. These are the approximate moisture contents at which the goods will tend to equalise with their surrounding atmosphere under normal conditions of storage. In the event of these moisture contents being appreciably exceeded, the goods will be incapable of lengthy, close and undisturbed storage periods without damage developing, e.g. timber develops mould, grain is subject to spontaneous heating.

In order to determine the dew point temperature of the air surrounding a certain commodity, which is in equilibrium with its ambient air, at that particular moisture content of the goods and the particular temperature prevailing at the time, it is necessary to determine what is referred to as "the equilibrium relative humidity" of the air surrounding the goods. With such information we would be able to forecast the environment which the timber would create inside an unventilated cargo hold on board ship. We would also be in a position to decide whether ventilating the timber could result in it adsorbing or desorbing moisture.

The moisture content of merchandise of vegetable origin is the actual weight of the moisture contained and held in the goods as opposed to the total weight of the material expressed as a percentage. Therefore, if we refer to timber having a moisture content of 15%, and one bundle of timber weighs two metric tons, then the actual fibres of which the wood is composed will weigh 1,700 kg and the moisture suspended in the bundle will weigh 300 kg.

The values of dew point and vapour pressure rely solely upon the quantity of moisture in the air. Therefore, if we wish to know what these values are for the ambient air surrounding any commodity it will be necessary to consult a moisture equilibrium chart (see Appendix, p. 319).

Timber is taken as a good example, because there have been, in the past, many combination cargoes of steel and timber from the Far East to northern Europe. Steel and timber associate admirably from a deadweight point of view and in the interests of utilising space. Unfortunately, there have been heavy claims entered against steel cargoes on these voyages for damages sustained through contact with cargo sweat. Invariably, the stowage has been condemned, as timber, stowed in the same compartment, is generally considered to be too great a source of moisture to be compatible with the steel, and with regard to the steel the timber is considered to be injurious. There is more than an element of truth in the above. However, there has been a considerable commotion made over the voyages where outturns have produced heavy damage, but similar voyages where the steel has been received without complaint have naturally attracted no attention whatsoever. The fact is that hygroscopic commodities such as timber can be successfully transported together with steel if certain conditions and requirements are satisfied.

Unseasoned Meranti timber, or what is referred to as air dried timber, usually has a moisture content of around 30% on weight and would be loaded at a temperature of, on average, 26°C. By consulting the "Moisture Content Curves for Wood" (see p. 319) it will be seen that timber with a moisture content of 14% and a temperature of 26°C has a relative humidity of 67%. Taking into consideration that bare steel will not rust at a RH less than 70%, it is obvious that any timber commingled with the coil stowage with a moisture content in excess of 14% is unacceptable as it will create a hostile environment for steel products inside the cargo hold. The same might be said of kiln dried timber where the certificates declare that the goods have been dried to a moisture level of 14%. From the drying plant to time of shipment the goods usually increase their moisture level to about 17% in hot and humid countries, which at a temperature of 26°C produces a RH of 79% (DPT 22°C), which is too high considering that the temperature of the steel will be below 22°C.

From the above it is evident that wrapped material cannot be stowed together with timber unless the dew point temperature of the timber is equivalent to, or lower than, the temperature of the steel. With regard to hot-rolled semi-finished products, providing they are of recent production, commingling stowage, as indicated above, may be a possibility, especially with hot-rolled steel coils, as the internal surfaces of the windings will be protected from rusting to a great extent by an intact surface layer of mill scale. However, such stowage is fraught with potential hazards, and in many areas such steel shipments are an easy target for unscrupulous interests. In the case of the shipment from Santos to Japan, as referred to in the early part of this report, the goods were kept under cover before shipment, were of recent production, and were therefore shipped in a factory blue condition. Loading the goods during rain and acceptance of a letter of indemnity did much to convince the court that the carrier was party to a fraudulent act. The claimant then managed to convince the court that the goods were imported for uses where a protective coating is not usually applied and the material had to be received in a factory blue condition, as it was when received by the carrier in Santos. No amount of explanation from the side of the carrier to the effect that the goods were semi-finished products for further processing, were unwrapped and exposed to a marine atmosphere for about six weeks at sea and could therefore not be delivered rust free, made any impression on the court. If, hypothetically speaking, the cargo had been shipped during dry weather the final outcome may very well have been the same. Bundles of plywood are a better proposition as they have a moisture content of only ±9%. Other incompatible cargoes might be bales of raw jute, wood pulp, chemical cargoes, bagged fertilizers and various bulk cargoes in the same hold but separated by a temporary athwartships bulkhead, etc.

Keeping ventilation records

The purpose of keeping proper ventilation records is to stay abreast of the situation and to know whether or not the action taken, of not ventilating, is actually working. If, for instance, it is seen that wet-bulb hold temperatures are keeping pace with wet-bulb temperatures recorded outside the compartment it will be obvious that the cargo hold "dew point temperature" is the same as that of the atmosphere surrounding the vessel. Such circumstances are proof that there are air leaks into the cargo hold that must be arrested by furthering sealing off the compartment from the outside air. A separate temperature record log should be kept as indicated for steel cargoes (see Appendix, p. 317).

Bulk carrier ventilation arrangements and air tightness of cargo holds

Nowadays most vessels carrying steel products on long international voyages are bulk carriers in the range of Handysize 10,000–34,000 DWT and Handymax 35,000–49,000 DWT. They are usually equipped with a natural draft type ventilation system, consisting of either four mushroom headed ventilators on deck, two at the forward end of each hatchway—one placed to the port side and the other to starboard—and two at the after end of the hatchway similarly placed, as for the forward end. Another arrangement is to have only two natural draft type, mushroom headed, ventilators installed, one at the forward end of each hatchway on the starboard side, and one at the after end on the port side. Each ventilator is usually fitted with a screw-down valve installed in the top of the ventilator pipe. This valve consists of a circular steel disk, the peripheral underside of which is fitted with a rubber seal, which engages with a circular compression bar at the upper extremity of the ventilator pipe. When closed this arrangement should form an airtight seal.

Another type of double-headed mushroom type ventilator is fitted and closed by means of a steel flap inside the ventilator pipe. This flap is maneuverable by being attached to a spindle, which extends outside the ventilator pipe where it is attached to a handle. The flap is supposed to be sufficiently airtight to seal off the compartment and prevent oxygen flow inwards in the event of fire. The airtightness of this arrangement is considered dubious and may have to be complemented. Ventilators fitted with stumps, that permit removal of the upper pipe, also the head of the ventilator, can be efficiently closed by bolting a circular steel plate, with a seal, to the circular coaming plate, or bolt-holed flange at the head of the stump. In most instances these ventilators may be short or tall; they usually terminate inside the compartment at deck head level, i.e. roof of the compartment level.

The moulded depth of a 40,000 DWT bulk carrier's hold is about 18 m (59 feet), whereas the cargo height of high density steel cargoes terminates at about 4 m (13 feet), thus leaving an empty space above the cargo of 46 feet. The ventilators on deck are usually of the same height and offer no aspirating effects whatsoever. Even with electrically operated fans, the forced air, once it enters the cargo hold, takes the least line of resistance and escapes to the atmosphere through the forward ventilators. With electrical ventilation there is some aspirating effect but it is too shallow at the top of the compartment to have any worthwhile effect upon the cargo in the bottom of the hold. The only ventilation arrangement that can be effective is sufficiently strong electrical ventilation, the air intakes of which are trunked to the bottom of the hold and introduced at tank top level. This was a method used with great effectiveness when cargo carrying was more of an art, in the days of the conventional liner trade cargo ships. It was during that period that much fieldwork was done with regard to the ventilation of cargoes carried by sea.

Another source of leakage is through the joints of steel hatch-closing appliances. Even with the best set of hatches possible there is always a miniscule gap between the rubber joints, through which air can leak, especially when a vessel is working in a seaway and the compensating qualities of the rubber joints are less effective through age and the wear-and-tear of continuous service. If the carrier decides to use masking tape, to complement the weathertightness of the joints, it is usually applied to the steel hatch panel transverse joints. If the rubber gaskets of the panels are in good condition and reasonably calm weather prevails throughout the voyage, sealing the transverse joints should be sufficient. However, if the gap between the wet and dry-bulb hold temperatures begins to narrow—the dew point temperature is beginning to increase—it may be necessary to complement the sealing of the transverse joint by applying masking tape to the peripheral joints of the hatch closing appliances.

Failure to prevent sweat

Most marine surveyors would agree that a vessel's cargo hold cannot be made absolutely airtight, and one of the arguments supporting this is given below. The theory involved in the proposed ventilation procedures is sound but often does not produce the desired results. If during the course of a voyage it is seen, when moving from a cold to a relatively warmer climate, that the dew point temperature within the cargo hold is keeping pace with a rising outside dew point temperature then clearly the exercise is not working. The reason for this is that a ship's cargo hold breathes, because, during the day, the air in the hold heats up and expands, whereas at night time air cools and contracts. Recalling what has been said with regard to relative humidity, air expanding and contracting under the influence of varying temperatures, the following should make matters clear:

Given

Volume of cargo space	200,000 ft^3
Daytime density of hold air at 90°F, 40% RH	14.2 ft^3/lb
Night-time density of hold air at 70°F, 85% RH	13.6 ft^3/lb
Night-time density of atmosphere of air 70°F, 90% RH	13.7 ft^3lb

Proof

Night 200,000 ft^3/13.6 ft^3/lb	14,705.9 lb of air
Day 200,000 ft^3/14.2 ft^3/lb	14,084.5 lb of air
Difference	621.4 lb of air

621.4 lb × 13.7 ft^3 = 8,513 ft^3 make up of air per day.

Most steel is such a high density cargo that it occupies relatively little space in the hold of a ship. If steel coils are stowed three high the height of the stow would generally be less than four metres above the level of the tank top. The depth of hold of a 30,000 ton deadweight bulk carrier, normally used in the trade, would be about 15 metres.

One of the problems associated with battening down a ship's cargo hold and supplying no ventilation, is to create a still air situation within the space. With sea temperatures and outside air temperatures rising some slight movement of air would be created as warm air rises—there would be some propensity to movement of air due to the movement of the ship. However, in the bottom of the ship, that part of the space bounded by the double bottom tanks and lower wing tanks would be virtually unaffected—experience dictates that this is so.

The storage of any merchandise under a still air situation is something to be avoided. The quality of the environment within the space deteriorates and can be directly responsible for the deterioration of both hygroscopic and non-hygroscopic materials. Where steel coils are concerned, during discharge a heavy crusty rust formation has been observed on the wrappers as the stow is broken down. The rust development, characteristic of cargo sweat, often terminates in a clearly defined line across the vertical face of the packing in the stow. The surface of the packing below this line is completely rusty, whereas all packing above the line is free of rust. Clearly this phenomenon has been caused by the still air situation brought about by one of the following or a combination of both (see Plate 19):

(1) The still air between the coils picks up moisture from the dunnage which is often too moist.
(2) Ship sweat from the ship's sides collecting upon the tank-top plating causes the dewpoint of the still air between the coils, or other cargo, to be raised above the temperature of the cargo.

Clearly in order to avoid such a situation developing a means of circulating the air in the ship's hold needs to be found and provided during the transport of steel products. See Plate 19 where the coils are affected by cargo sweat, whereas the overstowing packages are unaffected. None of the goods showed rust when loaded.

Experience has proved that the ventilation systems installed in most bulk carriers are incapable of dealing with a ship sweat situation in the hold of the ship. Cargo sweat can be controlled by taking measures already referred to in this article, i.e. properly sealing off the compartments, efficiently observing and recording and evaluating respectively temperatures, etc., but for this to be achieved, or even to be partially effective, in many instances seems to be, to a greater or lesser degree, beyond the capacities of present-day vessels' staff.

Because the cargo holds are not properly sealed off and the necessary isolation maintained, the effects of contrast between inside hold and outside atmosphere vapour pressures result in the hold dew point following closely increasing atmospheric dew points as the ship moves into warmer climates. Further to this, the instruments used to obtain the necessary data for compiling records necessary to control the situation are often badly maintained and incorrectly positioned for obtaining accurate readings (see Fig. 75). With the benefit of hindsight it is considered that a more modern and progressive approach to the age-old sweat problem is called for.

From the above it does appear that a ship's hold must breath and normally it does without any unfavorable consequences as far as the cargo is concerned, if the cargo holds are equipped with an efficient dehumidification system, when the phenomena is catered for by drying out the make-up air, as it enters the cargo hold. The make-up air is fully controlled by the dew point temperature of the ambient air being kept at a moisture margin as low as 5°C below the temperature of the cargo.

The situation is less easily overcome when available technology is not used for cost reasons.

Reasons for airtightness of cargo holds

Situations have arisen which do not equate with the popular misconception that a vessel's cargo holds cannot be made airtight. What has been said earlier is proof enough that any imbalance in the atmosphere, between that outside a cargo compartment and that prevailing inside, will strive to equalise. If airtightness of cargo compartments was not achievable, how do we explain the many incidents involving partly-filled cargo compartments lacking oxygen, when opened and entered at sea via the escape hatch, resulting in death of the entrant? If such a cargo hold is not airtight, it is certainly as near to airtightness to effectively produce, within the cargo hold, a different environment to that prevailing in the outside atmosphere. In another case, when a vessel loaded with shredded scrap was spontaneously heating, and the temperature of the scrap was rising, quantities of CO_2 were introduced into the cargo holds, and the decks were cooled with seawater, but to no avail. After extra special measures were taken to seal the cargo holds with masking tape and cement, and the removal of ventilators and sealing of the ventilator stumps, the temperature of the cargo stabilised and slowly reduced to just above normal. Using the simple psychrometric ventilation procedures mentioned above has resulted in the successful carriage of many hygroscopic and non-hygroscopic cargoes previously associated with moisture damage problems on voyages in fixed trades.

It will now be understood that vapour pressure is the pressure exerted by the water vapour contained in the air; it is measured in millibars of mercury equivalent to the pressure as indicated on the mercurial barometer scale. When air expands due to increase of temperature it increases its moisture content, i.e. its RH. And its vapour pressure increases (see Appendix, p. 318, where this will become even clearer). Therefore, at high temperatures VP becomes very powerful; conversely at zero temperatures, and below, the pressure becomes feeble. This fact has great importance in avoiding damage to cargo, especially where wrapped, sophisticated, steel products are concerned.

Some years ago the Stanford Research Institute followed a number of voyages from the West Coast of the USA to the East Coast via the Panama Canal during the northern winter. They eventually produced a report called "Evaluation and Control of Sweat Damage". The vessels involved were conventional cargo ships which had open shelter decks, and in the upper tween decks, which were the open shelter deck areas, the bulkheads did not have to be watertight. These bulkheads were pierced to allow the passage of electric cables and pipes, etc. The cargoes concerned were steel canned goods. One tween deck fully loaded, closed down and sealed at Seattle remained undisturbed, when loading took place in an adjacent compartment at Los Angeles. There was quite a temperature difference between the two ports and it was determined that after opening the adjacent tween decks the dew point temperature in the undisturbed compartment immediately equalised with the open tween deck areas. It was finally discovered that the holes in the bulkheads allowing the passage of electrical wires was the cause of the environments equalising in the two compartments. This serves to demonstrate the power of vapour pressure at high temperatures.

The ventilating procedure in practice

The various P & I Associations encourage their members to appoint a surveyor, to carry out a preshipment survey, when loading steel products. Assuming the surveyor is also a competent marine surveyor, it would be in the owner's interest for the surveyor to discuss with the master of the vessel the measures he intends to take with regard to ventilation of the cargo during the voyage. The surveyor could inspect the instruments to be used for determining dew point temperatures and include associated remarks in his report. The master would make an entry in his log book to the effect that the surveyor had examined the equipment in question and approved it as being appropriate and well maintained. The surveyor could check the moisture level of the dunnage and the timbers used to secure the cargo.

During the course of loading operations wet and dry-bulb atmospheric temperatures should be recorded. The apparent condition of the cargo shipped should be accurately described in the bill(s) of lading At commencement of the voyage, before atmospheric temperatures begin to rise, the cargo holds must be sealed and made as airtight as possible. Complete ventilation records will now be commenced as per the suggested formula (see Appendix) and be continued throughout the voyage, i.e. from commencement of loading to completion of discharge. During discharge wet and dry-bulb temperatures of the atmosphere must be recorded when cargo is being worked. If cargo is seen to be affected by humidity or sweat, at any time from when hatches are opened and throughout the course of discharge, frequent temperature checks, i.e wet and dry-bulb temperatures, dew point temperature, cargo surface temperatures, should be made and brought to the notice of any attending surveyors, shippers and/or Charterer's representatives; without delay.

The instruments to be used

Preference must be given to the "whirling psychomotor" as being the easiest, handiest and most accurate instrument that can be used on board a sea-going vessel. There will of course be no difficulty involved in using this instrument on the vessel's bridge. The cargo hold will be entered through the escape hatch, which on some vessels is situated on the weather deck; on others is may be inside a mast house, which makes matters easier, as the mast house door can be closed. What must be borne in mind is that the escape hatch lid will have to be opened six times a day. When opened to admit a person it must be opened and closed as rapidly as possible. An arclight can be provided and, as soon as the person has entered the cargo hold, the lid must be closed until he is ready to come back on deck. The person using the psychrometer can take a temperature of the top platform of the Australian ladder and another at the bottom of the ladder near to the surface of the cargo. He may be able to move across the surface of the cargo and take more readings, eventually taking the mean of the two lots of temperatures but with separate evaluation. Naturally, the exercise will have to cease if weather conditions prevent men going on deck. It may be argued that the danger of depleted oxygen in closed-in spaces may prevent temperature readings being taken on the surface of the cargo. Steel cargoes are not the same as cereal cargoes, nor some mineral cargoes that mop up oxygen; further to this they do not occupy so much space. In any case, Chapter Vl, Regulation 3, of SOLAS under the heading of "oxygen analysis and gas detection equipment" should ensure that every vessel is equipped to test for depleted oxygen supply. On long voyages when cargoes have shifted and/or need re-securing, crew members have inadvertently (with regard to depleted oxygen supply) entered the holds and spent considerable time below without any dire results.

What is stated above is, of course, all that can be done with the equipment generally available on board the average merchant ship. On the other hand, keeping in mind that in the context of being expected to take the maximum precautions humanly possible, to carefully carry and care for the cargo, one will be conscious of the fact that what is required under present-day conditions forces what can be achievable to the very limit. A few decades ago a 3,000 DWT cargo ship carried, on deck, a minimum of six able seamen, three ordinary seamen, a bo'sun and a carpenter; there were always three men in a deck watch. Today a 30,000 DWT bulk carrier vessel may have on deck only three able seamen, one ordinary seaman and a bo'sun. Considering that, every four hours of the voyage, at least two men are required to take temperatures in at least five cargo holds (and often seven), doubts can be raised as to how enthusiastically and efficiently such an exercise can be maintained over an average voyage of about 15 days. The temptation to "flog" the figures must be overwhelming, and this is, of course, a complete waste of time. It serves no useful purpose, and the figures produced are usually unconvincing and may be contributory to destroying the carrier's defence in the event of a cargo claim. What can be done to remedy this situation?

Measures that could be taken

The constant opening and closing of escape hatches to admit an observer to take temperatures in the cargo holds is unsatisfactory. The ideal situation would be to seal the hatches before the voyage commences and only open them again to allow discharge of the cargo at the final port of destination. Naturally, there would be occasions, as few as possible, when the cargo holds would have to be entered to inspect the stowage. Here again, escape hatches must not be left open whilst this inspection tales place. The present-day method of taking hold temperatures with a whirling psychrometer is archaic, risky, and, in view of the small number of crew available and the circumstances in general, is highly subject to the ventilation records being erroneous. Of course, in the event of a cargo claim, the carrier cannot plead lack of personnel available on board, lack of the right type of equipment being available, nor that it was unavailable (due to the vessel's position) when notified of the next charter and port of loading, nor inexperience of the vessel's staff to perform the task. The cargo interest's answer to this would be demonstrated by production of a clean bill of lading; the passage in the Carriage of Goods by Sea Act (COGSA), whereby the Master is obliged to carefully load, stow, carry, care for, and discharge the cargo, and finally the fact that his cargo was received on board in apparent good order and condition and was not so received at the final port of destination, to which the only answer often is near enough to amen.

In the 21st century, the atomic age, there is ample technology available which can be used to compile temperature records without having to enter cargo holds every four hours. Since the early part of the 20th century, dew cells and distant temperature recorders have been available. The most modern and satisfactory arrangement would be to use thermocouples wired up to recorders on the vessel's navigation bridge or, failing that, inside a mast house. The recorder would produce a continuous graph providing surface temperature of the cargo, dew point temperature within the cargo hold and dew point temperature of the atmosphere. This would rule out any accusation of falsification of records. Further to this, a fringe benefit of the system would be early warning of any leakage of hatches during boisterous or heavy weather being experienced: such suspicions would be raised by a sudden appreciable rise in the relative humidity within any cargo hold. Dehumidification systems are very effective, and portable units are used in the liner, bulk carrier, steel out and forest products back, Europe to the North American West Coast.

Incompatible cargoes

Mixed stowages often present problems, when for instance semi-finished steel products are shipped in a wet condition, or covered with snow and ice, due to outside preshipment storage. These goods are often stowed in the same compartment as wrapped material such as cold-rolled and galvanised steel coils. Such an arrangement constitutes a case of bad stowage for which initially the Master/Owner is in the front of the claims firing line. The difficulties arise when the hatches are closed down. The excess moisture brought on board with the wet cargo causes the relative humidity of the ambient air within the cargo hold to increase, often dramatically. The due point temperature of the ambient air increases accordingly, to exceed the temperature of the dry cargo which commences to sustain damage from ship and cargo sweat.

These mixed cargoes are booked in order to fill up one vessel instead of having to charter two vessels, and/or whatever cargo is on offer is accepted. Planning both the loading and discharging operations is done in such a way as to, as much as possible, have all hatches being worked and finishing at the same time, in addition to which the vessel's draft has to be taken into consideration when placing weights in, and removing weights from, the various cargo holds, as well as port rotation. In such circumstances, in order to gain the maximum from the venture, separating out, and segregating wet and dry cargo into their own compartments, is not only a viably uninteresting proposition but often an impossibility.

The question arises as to what can be done with the ventilation to expedite the above-mentioned situation? Can anything be done? Usually nothing is done and the vessel faces claims for moisture damage to the wrapped cargo at the port of discharge. The inadequacy of bulk carrier ventilation has already been mentioned. When such a situation arises there are measures that can be taken, and the ship's Master must discuss these with the charterers. When contemplating a voyage where atmospheric temperatures will be rising in comparison to loading temperatures, the main objective will be to remove the moisture from the cargo holds in order to lower the ambient dew point temperature. This may be accomplished by leaving the ventilators open and increasing the heat of the heavy oil bunkers in the double bottom tanks beneath the various cargo holds, in an effort to evaporate the moisture, which should pass to the atmosphere. The owner's answer to this may well be "why should we spend money for the purpose of reconditioning cargo"; perhaps this might be a point to be taken up with charterers prior to the cargo being shipped. The fact that something of a positive nature is done, rather than nothing, providing it is not seen to exacerbate an already deteriorating situation, is always looked upon favorably. As long as the temperature of the air in the cargo hold is heated up and warm air is rising out through the vents, the dew point temperature of the outside atmosphere surrounding the vessel can be disregarded, although it should be checked and recorded. Once the excess moisture has been removed from the cargo compartment (if it can be removed), it can be battened down and normal measures taken to protect the cargo for the rest of the voyage. The entire situation should always be monitored in accordance with the wet and dry-bulb temperature observations.

With such a stowage, when moving into a relatively colder climate, heating of the double bottom tanks when atmospheric temperatures are falling will promote copious ship sweat, which may rain down from above, also stream down the ship's sides to end up on the tank-top, to be soaked up by the dunnage and create a humid high dew point temperature zone at the base of the cargo. This can result in cargo sweating of the steel coils up to a defined level above the tank-top and eventual moisture damage to the cargo. Such a situation defies the justification of any action taken in consideration of the fact that bulk carrier ventilation is unable to cope with a ship sweat situation. If in the early part of the voyage there is a period when atmospheric temperatures are to some extent static, extra heating of the double bottom tanks may be a consideration. Once temperatures begin to fall normally, full ventilation would be required in order to flush out the relatively higher DPT air from the loading port, to be continuously replaced with the progressively cooler air as temperatures keep falling. There is not much that can be done, but to be seen to do something positive is better than doing nothing. When temperatures begin to fall, if the dew point temperature of the air in the cargo hold exceeds the dew point temperature of the atmosphere, what ventilation is available should be used.

Still air storage

Any type of material/cargo will deteriorate if residing too long in a still air storage situation. The air in a room, closed down and not ventilated, will absorb moisture from the furniture, etc; air films develop and, given time, even the wallpaper may start to peel off the walls. Where steel cargoes are concerned, in most instances, such a situation is not so important with semi-finished products. However, where wrapped goods and galvanised material are concerned, it can result in condensation damage developing. Therefore, the ideal situation would be to devise a method of keeping the air circulating within a closed-down compartment.

The practice of covering the cargo with plastic is not recommended. This is usually done to prevent cargo sweat or water from leaking hatches contacting the cargo. The sensible thing to do would be to avoid cargo sweat and the leaking of hatches. The ship sweat situation usually develops when the vessel is voyaging from a warmer to a relatively cooler climate. The cargo does not adjust readily to the falling air temperatures, especially with blocks of steel cargo protected from changing sea temperatures by double bottom tanks and hopper tanks (whether the cargo hold is ventilated or not) and encapsulated in plastic sheeting.

Whether the hold is ventilated or not, the temperature of the air above the cargo/plastic will cool. The surface of the plastic will cool and the warm air beneath the plastic may condense on the underside of the plastic and rain back onto the cargo in the form of ship sweat. It may be argued that the amount of air beneath the plastic and within the cargo is minimal and carries insufficient moisture to cause any appreciable damage, if any at all. Theoretically this is true. However, in practice the small amount of trapped air may develop a high relative humidity through moisture picked up from the usually too-moist dunnage. Further to this, sweat water forming on the inside of the shell plating (ship sweat) drips down onto the tank-top beneath the cargo and charges the air with further moisture and saturates the dunnage beneath the cargo. The cooler plastic causes the moisture to reach the point of super saturation and commences to sweat upon the plastic. Further to this, the heightened DPT of the ambient air under the tarpaulins can exceed the temperature of the cargo giving rise to cargo sweat. Blocks of steel cargo tend to control their own environment. Covering the cargo with plastic sheets or tarpaulins retards the warm-up or fall-off of the cargo temperature.

The fact is that when temperatures are falling the outside air cools and its RH increases. The air in the hold is warmer and has a lower RH. To equalise moisture, vapour flows from outside towards the cargo holds if the cargo hold is not properly isolated; this is where the ventilation records can be of great assistance. The conclusions are that, if it is decided to ventilate a steel cargo, under normal circumstances, it should only be done when temperatures are falling. No ventilation should be given under any circumstances without the guidance of the temperature records. Bulk carrier ventilation is incapable of combating the development of ship sweat. Electrical ventilation installed in bulk carriers is incorrectly placed and invariably too weak to be of any use. From experience, the electrical ventilation installed in the break bulk ships of the liner companies was capable and more-or-less immediately effective, when turned on, of arresting the formation of ship sweat on deck heads and ship's sides. It is considered that if the rules on cargo hold ventilation are properly adhered to, and where, under closed-down ventilation, the air in the hold is kept circulating, localised concentrations, zones and areas of high humidity will be prevented from forming. It is such humidity developments that result in coil cargoes outturning with part of a block of cargo being moisture damaged while the balance is received without complaint.

WIRE RODS

Wire rods are manufactured from billets. This is accomplished by an extrusion process, whereby the red-hot metal bar is pressed into the orifice of a die. The metal is forced against the die by hydraulic rams and the continuous form which emerges takes the shape of the die opening. Sizes vary but are often between 5 and 9 mm in diameter.

This semi-finished—mainly hot rolled—product consists of steel wire wound into coils, each being secured with wire bands, the ends of which are merely twisted together; although a customary method of securing, this is open to criticism where claims for loose bundles and tangling are involved upon delivery. Some coils in shipments are consolidated into units of four coils bound together by metal strapping bands. Each unit is referred to as a "unitised bundle of wire rods". Sometimes the coils are the same size as unitised bundles but the wire in the bundles is continuous. Approximate weights are between 1.5 to 2.0 m/tons each, and the bundles on average measure about 1m in diameter and 1.25 m in breadth.

Wire rods are in most instances shipped in an unwrapped condition, and are therefore usually affected by rust (see Plate 20). They are invariably stored in the open prior to shipment, so that it is not uncommon to observe water dripping out of the bundles when they are shipped. This is a hot-rolled product which is subjected to further processing in order that it might be directed to a large range of end uses, such as the manufacture of nails, galvanised wire for fencing (including barbed wire), road mesh, to mention a few applications. When the wire is wrapped, this is an indication that the goods are destined for a fabrication of a more delicate nature, e.g. high tensile steel wire rod, tyre cord, wire for musical instruments; the type of wrapping is usually a single plastic sheet. Defects which develop during the time that the goods are in the care of the carrier will ultimately provoke claims.

Eventual uses of wire rods

Many shipments of wire rods, in their condition as transported by sea, are eventually destined to be cold drawn. During this process the wire is forced through dies which reduce its gauge, and cause it to increase in length. Because of this, kinks and nicks in the wire are inadmissible, as when being drawn through the dies the wire can break; even if it does not break defects elongate and remain a defect. Even if the wire is not for redrawing such defects are undesirable, e.g. in the manufacture of road mesh, as these imperfections show up in the finished product. Disintegration of bundles during the voyage, caused by bad stowage, crushing and breakage of the strapping bands, is to be avoided as this leads to loose turns of wire, which develop into tangling, intertwining and twisting of the wire (see Plate 21). As a result of this, parts of the coils may have to be cut off and scrapped. If this is not the ultimate solution, depending upon the uses for which the wire is intended, tangling and twisting of the turns in the bundles results in loss of time on the production line and eventual insurance claims.

Stowage of board ship

Compared with other steel products wire rods have a higher stowage factor. They can be stowed in lower holds or tweendecks. Usually deep stowages are involved and, with the present-day lack of tweendeckers, such stowages commence directly on the vessel's tank tops, in bulk carriers.

Compact, rigid bundles are necessary (as shown in Fig. 76) in order to effect a good stow. The bundles or coils are stowed with their cylindrical centre cores placed in a fore and aft direction in the ship, in much the same way as coils of steel sheeting are placed in stow. The stowage of the individual coils or bundles must be uniform and compactly arranged in order to avoid breakdowns of the stow with accompanying crushing damage and disintegration of the bundles. Slackness in bundles,

Fig. 76. Wire rods—correct bundling.

Fig. 77. Wire rods—slack bundling.

when offered for shipment (as shown in Fig. 77) is usually attributable to failure to compress the bundles sufficiently when the securing wires, or metal strapping bands, are applied. Slack bundles invariably suffer from convergence of the securing bands, which results in the bundle windings spraying open (see Plate 19). Such bundles or coils cannot be properly stowed on board ship and also cannot be properly handled when being loaded; damage inevitably results from this. Dump stowage of wire rods should be avoided at all costs.

When stowages terminate at a half hatch position, the face of the stow should be secured as indicated in Fig. 78, in order to prevent bulging and eventual collapse of the stow. If the bundles are well formed and rigid, a wire rod stowage can usually be stowed to eight tiers high; beyond this there is a risk of crushing of lower tier bundles, and for further loading shippers conformation of tier limitation should be obtained in writing. Stowages must be tight and well secured against movement, to avoid tripping of bundles.

Care in handling

Wire rods must be handled with care to avoid nicking, scoring, scratching and localised sharp bends and twists to the windings. In the circumstances, the forks on the fork-lift trucks need to be protected or of cylindrical construction. The bundles or coils should be hoisted by means of "Cobra" or braided wire slings; "C" hooks are often used. When only part of the cargo space is filled with wire rods the ideal stowage position is against the after bulkhead, with the stow projecting forward of this. Adequate dunnage must be used to keep the goods clear of all metal parts comprising the vessel's structure. No dunnage is used between tiers.

All unwrapped mild steel wire is usually affected by rust at time of shipment, prior to which it is often stored outside and exposed to the elements for some considerable time. In most instances, fresh-water rust development is accepted as normal for these goods and is therefore of no consequence. It is not unusual for these goods to be loaded during light rain showers. However, the remarks made referring to loading of hot rolled steel products during rainy weather should be kept in mind. See under Hot rolled steel sheeting.

Points to be considered

The delivery of loose, or partly loose, bundles will result in claims being entered to cover the costs of restrapping to facilitate carriage of the goods to final destination. Physical damage may be claimed on account of disintegration of the coils or bundles.

Contact with salt water will impair the quality of the goods, which may have to undergo a pickling process in order to restore them to their original good condition. The development of rust pitting can permanently damage wire rods. This can also be caused through contact with salt water, but long exposure to fresh water may also result in severe pitting under certain conditions.

When the various windings in the bundles are affected by severe bending, kinking, twisting, nicks, scoring and heavy scratch marks, claims against the carrier can be forthcoming when the goods are delivered.

It cannot be too strongly emphasised that wire rods must be handled with care.

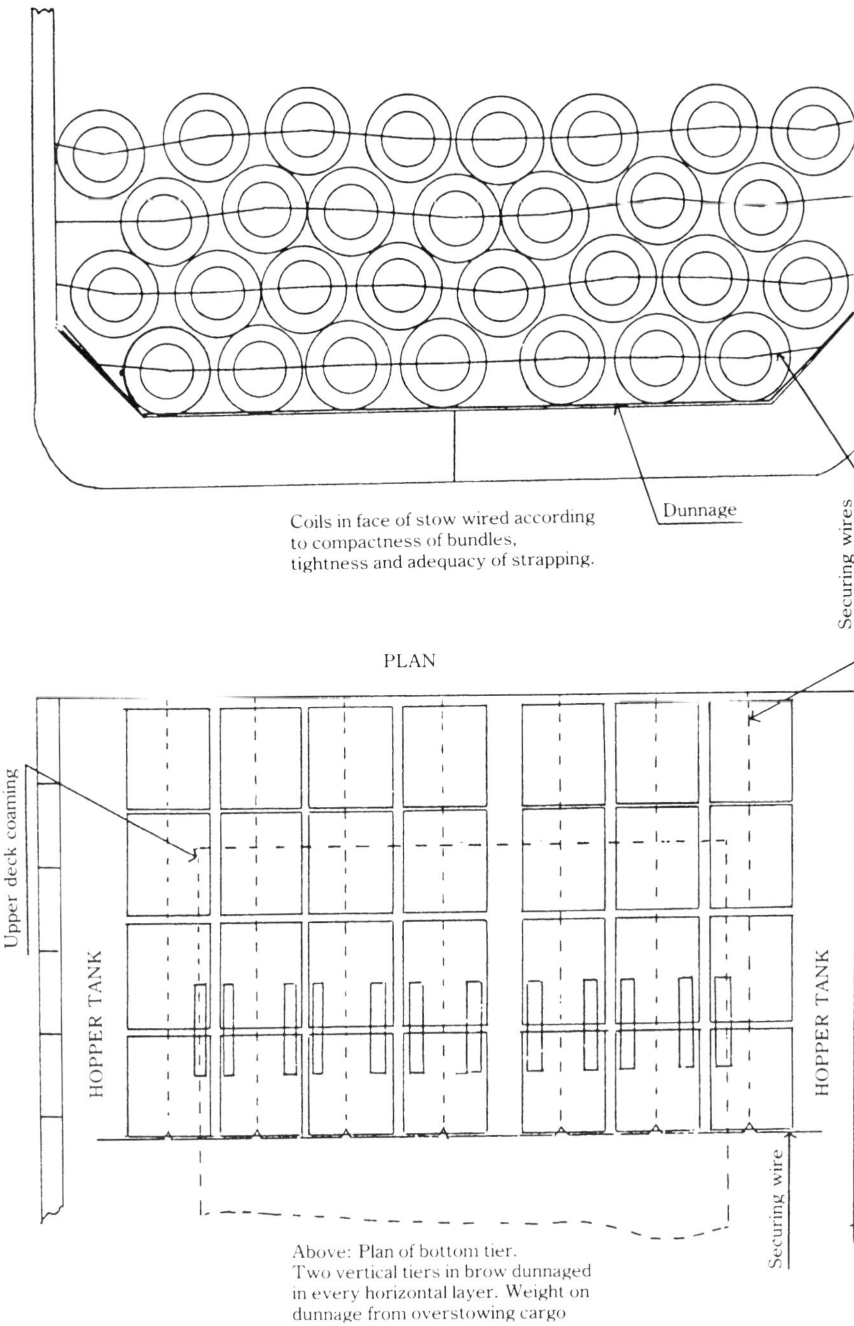

Fig. 78. Securing of wire rod cargo (open brow stowage).

Dropping the bundles or coils into places considered inaccessible, i.e. dump stowage, on board the ship, at time of loading may cause serious distortion—bundles or coils become out of round shape. Haphazard dumping of bundles into the stowage as shown in Fig. 79 leads to damage. A more proper and correct type of stowage is shown in Fig. 80.

Defects which develop as indicated above invariably develop into claims consisting of extra expenses for rebundling, loss of production by the factory owing to loss of time, and occasionally a certain degree of reprocessing is necessary in order to rectify some serious defects caused through heavy rusting.

In view of the above, wire rods should be examined carefully at time of shipment, and their condition noted accurately.

The following is a reproduction of the details entered in a bill of lading for a wire rod shipment from Australia.

358 bundles (1,432 coils) 5.5 mm wire rods 1015.5 T.

Surface rust on all bundles.

Description of goods

Notes:

(1) Description: wire drawing rods, in coils commercial quality, manufactured from open hearth and/or basic oxygen furnace steel at seller's option.

Specification:

SAE1006 semi-killed, carbon—0.08% maximum Manganese—aim 0.40%.

Fig. 79. Wire rods—bad stowage.

Fig. 80. Wire rods—reasonably good stowage.

Tolerance:

According to A.I.S.I.

Size:

5.5 mm.

Surface:

Rods are commercially free from harmful surface defects such as seams, cracks, slivers and overfills.

Banding and tagging:

(a) Red colour strip is placed on bands.
(b) Coils are labelled with two red labels per bundle. Labels show: specification, size, weight, destination.

Bundling/Stowage:

(a) Wire rod is shipped in master bundle consisting of four coils per bundle. Each coil with two wire ties and each bundle with four bands of steel strapping. The wire ties are 9–11 ISWG diameters.
Coil I.D. approx 850 mm
Coil O.D. approx 1,250 mm.
(b) Material is block stowed.

The goods were discharged in a north European port and claims were entered against the carrier in respect of:

—Broken strapping bands which had to be replaced if damage was to be avoided in further handling.
—The forks of the fork-lift trucks damaged the windings owing to rough handling by the stevedores.
—Bundles dropped into the wings in the ship's holds had become deformed and windings were heavily chafed when the bundles struck the ship's sides.
—Windings of bundles chafed on the metal parts of the ship's structure during the course of the voyage caused by vibration of the ship.

The loss sustained owing to this damage was eventually agreed by the surveyors as amounting to 11% of the sound value of the cargo.

Claims are regularly entered against this material as mentioned above, so proving the vulnerability of the material to sustain damage.

WIRE RODS—FOR PRE-STRESSING

High tensile wire, free from mill scale and must be rust free. The packing is usually consisting of a complete coverage with:

—1st layer of VCI (Volatile Corrosion Inhibiter) paper, overlaps secured with tape;
—2nd layer: plastified kraft paper, secured with tape;
—3rd layer: woven polypropylene sheeting;
—4th layer: bandaged with strip burlap.

The goods are used in the pre-stressing of concrete, railway sleepers, flooring, roof beams, wall units, industrial panelling, bridge building, etc. No damage of any description is acceptable. Extra care should be taken to ensure that these goods are protected in every respect and provided with a top stowage position.

Appendix

TEXT OF THE HAGUE RULES 1924

International convention for the unification of certain rules of law relating to bills of lading signed at Brussels on 26 August 1924

Article I

In this Convention the following words are employed with the meanings set out below:

(a) "Carrier" includes the owner or the charterer who enters into a contract of carriage with a shipper.
(b) "Contract of carriage" applies only to contracts of carriage covered by a bill of lading or any similar document of title, in so far as such document relates to the carriage of goods by sea, including any bill of lading or any similar document as aforesaid issued under or pursuant to a charterparty from the moment at which such bill of lading or similar document of title regulates the relations between a carrier and a holder of the same.
(c) "Goods" includes goods, wares, merchandise and articles of every kind whatsoever except live animals and cargo which by the contract of carriage is stated as being carried on deck and is so carried.
(d) "Ship" means any vessel used for the carriage of goods by sea.
(e) "Carriage of goods" covers the period from the time when the goods are loaded on to the time they are discharged from the ship.

Article II

Subject to the provisions of Article 6, under every contract of carriage of goods by sea the carrier, in relation to the loading, handling, stowage, carriage, custody, care and discharge of such goods, shall be subject to the responsibilities and liabilities, and entitled to the rights and immunities hereinafter set forth.

Article III

1. The carrier shall be bound before and at the beginning of the voyage to exercise due diligence to:

(a) Make the ship seaworthy;
(b) Properly man, equip and supply the ship;
(c) Make the holds, refrigerating and cool chambers, and all other parts of the ship in which goods are carried, fit and safe for their reception, carriage and preservation.

2. Subject to the provisions of Article 4, the carrier shall properly and carefully load, handle, stow, carry, keep, care for, and discharge the goods carried.

3. After receiving the goods into his charge the carrier or the master or agent of the carrier shall, on demand of the shipper, issue to the shipper a bill of lading showing among other things:

(a) The leading marks necessary for identification of the goods as the same are furnished in writing by the shipper before the loading of such goods starts, provided such marks are stamped or otherwise shown clearly upon the goods if uncovered, or on the cases or coverings in which such goods are contained, in such a manner, as should ordinarily remain legible until the end of the voyage.

(b) Either the number of packages or pieces, or the quantity, or weight, as the case may be, as furnished in writing by the shipper.

(c) The apparent order and condition of the goods.

Provided that no carrier, master or agent of the carrier shall be bound to state or show in the bill of lading any marks, number, quantity, or weight which he has reasonable ground for suspecting not accurately to represent the goods actually received, or which he has had no reasonable means of checking.

4. Such a bill of lading shall be *prima facie* evidence of the receipt by the carrier of the goods as therein described in accordance with paragraph 3(a), (b) and (c).

5. The shipper shall be deemed to have guaranteed to the carrier the accuracy at the time of shipment of the marks, number, quantity and weight, as furnished by him, and the shipper shall indemnify the carrier against all loss, damages and expenses arising or resulting from inaccuracies in such particulars. The right of the carrier to such indemnity shall in no way limit his responsibility and liability under the contract of carriage to any person other than the shipper.

6. Unless notice of loss or damage and the general nature of such loss or damage be given in writing to the carrier or his agent at the port of discharge before or at the time of the removal of the goods into the custody of the person entitled to delivery thereof under the contract of carriage, such removal shall be *prima facie* evidence of the delivery by the carrier of the goods as described in the bill of lading.

If the loss or damage is not apparent, the notice must be given within three days of the delivery of the goods.

The notice in writing need not be given if the state of the goods has, at the time of their receipt, been the subject of joint survey or inspection.

In any event the carrier and the ship shall be discharged from all liability in respect of loss or damage unless suit is brought within one year after delivery of the goods or the date when the goods should have been delivered.

In the case of any actual or apprehended loss or damage the carrier and the receiver shall give all reasonable facilities to each other for inspecting and tallying the goods.

7. After the goods are loaded the bill of lading to be issued by carrier, master, or agent of the carrier, to the shipper shall, if the shipper so demands, be a "shipped" bill of lading, provided that if the shipper shall have previously taken up any document of title to such goods, he shall surrender the same as against the issue of the "shipped" bill of lading, but at the option of the carrier such document of title may be noted at the port of shipment by the carrier, master, or agent with the name or names of the ship or ships upon which the goods have been shipped and the date or dates of shipment, and when so noted, if it shows the particulars mentioned in paragraph 3 of Article 3, shall for the purpose of this Article be deemed to constitute a "shipped" bill of lading.

8. Any clause, covenant, or agreement in a contract of carriage relieving the

carrier or the ship from liability for loss or damage to, or in connection with, goods arising from negligence, fault, or failure in the duties and obligations provided in this Article or lessening such liability otherwise than as provided in this Convention, shall be null and void and of no effect. A benefit of insurance clause in favour of the carrier or similar clauses shall be deemed to be a clause relieving the carrier from liability.

Article IV

1. Neither the carrier nor the ship shall be liable for loss or damage arising or resulting from unseaworthiness unless caused by want of due diligence on the part of the carrier to make the ship seaworthy, and to secure that the ship is properly manned, equipped and supplied, and to make the holds, refrigerating and cool chambers and all other parts of the ship in which goods are carried fit and safe for their reception, carriage and preservation in accordance with the provisions of paragraph 1 of Article 3. Whenever loss or damage has resulted from unseaworthiness the burden of proving the exercise of due diligence shall be on the carrier or other person claiming exemption under this Article.

2. Neither the carrier nor the ship shall be responsible for loss or damage arising or resulting from:

(a) Act, neglect, or default of the master, mariner, pilot, or the servants of the carrier in the navigation or in the management of the ship;
(b) Fire, unless caused by the actual fault or privity of the carrier;
(c) Perils, dangers and accidents of the sea or other navigable waters;
(d) Act of God;
(e) Act of war;
(f) Act of public enemies;
(g) Arrest or restraint of princes, rulers or people, or seizure under legal process;
(h) Quarantine restrictions;
(i) Act or omission of the shipper or owner of the goods, his agent or representative;
(j) Strikes or lockouts or stoppage or restraint of labour from whatever cause, whether partial or general;
(k) Riots and civil commotions;
(l) Saving or attempting to save life or property at sea;
(m) Wastage in bulk or weight or any other loss or damage arising from inherent defect, quality or vice of the goods;
(n) Insufficiency of packing;
(o) Insufficiency or inadequacy of marks;
(p) Latent defects not discoverable by due diligence;
(q) Any other cause arising without the actual fault or privity of the carrier, or without the fault or neglect of the agents or servants of the carrier, but the burden of proof shall be on the person claiming the benefit of this exception to show that neither the actual fault or privity of the carrier nor the fault or neglect of the agents or servants of the carrier contributed to the loss or damage.

3. The shipper shall not be responsible for loss or damage sustained by the carrier or the ship arising or resulting from any cause without the act, fault or neglect of the shipper, his agents or his servants.

4. Any deviation in saving or attempting to save life or property at sea or any reasonable deviation shall not be deemed to be an infringement or breach of this Convention or of the contract of carriage, and the carrier shall not be liable for any loss or damage resulting therefrom.

5. Neither the carrier nor the ship shall in any event be or become liable for any loss or damage to or in connection with goods in an amount exceeding 100 pounds sterling per package or unit, or the equivalent of that sum in other currency unless the nature and value of such goods have been declared by the shipper before shipment and inserted in the bill of lading.

This declaration if embodied in the bill of lading, shall be *prima facie* evidence, but shall not be binding or conclusive on the carrier.

By agreement between the carrier, master or agent of the carrier and the shipper another maximum amount than that mentioned in this paragraph may be fixed, provided that such maximum shall not be less than the figure above named.

Neither the carrier nor the ship shall be responsible in any event for loss or damage to, or in connection with, goods if the nature or value thereof has been knowingly misstated by the shipper in the bill of lading.

6. Goods of an inflammable, explosive or dangerous nature to the shipment whereof the carrier, master or agent of the carrier has not consented with knowledge of their nature and character, may at any time before discharge be landed at any place, or destroyed or rendered innocuous by the carrier without compensation and the shipper of such goods shall be liable for all damages and expenses directly or indirectly arising out of or resulting from such shipment. If any such goods shipped with such knowledge and consent shall become a danger to the ship or cargo, they may in like manner be landed at any place, or destroyed or rendered innocuous by the carrier without liability on the part of the carrier except to general average, if any.

Article V

A carrier shall be at liberty to surrender in whole or in part all or any of his rights and immunities or to increase any of his responsibilities and obligations under this Convention, provided such surrender or increase shall be embodied in the bill of lading issued to the shipper.

The provisions of this Convention shall not be applicable to charterparties, but if bills of lading are issued in the case of a ship under a charterparty they shall comply with the terms of this Convention. Nothing in these rules shall be held to prevent the insertion in a bill of lading of any lawful provision regarding general average.

Article VI

Notwithstanding the provisions of the preceding Articles, a carrier, master or agent of the carrier and a shipper shall in regard to any particular goods be at liberty to enter into any agreement in any terms as to the responsibility and liability of the carrier for such goods, and as to the rights and immunities of the carrier in respect of such goods, or his obligation as to seaworthiness, so far as this stipulation is not contrary to public policy, or the care or diligence of his servants or agents in regard to the loading, handling, stowage, carriage, custody, care and discharge of the goods carried by sea, provided that in this case no bill of lading has been or shall be issued and that the terms agreed shall be embodied in a receipt which shall be a non-negotiable document and shall be marked as such.

Any agreement so entered into shall have full legal effect.

Provided that this Article shall not apply to ordinary commercial shipments made in the ordinary course of trade, but only to other shipments where the character or condition of the property to be carried or the circumstances, terms and conditions under which the carriage is to be performed are such as reasonably to justify a special agreement.

Article VII

Nothing herein contained shall prevent a carrier or a shipper from entering into any agreement, stipulation, condition, reservation or exemption as to the responsibility and liability of the carrier or the ship for the loss or damage to, or in connection with, the custody and care and handling of goods prior to the loading on, and subsequent to, the discharge from the ship on which the goods are carried by sea.

Article VIII

The provisions of this Convention shall not affect the rights and obligations of the carrier under any statute for the time being in force relating to the limitation of the liability of owners of seagoing vessels.

Article IX

The monetary units mentioned in this Convention are to be taken to be gold value.

Those contracting States in which the pound sterling is not a monetary unit reserve to themselves the right of translating the sums indicated in this Convention in terms of pound sterling into terms of their own monetary system in round figures.

The national laws may reserve to the debtor the right of discharging his debt in national currency according to the rate of exchange prevailing on the day of the arrival of the ship at the port of discharge of the goods concerned.

Article X

The provisions of this Convention shall apply to all bills of lading issued in any of the contracting States.

Article XI

After an interval of not more than two years from the day on which the Convention is signed the Belgian Government shall place itself in communication with the Governments of the High Contracting Parties which have declared themselves prepared to ratify the Convention, with a view to deciding whether it shall be put into force. The ratification shall be deposited at Brussels at a date to be fixed by agreement among the said Governments. The first deposit of ratifications shall be recorded in a procès-verbal signed by the representatives of the Powers which take part therein and by the Belgian Minister for Foreign Affairs.

The subsequent deposit of ratifications shall be made by means of a written notification, addressed to the Belgian Government and accompanied by the instrument of ratification.

A duly certified copy of the procès-verbal relating to the first deposit of ratifications, of the notifications referred to in the previous paragraph, and also of the instruments of ratification accompanying them, shall be immediately sent by the Belgian Government through the diplomatic channel to the Powers who have signed this Convention or who have acceded to it. In the cases contemplated in the preceding paragraph, the said Government shall inform them at the same time of the date on which it received the notification.

Article XII

Non-signatory States may accede to the present Convention whether or not they have been represented at the International Conference at Brussels.

A State which desires to accede shall notify its intention in writing to the Belgian Government, forwarding to it the document of accession, which shall be deposited in the archives of the said Government.

The Belgian Government shall immediately forward to all the States which have signed or acceded to the Convention a duly certified copy of the notification and of the act of accession, mentioning the date on which it received the notification.

Article XIII

The High Contracting Parties may at the time of signature, ratification or accession declare that their acceptance of the present Convention does not include any or all of the self-governing dominions, or of the colonies, overseas possessions, protectorates or territories under their sovereignty or authority, and they may subsequently accede separately on behalf of any self-governing dominion, colony, overseas possession, protectorate or territory excluded in their declaration. They may also denounce the Convention separately in accordance with its provisions in respect of any self-governing dominion, or any colony, overseas possession, protectorate or territory under their sovereignty or authority.

Article XIV

The present Convention shall take effect, in the case of the States which have taken part in the first deposit of ratifications, one year after the date of the protocol recording such deposit.

As respects the States which ratify subsequently or which accede, and also in cases in which the Convention is subsequently put into effect in accordance with Article 13, it shall take effect six months after the notifications specified in paragraph 2 of Article 11 and paragraph 2 of Article 12 have been received by the Belgian Government.

Article XV

In the event of one of the contracting States wishing to denounce the present Convention, the denunciation shall be notified in writing to the Belgian Government, which shall immediately communicate a duly certified copy of the notification to all the other States, informing them of the date on which it was received.

The denunciation shall only operate in respect of the State which made the notification, and on the expiry of one year after the notification has reached the Belgian Government.

Article XVI

Any one of the contracting States shall have the right to call for a fresh conference with a view to considering possible amendments.

A State which would exercise this right should notify its intention to the other

States through the Belgian Government, which would make arrangements for convening the conference.

Done at Brussels, in a single copy, 25 August 1924.

(Follow the signatures)

Protocol of signature

At the time of signing the International Convention for the unification of certain rules of law relating to bills of lading the Plenipotentiaries whose signatures appear below have adopted this Protocol, which will have the same force and the same value as if its provisions were inserted in the text of the Convention to which it relates.

The High Contracting Parties may give effect to this Convection either by giving it the force of law or by including in their national legislation in a form appropriate to that legislation the rules adopted under this Convention.

They may reserve the right:

1. To prescribe that in the cases referred to in paragraph 2(c) to (p) of Article 4 the holder of a bill of lading shall be entitled to establish responsibility for loss or damage arising from the personal fault of the carrier or the fault of his servants which are not covered by paragraph (a).

2. To apply Article 6 in so far as the national coasting trade is concerned to all classes of goods without taking account of the restriction set out in the last paragraph of that Article.

Done at Brussels, in a single copy, 25 August 1924.

(Follow the signatures)

TEXT OF THE HAGUE-VISBY RULES

The Hague Rules as Amended by the Brussels Protocol 1968

Article I

In these Rules the following words are employed, with the meanings set out below:

(a) "Carrier" includes the owner or the charterer who enters into a contract of carriage with a shipper.

(b) "Contract of carriage" applies only to contracts of carriage covered by a bill of lading or any similar document of title, in so far as such document relates to the carriage of goods by sea, including any bill of lading or any similar document as aforesaid issued under or pursuant to a charterparty from the moment at which such bill of lading or similar document of title regulates the relations between a carrier and a holder of the same.

(c) "Goods" includes goods, wares, merchandise, and articles of every kind whatsoever except live animals and cargo which by the contract of carriage is stated as being carried on deck and is so carried.

(d) "Ship" means any vessel used for the carriage of goods by sea.

(e) "Carriage of goods" covers the period from the time when the goods are loaded on to the time they are discharged from the ship.

Article II

Subject to the provisions of Article VI, under every contract of carriage of goods by sea the carrier, in relation to the loading, handling, stowage, carriage, custody, care and discharge of such goods, shall be subject to the responsibilities and liabilities, and entitled to the rights and immunities hereinafter set forth.

Article III

1. The carrier shall be bound before and at the beginning of the voyage to exercise due diligence to:

(a) Make the ship seaworthy.
(b) Properly man, equip and supply the ship.
(c) Make the holds, refrigerating and cool chambers, and all other parts of the ship in which goods are carried, fit and safe for their reception, carriage and preservation.

2. Subject to the provisions of Article IV, the carrier shall properly and carefully load, handle, stow, carry, keep, care for, and discharge the goods carried.

3. After receiving the goods into his charge the carrier or the master or agent of the carrier shall, on demand of the shipper, issue to the shipper a bill of lading showing among other things:

(a) The leading marks necessary for identification of the goods as the same are furnished in writing by the shipper before the loading of such goods starts, provided such marks are stamped or otherwise shown clearly upon the goods if uncovered, or on the cases or coverings in which such goods are contained, in such a manner as should ordinarily remain legible until the end of the voyage.
(b) Either the number of packages or pieces, or the quantity, or weight, as the case may be, as furnished in writing by the shipper.
(c) The apparent order and condition of the goods.

Provided that no carrier, master or agent of the carrier shall be bound to state or show in the bill of lading any marks, number, quantity, or weight which he has reasonable ground for suspecting not accurately to represent the goods actually received, or which he has had no reasonable means of checking.

4. Such a bill of lading shall be *prima facie* evidence of the receipt by the carrier of the goods as therein described in accordance with paragraph 3(*a*), (*b*) and (*c*). However, proof to the contrary shall not be admissible when the bill of lading has been transferred to a third party acting in good faith.

5. The shipper shall be deemed to have guaranteed to the carrier the accuracy at the time of shipment of the marks, number, quantity and weight, as furnished by him, and the shipper shall indemnify the carrier against all loss, damages and expenses arising or resulting from inaccuracies in such particulars. The right of the carrier to such indemnity shall in no way limit his responsibility and liability under the contract of carriage to any person other than the shipper.

6. Unless notice of loss or damage and the general nature of such loss or damage be given in writing to the carrier or his agent at the port of discharge before or at the time of the removal of the goods into the custody of the person entitled to delivery thereof under the contract of carriage, or, if the loss or damage be not apparent, within three days, such removal shall be *prima facie* evidence of the delivery by the carrier of the goods as described in the bill of lading.

The notice in writing need not be given if the state of the goods has, at the time of their receipt, been the subject of joint survey or inspection.

Subject to paragraph 6*bis* the carrier and the ship shall in any event be discharged from all liability whatsoever in respect of the goods, unless suit is brought within one year of their delivery or of the date when they should have been delivered. This period may, however, be extended if the parties so agree after the cause of action has arisen.

In the case of any actual or apprehended loss or damage the carrier and the receiver shall give all reasonable facilities to each other for inspecting and tallying the goods.

6*bis*. An action for indemnity against a third person may be brought even after the expiration of the year provided for in the preceding paragraph if brought within the time allowed by the law of the Court seized of the case. However, the time allowed shall be not less than three months, commencing from the day when the person bringing such action for indemnity has settled the claim or has been served with process in the action against himself.

7. After the goods are loaded the bill of lading to be issued by the carrier, master, or agent of the carrier, to the shipper shall, if the shipper so demands, be a "shipped" bill of lading, provided that if the shipper shall have previously taken up any document of title to such goods, he shall surrender the same as against the issue of the "shipped" bill of lading, but at the option of the carrier such document of title may be noted at the port of shipment by the carrier, master, or agent with the name or names of the ship or ships upon which the goods have been shipped and the date or dates of shipment, and when so noted, if it shows the particulars mentioned in paragraph 3 of Article III, shall for the purpose of this article be deemed to constitute a "shipped" bill of lading.

8. Any clause, covenant, or agreement in a contract of carriage relieving the carrier or the ship from liability for loss or damage to, or in connection with, goods arising from negligence, fault, or failure in the duties and obligations provided in this article or lessening such liability otherwise than as provided in these Rules, shall be null and void and of no effect. A benefit of insurance in favour of the carrier or similar clause shall be deemed to be a clause relieving the carrier from liability.

Article IV

1. Neither the carrier nor the ship shall be liable for loss or damage arising or resulting from unseaworthiness unless caused by want of due diligence on the part of the carrier to make the ship seaworthy, and to secure that the ship is properly manned, equipped and supplied, and to make the holds, refrigerating and cool chambers and all other parts of the ship in which goods are carried fit and safe for their reception, carriage and preservation in accordance with the provisions of paragraph 1 of Article III. Whenever loss or damage has resulted from unseaworthiness the burden of proving the exercise of due diligence shall be on the carrier or other person claiming exemption under this article.

2. Neither the carrier nor the ship shall be responsible for loss or damage arising or resulting from:

(a) Act, neglect, or default of the master, mariner, pilot, or the servants of the carrier in the navigation or in the management of the ship.
(b) Fire, unless caused by the actual fault or privity of the carrier.
(c) Perils, dangers and accidents of the sea or other navigable waters.
(d) Act of God.
(e) Act of war.
(f) Act of public enemies.
(g) Arrest or restraint of princes, rulers or people, or seizure under legal process.
(h) Quarantine restrictions.
(i) Act or omission of the shipper or owner of the goods, his agent or representative.

(j) Strikes or lockouts or stoppage or restraint of labour from whatever cause, whether partial or general.
(k) Riots and civil commotions.
(l) Saving or attempting to save life or property at sea.
(m) Wastage in bulk or weight or any other loss or damage arising from inherent defect, quality or vice of the goods.
(n) Insufficiency of packing.
(o) Insufficiency or inadequacy of marks.
(p) Latent defects not discoverable by due diligence.
(q) Any other cause arising without the actual fault or privity of the carrier, or without the fault or neglect of the agents or servants of the carrier, but the burden of proof shall be on the person claiming the benefit of this exception to show that neither the actual fault or privity of the carrier nor the fault or neglect of the agents or servants of the carrier contributed to the loss or damage.

3. The shipper shall not be responsible for loss or damage sustained by the carrier or the ship arising or resulting from any cause without the act, fault or neglect of the shipper, his agents or his servants.

4. Any deviation in saving or attempting to save life or property at sea or any reasonable deviation shall not be deemed to be an infringement or breach of these Rules or of the contract of carriage, and the carrier shall not be liable for any loss or damage resulting therefrom.

5. (*a*) Unless the nature and value of such goods have been declared by the shipper before shipment and inserted in the bill of lading, neither the carrier nor the ship shall in any event be or become liable for any loss or damage to or in connection with the goods in an amount exceeding the equivalent of 10,000 francs per package or unit or 30 francs per kilo of gross weight of the goods lost or damaged, whichever is the higher.

(*b*) The total amount recoverable shall be calculated by reference to the value of such goods at the place and time at which the goods are discharged from the ship in accordance with the contract or should have been so discharged.

The value of the goods shall be fixed according to the commodity exchange price, or if there be no such price according to the current market price, or, if there be no commodity exchange price or current market price, by reference to the normal value of goods of the same kind and quality.

(*c*) Where a container, pallet or similar article of transport is used to consolidate goods, the number of packages or units enumerated in the bill of lading as packed in such article of transport shall be deemed the number of packages or units for the purpose of this paragraph as far as these packages or units are concerned. Except as aforesaid such article of transport shall be considered the package or unit.

(*d*) A franc means a unit consisting of 65.5 milligrammes of gold of millesimal fineness 900. The date of conversion of the sum awarded into national currencies shall be governed by the law of the Court seized of the case.

(*e*) Neither the carrier nor the ship shall be entitled to the benefit of the limitation of liability provided for in this paragraph if it is proved that the damage resulted from an act or omission of the carrier done with intent to cause damage, or recklessly and with knowledge that damage would probably result.

(*f*) The declaration mentioned in sub-paragraph (*a*) of this paragraph, if embodied in the bill of lading, shall be *prima facie* evidence, but shall not be binding or conclusive on the carrier.

(*g*) By agreement between the carrier, master or agent of the carrier and the shipper other maximum amounts than those mentioned in sub-paragraph (*a*) of this paragraph may be fixed, provided that no maximum amount so fixed shall be less than the appropriate maximum mentioned in that sub-paragraph.

(*h*) Neither the carrier nor the ship shall be responsible in any event for loss or damage to, or in connection with, goods if the nature or value thereof has been knowingly mis-stated by the shipper in the bill of lading.

6. Goods of an inflammable, explosive or dangerous nature to the shipment whereof the carrier, master or agent of the carrier has not consented with knowledge of their nature and character, may at any time before discharge be landed at any place, or destroyed or rendered innocuous by the carrier without compensation and the shipper of such goods shall be liable for all damages and expenses directly or indirectly arising out of or resulting from such shipment. If any such goods shipped with such knowledge and consent shall become a danger to the ship or cargo, they may in like manner be landed at any place, or destroyed or rendered innocuous by the carrier without liability on the part of the carrier except to general average, if any.

Article IV bis

1. The defences and limits of liability provided for in these Rules shall apply in any action against the carrier in respect of loss or damage to goods covered by a contract of carriage whether the action be founded in contract or in tort.

2. If such an action is brought against a servant or agent of the carrier (such servant or agent not being an independent contractor), such servant or agent shall be entitled to avail himself of the defences and limits of liability which the carrier is entitled to invoke under these Rules.

3. The aggregate of the amounts recoverable from the carrier, and such servants and agents, shall in no case exceed the limit provided for in these Rules.

4. Nevertheless, a servant or agent of the carrier shall not be entitled to avail himself of the provisions of this article, if it is proved that the damage resulted from an act or omission of the servant or agent done with intent to cause damage or recklessly and with knowledge that damage would probably result.

Article V

A carrier shall be at liberty to surrender in whole or in part all or any of his rights and immunities or to increase any of his responsibilities and obligations under these Rules, provided such surrender or increase shall be embodied in the bill of lading issued to the shipper. The provisions of these Rules shall not be applicable to charterparties, but if bills of lading are issued in the case of a ship under a charter-party they shall comply with the terms of these Rules. Nothing in these Rules shall be held to prevent the insertion in a bill of lading of any lawful provision regarding general average.

Article VI

Notwithstanding the provisions of the preceding articles, a carrier, master or agent of the carrier and a shipper shall in regard to any particular goods be at liberty to enter into any agreement in any terms as to the responsibility and liability of the carrier for such goods, and as to the rights and immunities of the carrier in respect of such goods, or his obligation as to seaworthiness, so far as this stipulation is not contrary to public policy, or the care or diligence of his servants or agents in regard to the loading, handling, stowage, carriage, custody, care and discharge of the goods carried by sea, provided that in this case no bill of lading has been or shall be issued and that the terms agreed shall be embodied in a receipt which shall be a non-negotiable document and shall be marked as such.

Any agreement so entered into shall have full legal effect.

Provided that this article shall not apply to ordinary commercial shipments made in the ordinary course of trade, but only to other shipments where the character or condition of the property to be carried or the circumstances, terms and conditions under which the carriage is to be performed are such as reasonably to justify a special agreement.

Article VII

Nothing herein contained shall prevent a carrier or a shipper from entering into any agreement, stipulation, condition reservation or exemption as to the responsibility and liability of the carrier or the ship for the loss or damage to, or in connection with, the custody and care and handling of goods prior to the loading on, and subsequent to, the discharge from the ship on which the goods are carried by sea.

Article VIII

The provisions of these Rules shall not affect the rights and obligations of the carrier under any statute for the time being in force relating to the limitation of the liability of owners of sea-going vessels.

Article IX

These Rules shall not affect the provisions of any international Convention or national law governing liability for nuclear damage.

Article X

The provisions of these Rules shall apply to every bill of lading relating to the carriage of goods between ports in two different States if:

(a) the bill of lading is issued in a contracting State, or
(b) the carriage is from a port in a contracting State, or
(c) the contract contained in or evidenced by the bill of lading provides that these Rules or legislation of any State giving effect to them are to govern the contract,

whatever may be the nationality of the ship, the carrier, the shipper, the consignee, or any other interested person.

(The last two paragraphs of this article are not reproduced. They require contracting States to apply the Rules to bills of lading mentioned in the article and authorise them to apply the Rules to other bills of lading.)

(Articles XI to XVI of the International Convention for the unification of certain rules of law relating to bills of lading signed at Brussels on 25 August 25 1974 are not reproduced. They deal with the coming into force of the Convention, procedure for ratification, accession and denunciation and the right to call for a fresh conference to consider amendments to the Rules contained in the Convention.)

TEXT OF THE HAMBURG RULES

Status of this Convention. The States Parties to this Convention,

Having recognised the desirability of determining by agreement certain rules relating to the carriage of goods by sea,

Have decided to conclude a Convention for this purpose and have thereto agreed as follows:

PART I—GENERAL PROVISIONS

Article 1—Definitions

In this Convention:

1. "Carrier" means any person by whom or in whose name a contract of carriage of goods by sea has been concluded with a shipper.

2. "Actual carrier" means any person to whom the performance of the carriage of the goods, or of part of the carriage, has been entrusted by the carrier, and includes any other person to whom such performance has been entrusted.

3. "Shipper" means any person by whom or in whose name or on whose behalf a contract of carriage of goods by sea has been concluded with a carrier, or any person by whom or in whose name or on whose behalf the goods are actually delivered to the carrier in relation to the contract of carriage by sea.

4. "Consignee" means the person entitled to take delivery of the goods.

5. "Goods" includes live animals; where the goods are consolidated in a container, pallet or similar article of transport or where they are packed, "goods" includes such article of transport or packaging if supplied by the shipper.

6. "Contract of carriage by sea" means any contract whereby the carrier undertakes against payment of freight to carry goods by sea from one port to another; however, a contract which involves carriage by sea and also carriage by some other means is deemed to be a contract of carriage by sea for the purposes of this Convention only in so far as it relates to the carriage by sea.

7. "Bill of lading" means a document which evidences a contract of carriage by sea and the taking over or loading of the goods by the carrier, and by which the carrier undertakes to deliver the goods against surrender of the document. A provision in the document that the goods are to be delivered to the order of a named person, or to order, or to bearer, constitutes such an undertaking.

8. "Writing" includes, *inter alia*, telegram and telex.

Article 2—Scope of application

1. The provisions of this Convention are applicable to all contracts of carriage by sea between two different States, if:

(a) The port of loading as provided for in the contract of carriage by sea is located in a Contracting State, or
(b) The port of discharge as provided for in the contract of carriage by sea is located in a Contracting State, or
(c) One of the optional ports of discharge provided for in the contract of carriage by sea is the actual port of discharge and such port is located in a Contracting State, or
(d) The bill of lading or other document evidencing the contract of carriage by sea is issued in a Contracting State, or
(e) The bill of lading or other document evidencing the contract of carriage by sea provides that the provisions of this Convention or the legislation of any State giving effect to them are to govern the contract.

2. The provisions of this Convention are applicable without regard to the nationality of the ship, the carrier, the actual carrier, the shipper, the consignee or any other interested person.

3. The provisions of this Convention are not applicable to charter-parties. However, where a bill of lading is issued pursuant to a charter-party, the provisions of the Convention apply to such a bill of lading if it governs the relation between the carrier and the holder of the bill of lading, not being the charterer.

4. If a contract provides for future carriage of goods in a series of shipments during an agreed period, the provisions of this Convention apply to each shipment. However, where a shipment is made under a charter-party, the provisions of paragraph 3 of this Article apply.

Article 3—Interpretation of the Convention

In the interpretation and application of the provisions of this Convention regard shall be had to its international character and to the need to promote uniformity.

PART II—LIABILITY OF THE CARRIER

Article 4—Period of responsibility

1. The responsibility of the carrier for the goods under this Convention covers the period during which the carrier is in charge of the goods at the port of loading, during the carriage and at the port of discharge.

2. For the purpose of paragraph 1 of this Article, the carrier is deemed to be in charge of the goods

(a) From the time he has taken over the goods from:
  (i) The shipper, or a person acting on his behalf; or
  (ii) An authority or other third party to whom, pursuant to law or regulations applicable at the port of loading, the goods must be handed over for shipment;
(b) Until the time he has delivered the goods:
  (i) By handing over the goods to the consignee; or
  (ii) In cases where the consignee does not receive the goods from the carrier, by placing them at the disposal of the consignee in accordance with the contract or with the law or with the usage of the particular trade, applicable at the port of discharge, or
  (iii) By handing over the goods to an authority or other third party to whom, pursuant to law or regulations applicable at the port of discharge, the goods must be handed over.

3. In paragraphs 1 and 2 of this Article, reference to the carrier or to the consignee means, in addition to the carrier or the consignee, the servants or agents, respectively of the carrier or the consignee.

Article 5—Basis of liability

1. The carrier is liable for loss resulting from loss of or damage to the goods, as well as from delay in delivery, if the occurrence which caused the loss, damage or delay took place while the goods were in his charge as defined in Article 4, unless the carrier proves that he, his servants or agents took all measures that could reasonably be required to avoid the occurrence and its consequences.

2. Delay in delivery occurs when the goods have not been delivered at the port of

discharge provided for in the contract of carriage by sea within the time expressly agreed upon or, in the absence of such agreement, within the time which it would be reasonable to require of a diligent carrier, having regard to the circumstances of the case.

3. The person entitled to make a claim for the loss of goods may treat the goods as lost if they have not been delivered as required by article 4 within 60 consecutive days following the expiry of the time for delivery according to paragraph 2 of this Article.

4. (a) The carrier is liable

(i) For loss or damage to the goods or delay in delivery caused by fire, if the claimant proves that the fire arose from fault or neglect on the part of the carrier, his servants or agents;

(ii) For such loss, damage or delay in delivery which is proved by the claimant to have resulted from the fault or neglect of the carrier, his servants or agents, in taking all measures that could reasonably be required to put out the fire and avoid or mitigate its consequences.

(b) In case of fire on board the ship affecting the goods, if the claimant or the carrier so desires, a survey in accordance with shipment practices must be held into the cause and circumstances of the fire, and a copy of the surveyor's report shall be made available on demand to the carrier and the claimant.

5. With respect to live animals, the carrier is not liable for loss, damage or delay in delivery resulting from any special risks inherent in that kind of carriage. If the carrier proves that he has complied with any special instructions given to him by the shipper respecting the animals and that, in the circumstances of the case, the loss, damage or delay in delivery could be attributed to such risks, it is presumed that the loss, damage or delay in delivery was so caused, unless there is proof that all or a part of the loss, damage or delay in delivery resulted from fault or neglect on the part of the carrier, his servants or agents.

6. The carrier is not liable, except in general average, where loss, damage or delay in delivery resulted from measures to save life or from reasonable measures to save property at sea.

7. Where fault or neglect on the part of the carrier, his servants or agents combines with another cause to produce loss, damage or delay in delivery the carrier is liable only to the extent that the loss, damage or delay in delivery is attributable to such fault or neglect, provided that the carrier proves the amount of the loss, damage or delay in delivery not attributable thereto.

Article 6—Limits of liability

1. (a) The liability of the carrier for loss resulting from loss of or damage to goods according to the provisions of Article 5 is limited to an amount equivalent to 835 units of account per package or other shipping unit or 2.5 units of account per kilogram of gross weight of the goods lost or damaged, whichever is the higher.

(b) The liability of the carrier for delay in delivery according to the provisions of Article 5 is limited to an amount equivalent to two and a half times the freight payable for the goods delayed, but no exceeding the total freight payable under the contract of carriage of goods by sea.

(c) In no case shall the aggregate liability of the carrier, under both subparagraphs (a) and (b) of this paragraph, exceed the limitation which would be established under subparagraph (a) of this paragraph for total loss of the goods with respect to which such liability was incurred.

2. For the purpose of calculating which amount is the higher in accordance with paragraph 1(a) of this Article, the following rules apply:

(a) Where a container, pallet or similar article of transport is used to consolidate goods, the package or other shipping units enumerated in the bill of lading, if issued, or otherwise in any other document evidencing the contract of carriage by sea, as packed in such article of transport are deemed packages or shipping units. Except as aforesaid the goods in such article of transport are deemed one shipping unit.

(b) In cases where the article of transport itself has been lost or damaged, that article of transport, if not owned or otherwise supplied by the carrier, is considered one separate shipping unit.

3. Unit of account means the unit of account mentioned in Article 26.

4. By agreement between the carrier and the shipper, limits of liability exceeding those provided for in paragraph 1 may be fixed.

Article 7—Application to non-contractual claims

1. The defences and limits of liability provided for in this Convention apply in any action against the carrier in respect of loss or damage to the goods covered by the contract of carriage by sea, as well as of delay in delivery whether the action is founded in contract, in tort or otherwise.

2. If such action is brought against a servant or agent of the carrier, such servant or agent, if he proves that he acted within the scope of his employment, is entitled to avail himself of the defences and limits of liability which the carrier is entitled to invoke under this Convention.

3. Except as provided in Article 8, the aggregate of the amounts recoverable from the carrier and from any persons referred to in paragraph 2 of this Article shall not exceed the limits of liability provided for in this Convention.

Article 8—Loss of right to limit responsibility

1. The carrier is not entitled to the benefit of the limitation of liability provided for in Article 6 if it is proved that the loss, damage or delay in delivery resulted from an act or omission of the carrier done with the intent to cause such loss, damage or delay, or recklessly and with knowledge that such loss, damage or delay would probably result.

2. Notwithstanding the provisions of paragraph 2 of Article 7, a servant or agent of the carrier is not entitled to the benefit of the limitation of liability provided for in Article 6 if it is proved that the loss, damage or delay in delivery resulted from an act or omission of such servant or agent, done with the intent to cause such loss, damage or delay, or recklessly and with knowledge that such loss, damage or delay would probably result.

Article 9—Deck cargo

1. The carrier is entitled to carry the goods on deck only if such carriage is in accordance with an agreement with the shipper or with the usage of the particular trade or is required by statutory rules or regulations.

2. If the carrier and the shipper have agreed that the goods shall or may be carried on deck, the carrier must insert in the bill of lading or other document evidencing the contract of carriage by sea a statement to that effect. In the absence of such statement the carrier has the burden of proving that an agreement for

carriage on deck has been entered into; however, the carrier is not entitled to invoke such an agreement against a third party, including a consignee, who has acquired the bill of lading in good faith.

3. Where the goods have been carried on deck contrary to the provisions of paragraph 1 of this Article or where the carrier may not under paragraph 2 of this Article invoke an agreement for carriage on deck, the carrier, notwithstanding the provisions of paragraph 1 of Article 5, is liable for loss of or damage to the goods, as well as for delay in delivery, resulting solely from the carriage on deck, and the extent of his liability is to be determined in accordance with the provisions of Article 6 or Article 8 of this Convention, as the case may be.

4. Carriage of goods on deck contrary to express agreement for carriage under deck is deemed to be an act or omission of the carrier within the meaning of Article 8.

Article 10—Liability of the carrier and actual carrier

1. Where the performance of the carriage or part thereof has been entrusted to an actual carrier, whether or not in pursuance of a liberty under the contract of carriage by sea to do so, the carrier nevertheless remains responsible for the entire carriage according to the provisions of this Convention. The carrier is responsible, in relation to the carriage performed by the actual carrier, for the acts and omissions of the actual carrier and of his servants and agents acting within the scope of their employment.

2. All the provisions of this Convention governing the responsibility of the carrier also apply to the responsibility of the actual carrier for the carriage performed by him. The provisions of paragraphs 2 and 3 of Article 7 and of paragraph 2 of Article 8 apply if an action is brought against a servant or agent of the actual carrier.

3. Any special agreement under which the carrier assumes obligations not imposed by this Convention or waives rights conferred by this Convention affects the actual carrier only if agreed to by him expressly and in writing. Whether or not the actual carrier has so agreed, the carrier nevertheless remains bound by the obligations or waivers resulting from such special agreement.

4. Where and to the extent that both the carrier and the actual carrier are liable, their liability is joint and several.

5. The aggregate of the amounts recoverable from the carrier, the actual carrier and their servants and agents shall not exceed the limits of liability provided for in this Convention.

6. Nothing in this Article shall prejudice any right of recourse as between the carrier and the actual carrier.

Article 11—Through carriage

1. Notwithstanding the provisions of paragraph 1 of Article 10, where a contract of carriage by sea provides explicitly that a specified part of the carriage covered by the said contract is to be performed by a named person other than the carrier, the contract may also provide that the carrier is not liable for loss, damage or delay in delivery caused by an occurrence which takes place while the goods are in the charge of the actual carrier during such part of the carriage. Nevertheless, any stipulation limiting or excluding such liability is without effect if no judicial proceedings can be instituted against the actual carrier in a court competent under paragraph 1 or 2 of Article 21. The burden of proving that any loss, damage or delay in delivery has been caused by such an occurrence rests upon the carrier.

2. The actual carrier is responsible in accordance with the provisions of

paragraph 2 of Article 10 for loss, damage or delay in delivery caused by an occurrence which takes place while the goods are in his charge.

PART III—LIABILITY OF THE SHIPPER

Article 12—General rule

The shipper is not liable for loss sustained by the carrier or the actual carrier, or for damage sustained by the ship, unless such loss or damage was caused by the fault or neglect of the shipper, his servants or agents. Nor is any servant or agent of the shipper liable for such loss or damage unless the loss or damage was caused by fault or neglect on his part.

Article 13—Special rules on dangerous goods

1. The shipper must mark or label in a suitable manner dangerous goods as dangerous.

2. Where the shipper hands over dangerous goods to the carrier or an actual carrier, as the case may be, the shipper must inform him of the dangerous character of the goods and, if necessary, of the precautions to be taken. If the shipper fails to do so and such carrier or actual carrier does not otherwise have knowledge of their dangerous character:

(a) The shipper is liable to the carrier and any actual carrier for the loss resulting from the shipment of such goods, and
(b) The goods may at any time be unloaded, destroyed or rendered innocuous, as the circumstances may require, without payment of compensation.

3. The provisions of paragraph 2 of this Article may not be invoked by any person if during the carriage he has taken the goods in his charge with knowledge of their dangerous character.

4. If, in cases where the provisions of paragraph 2, subparagraph (b), of this Article do not apply or may not be invoked, dangerous goods become an actual danger to life or property, they may be unloaded, destroyed or rendered innocuous, as the circumstances may require, without payment of compensation except where there is an obligation to contribute in general average or where the carrier is liable in accordance with the provisions of Article 5.

PART IV—TRANSPORT DOCUMENTS

Article 14—Issue of bill of lading

1. When the carrier or the actual carrier takes the goods in his charge, the carrier must, on demand of the shipper, issue to the shipper a bill of lading.

2. The bill of lading may be signed by a person having authority from the carrier. A bill of lading signed by the master of the ship carrying the goods is deemed to have been signed on behalf of the carrier.

3. The signature on the bill of lading may be in handwriting, printed in facsimile, perforated, stamped, in symbols, or made by any other mechanical or electronic means, if not inconsistent with the law of the country where the bill of lading is issued.

Article 15—Contents of bill of lading

1. The bill of lading must include, *inter alia*, the following particulars:

(a) The general nature of the goods, the leading marks necessary for identification of the goods, an express statement, if applicable, as to the dangerous character of the goods, the number of packages or pieces, and the weight of the goods or their quantity otherwise expressed, all such particulars as furnished by the shipper;
(b) the apparent condition of the goods;
(c) the name and principal place of business of the carrier;
(d) the name of the shipper;
(e) the consignee if named by the shipper;
(f) the port of loading under the contract of carriage by sea and the date on which the goods were taken over by the carrier at the port of loading;
(g) the port of discharge under the contract of carriage by sea;
(h) the number of originals of the bill of lading, if more than one;
(i) the place of issuance of the bill of lading;
(j) the signature of the carrier or a person acting on his behalf;
(k) the freight to the extent payable by the consignee or other indication that freight is payable by him;
(l) the statement referred to in paragraph 3 of Article 23;
(m) the statement, if applicable, that the goods shall or may be carried on deck;
(n) the date or the period of delivery of the goods at the port of discharge if expressly agreed upon between the parties; and
(o) any increased limit or limits of liability where agreed in accordance with paragraph 4 of Article 6.

2. After the goods have been loaded on board, if the shipper so demands, the carrier must issue to the shipper a "shipped" bill of lading which, in addition to the particulars required under paragraph 1 of this Article, must state that the goods are on board a named ship or ships, and the date or dates of loading. If the carrier has previously issued to the shipper a bill of lading or other document of title with respect to any of such goods, on request of the carrier, the shipper must surrender suchdocument in exchange for a "shipped" bill of lading. The carrier may amend any previously issued document in order to meet the shipper's demand for a "shipped" bill of lading if, as amended, such document includes all the information required to be contained in a "shipped" bill of lading.

3. The absence in the bill of lading of one or more particulars referred to in this Article does not affect the legal character of the document as a bill of lading provided that it nevertheless meets the requirements set out in paragraph 7 of Article 1.

Article 16—Bills of lading: reservations and evidentiary effect

1. If the bill of lading contains particulars concerning the general nature, leading marks, number of packages or pieces, weight or quantity of the goods which the carrier or other person issuing the bill of lading on his behalf knows or has reasonable grounds to suspect do not accurately represent the goods actually taken over or, where a "shipped" bill of lading is issued, loaded, or if he had no reasonable means of checking such particulars, the carrier or such other person must insert in the bill of lading a reservation specifying these inaccuracies, grounds of suspicion or the absence of reasonable means of checking.

2. If the carrier or other person issuing the bill of lading on his behalf fails to note

on the bill of lading the apparent condition of the goods, he is deemed to have noted on the bill of lading that the goods were in apparent good condition.

3. Except for particulars in respect of which and to the extent to which a reservation permitted under paragraph 1 of this Article has been entered:

(a) The bill of lading is *prima facie* evidence of the taking over or, where a "shipped" bill of lading is issued, loading, by the carrier of the goods as described in the bill of lading; and
(b) Proof to the contrary by the carrier is not admissible if the bill of lading has been transferred to a third party, including a consignee, who in good faith has acted in reliance on the description of the goods therein.

4. A bill of lading which does not, as provided in paragraph 1, subparagraph (h) of Article 15, set forth the freight or otherwise indicate that freight is payable by the consignee or does not set forth demurrage incurred at the port of loading payable by the consignee, is *prima facie* evidence that no freight or such demurrage is payable by him. However, proof to the contrary by the carrier is not admissible when the bill of lading has been transferred to a third party, including a consignee, who in good faith has acted in reliance on the absence in the bill of lading of any such indication.

Article 17—Guarantees by the shipper

1. The shipper is deemed to have guaranteed to the carrier the accuracy of particulars relating to the general nature of the goods, their marks, number, weight and quantity as furnished by him for insertion in the bill of lading. The shipper must indemnify the carrier against the loss resulting from inaccuracies in such particulars.

The shipper remains liable even if the bill of lading has been transferred by him. The right of the carrier to such indemnity in no way limits his liability under the contract of carriage by sea to any person other than the shipper.

2. Any letter of guarantee or agreement by which the shipper undertakes to indemnify the carrier against loss resulting from the issuance of the bill of lading by the carrier, or by a person acting on his behalf, without entering a reservation relating to particulars furnished by the shipper for insertion in the bill of lading, or to the apparent condition of the goods, is void and of no effect as against any third party, including a consignee, to whom the bill of lading has been transferred.

3. Such letter of guarantee or agreement is valid as against the shipper unless the carrier or the person acting on his behalf, by omitting the reservation referred to in paragraph 2 of this Article, intends to defraud a third party, including a consignee, who acts in reliance on the description of the goods in the bill of lading. In the latter case, if the reservation omitted relates to particulars furnished by the shipper for insertion in the bill of lading, the carrier has no right of indemnity from the shipper pursuant to paragraph 1 of this Article.

4. In the case of intended fraud referred to in paragraph 3 of this Article the carrier is liable, without the benefit of the limitation of liability provided for in this Convention, for the loss incurred by a third party, including a consignee, because he has acted in reliance on the description of the goods in the bill of lading.

Article 18—Documents other than bills of lading

Where a carrier issues a document other than a bill of lading to evidence the receipt of the goods to be carried, such a document is *prima facie* evidence of the conclusion of the contract of carriage by sea and the taking over by the carrier of the goods as therein described.

PART V—CLAIMS AND ACTIONS

Article 19—Notice of loss, damage or delay

1. Unless notice of loss or damage, specifying the general nature of such loss or damage, is given in writing by the consignee to the carrier not later than the working day after the day when the goods were handed over to the consignee, such handing over is *prima facie* evidence of the delivery by the carrier of the goods as described in the document of transport or, if no such document has been issued, in good condition.

2. Where the loss or damage is not apparent, the provisions of paragraph 1 of this Article apply correspondingly if notice in writing is not given within 15 consecutive days after the day when the goods were handed over to the consignee.

3. If the state of the goods at the time they were handed over to the consignee has been the subject of a joint survey or inspection by the parties, notice in writing need not be given of loss or damage ascertained during such survey or inspection.

4. In the case of any actual or apprehended loss or damage the carrier and the consignee must give all reasonable facilities to each other for inspecting and tallying the goods.

5. No compensation shall be payable for loss resulting from delay in delivery unless a notice has been given in writing to the carrier within 60 consecutive days after the day when the goods were handed over to the consignee.

6. If the goods have been delivered by an actual carrier, any notice given under this Article to him shall have the same effect as if it had been given to the carrier, and any notice given to the carrier shall have effect as if given to such actual carrier.

7. Unless notice of loss or damage, specifying the general nature of the loss or damage, is given in writing by the carrier or actual carrier to the shipper not later than 90 consecutive days after the occurrence of such loss or damage or after the delivery of the goods in accordance with paragraph 2 of Article 4, whichever is later, the failure to give such notice is *prima facie* evidence that the carrier or the actual carrier has sustained no loss or damage due to the fault or neglect of the shipper, his servants or agents.

8. For the purpose of this Article, notice given to a person acting on the carrier's or the actual carriers' behalf, including the master or the officer in charge of the ship, or to a person acting on the shipper's behalf is deemed to have been given to the carrier, to the actual carrier or to the shipper, respectively.

Article 20—Limitation of actions

1. Any action relating to carriage of goods under this Convention is time-barred if judicial or arbitral proceedings have not been instituted within a period of two years.

2. The limitation period commences on the day on which the carrier has delivered the goods or part thereof or, in cases where no goods have been delivered, on the last day on which the goods should have been delivered.

3. The day on which the limitation period commences is not included in the period.

4. The person against whom a claim is made may at any time during the running of the limitation period extend that period by a declaration in writing to the claimant. This period may be further extended by another declaration or declarations.

5. An action for indemnity by a person held liable may be instituted even after the expiration of the limitation period provided for in the preceding paragraphs if

instituted within the time allowed by the law of the State where proceedings are instituted. However, the time allowed shall not be less than 90 days commencing from the day when the person instituting such action for indemnity has settled the claim or has been served with process in the action against himself.

Article 21—Jurisdiction

1. In judicial proceedings relating to carriage of goods under this Convention the plaintiff, at his option, may institute an action in a court which, according to the law of the State where the court is situated, is competent and within the jurisdiction of which is situated one of the following places:

(a) The principal place of business or, in the absence thereof, the habitual residence of the defendant; or
(b) The place where the contract was made provided that the defendant has there a place of business, branch or agency through which the contract was made; or
(c) The port of loading or the port of discharge; or
(d) Any additional place designated for that purpose in the contract of carriage by sea.

2.

(a) Notwithstanding the preceding provisions of this Article, an action may be instituted in the courts of any port or place in a Contracting State at which the carrying vessel or any other vessel of the same ownership may have been arrested in accordance with applicable rules of the law of that State and of international law. However, in such a case, at the petition of the defendant, the claimant must remove the action, at his choice, to one of the jurisdictions referred to in paragraph 1 of this Article for the determination of the claim, but before such removal the defendant must furnish security sufficient to ensure payment of any judgement that may subsequently be awarded to the claimant in the action.
(b) All questions relating to the sufficiency or otherwise of the security shall be determined by the court of the port or place of the arrest.

3. No judicial proceedings relating to carriage of goods under this Convention may be instituted in a place not specified in paragraph 1 or 2 of this Article. The provisions of this paragraph do not constitute an obstacle to the jurisdiction of the Contracting States for provisional or protective measures.

4.

(a) Where an action has been instituted in a court competent under paragraph 1 or 2 of this Article or where judgement has been delivered by such a court, no new action may be started between the same parties on the same grounds unless the judgement of the court before which the first action instituted is not enforceable in the country in which the new proceedings are instituted.
(b) For the purpose of this Article the institution of measures with a view to obtaining enforcement of a judgement is not to be considered as the starting of a new action;
(c) For the purpose of this Article, the removal of an action to a different court within the same country, or to a court in another country, in accordance with paragraph 2 (a) of this Article, is not to be considered as the starting of a new action.

5. Notwithstanding the provisions of the preceding paragraphs, an agreement

made by the parties, after a claim under the contract of carriage by sea has arisen, which designates the place where the claimant may institute an action, is effective.

Article 22—Arbitration

1. Subject to the provisions of this Article, parties may provide by agreement evidenced in writing that any dispute that may arise relating to carriage of goods under this Convention shall be referred to arbitration.

2. Where a charter-party contains a provision that disputes arising thereunder shall be referred to arbitration and a bill of lading issued pursuant to the charterparty does not contain a special annotation providing that such provision shall be binding upon the holder of the bill of lading, the carrier may not invoke such provision as against a holder having acquired the bill of lading in good faith.

3. The arbitration proceedings shall, at the option of the claimant, be instituted at one of the following places:

(a) A place in a State within whose territory is situated:
  (i) The principal place of business of the defendant or, in the absence thereof, the habitual residence of the defendant; or
  (ii) The place where the contract was made, provided that the defendant has there a place of business, branch or agency through which the contract was made; or
  (iii) The port of loading or the port of discharge; or
(b) any place designated for that purpose in the arbitration clause or agreement.

4. The arbitrator or arbitration tribunal shall apply the rules of this Convention.

5. The provisions of paragraph 3 and 4 of this Article are deemed to be part of every arbitration clause or agreement, and any term of such clause or agreement which is inconsistent therewith is null and void.

6. Nothing in this Article affects the validity of an agreement relating to arbitration made by the parties after the claim under the contract of carriage by sea has arisen.

PART VI—SUPPLEMENTARY PROVISIONS

Article 23—Contractual stipulations

1. Any stipulation in a contract of carriage by sea, in a bill of lading, or in any other document evidencing the contract of carriage by sea is null and void to the extent that it derogates, directly or indirectly, from the provisions of this Convention. The nullity of such a stipulation does not affect the validity of the other provisions of the contract or document of which it forms a part. A clause assigning benefit of insurance of the goods in favour of the carrier, or any similar clause, is null and void.

2. Notwithstanding the provisions of paragraph 1 of this Article, a carrier may increase his responsibilities and obligations under this Convention.

3. Where a bill of lading or any other document evidencing the contract of carriage by sea is issued, it must contain a statement that the carriage is subject to the provisions of this Convention which nullify any stipulation derogating therefrom to the detriment of the shipper or the consignee.

4. Where the claimant in respect of the goods has incurred loss as a result of a stipulation which is null and void by virtue of the present Article, or as a result of the omission of the statement referred to in paragraph 3 of this Article, the carrier

must pay compensation to the extent required in order to give the claimant compensation in accordance with the provisions of this Convention for any loss of or damage to the goods as well as for delay in delivery. The carrier must, in addition pay compensation for costs incurred by the claimant for the purpose of exercising his right, provided that costs incurred in the action where the foregoing provision is invoked are to be determined in accordance with the law of the State where proceedings are instituted.

Article 24—General average

1. Nothing in this Convention shall prevent the application of provisions in the contract of carriage by sea or national law regarding the adjustment of general average.

2. With the exception of Article 20, the provisions of this Convention relating to the liability of the carrier for loss of or damage to the goods also determine whether the consignee may refuse contribution in general average and the liability of the carrier to indemnify the consignee in respect of any such contribution made or any salvage paid.

Article 25—Other conventions

1. This Convention does not modify the rights or duties of the carrier, the actual carrier and their servants and agents, provided for in international conventions or national law relating to the limitation of liability of owners of seagoing ships.

2. The provisions of Articles 21 and 22 of this Convention do not prevent the application of the mandatory provisions of any other multilateral convention already in force at the date of this Convention relating to matters dealt with in the said Articles, provided that the dispute arises exclusively between parties having their principal place of business in States members of such other convention. However, this paragraph does not affect the application of paragraph 4 of Article 22 of this Convention.

3. No liability shall arise under the provisions of this Convention for damage caused by a nuclear incident if the operator of a nuclear installation is liable for such damage:

(a) Under either the Paris Convention of 29 July 1960 on Third Party Liability in the Field of Nuclear Energy as amended by the Additional Protocol of 28 January 1964 or the Vienna Convention of 21 May 1963 on Civil Liability for Nuclear Damage, or
(b) By virtue of national law governing the liability for such damage, provided that such law is in all respects as favourable to persons who may suffer damage as either the Paris or Vienna Conventions.

4. No liability shall arise under the provisions of this Convention for any loss of or damage to or delay in delivery of luggage for which the carrier is responsible under any international convention or national law relating to the carriage of passengers and their luggage by sea.

5. Nothing contained in this Convention prevents a Contracting State from applying any other international convention which is already in force at the date of this Convention and which applies mandatorily to contracts of carriage of goods primarily by a mode of transport other than transport by sea. This provision also applies to any subsequent revision or amendment of such international convention.

Article 26—Unit of account

1. The unit of account referred to in Article 6 of this Convention is the Special Drawing Right as defined by the International Monetary Fund. The amounts mentioned in Article 6 are to be converted into the national currency of a State according to the value of such currency at the date of judgement or the date agreed upon by the parties. The value of a national currency, in terms of the Special Drawing Right, of a Contracting State which is a member of the International Monetary Fund is to be calculated in accordance with the method of valuation applied by the International Monetary Fund in effect at the date in question for its operations and transactions. The value of a national currency in terms of the Special Drawing Right of a Contracting State which is not a member of the International Monetary Fund is to be calculated in a manner determined by that State.

2. Nevertheless, those States which are not members of the International Monetary Fund and whose law does not permit the application of the provisions of paragraph 1 of this Article may, at the time of signature, or at the time of ratification, acceptance, approval or accession or at any time thereafter, declare that the limits of liability provided for in this Convention to be applied in their territories shall be fixed as: 12,500 monetary units per package or other shipping unit or 37.5 monetary units per kilogram of gross weight of the goods.

3. The monetary unit referred to in paragraph 2 of this Article corresponds to sixty-five and a half milligrams of gold of millesimal fineness nine hundred. The conversion of the amounts referred to in paragraph 2 into the national currency is to be made according to the law of the State concerned.

4. The calculation mentioned in the last sentence of paragraph 1 and the conversion mentioned in paragraph 3 of this Article is to be made in such a manner as to express in the national currency of the Contracting State as far as possible the same real value for the amounts in Article 6 as is expressed there in units of account. Contracting States must communicate to the depositary the manner of calculation pursuant to paragraph 1 of this Article, or the result of the conversion mentioned in paragraph 3 of this Article, as the case may be, at the time of signature or when depositing their instruments of ratification, acceptance, approval or accession, or when availing themselves of the option provided for in paragraph 2 of this Article and whenever there is a change in the manner of such calculation or in the result of such conversion.

PART VII—FINAL CLAUSES

Article 27—Depositary

The Secretary-General of the United Nations is hereby designated as the depositary of this Convention.

Article 28—Signature, ratification, acceptance, approval, accession

1. This Convention is open for signature by all States until 30 April 1979 at the Headquarters of the United Nations, New York.

2. This Convention is subject to ratification, acceptance or approval by the signatory States.

3. After 30 April 1979, this Convention will be open for accession by all States which are not signatory States.

4. Instruments of ratification, acceptance, approval and accession are to be deposited with the Secretary-General of the United Nations.

Article 29—Reservations

No reservations may be made to this Convention.

Article 30—Entry into force

1. This Convention enters into force on the first day of the month following the expiration of one year from the date of deposit of the 20th instrument of ratification, acceptance, approval or accession.

2. For each State which becomes a Contracting State to this Convention after the date of the deposit of the 20th instrument of ratification, acceptance, approval or accession, this Convention enters into force on the first day of the month following the expiration of one year after the deposit of the appropriate instrument on behalf of that State.

3. Each Contracting State shall apply the provisions of this Convention to contracts of carriage by sea concluded on or after the date of the entry into force of this Convention in respect of that State.

Article 31—Denunciation of other conventions

1. Upon becoming a Contracting State to this Convention, any State party to the International Convention for the Unification of Certain Rules relating to Bills of Lading signed at Brussels on 25 August 1924 (1924 Convention) must notify the Government of Belgium as the depositary of the 1924 Convention of its denunciation of the said Convention with a declaration that the denunciation is to take effect as from the date when this Convention enters into force in respect of that State.

2. Upon the entry into force of this Convention under paragraph 1 of Article 30, the depositary of this Convention must notify the Government of Belgium as the depositary of the 1924 Convention of the date of such entry into force, and of the names of the Contracting States in respect of which the Convention has entered into force.

3. The provisions of paragraphs 1 and 2 of this Article apply correspondingly in respect of States parties to the Protocol signed on 23 February 1968 to amend the International Convention for the Unification of Certain Rules relating to Bills of Lading signed at Brussels on 25 August 1924.

4. Notwithstanding Article 2 of this Convention, for the purposes of paragraph 1 of this Article, a Contracting State may, if it deems it desirable, defer the denunciation of the 1924 Convention and of the 1924 Convention as modified by the 1968 Protocol for a maximum period of five years from the entry into force of this Convention. It will then notify the Government of Belgium of its intention. During this transitory period, it must apply to the Contracting States this Convention to the exclusion of any other one.

Article 32—Revision and amendment

1. At the request of not less than one-third of the Contracting States to this Convention, the depositary shall convene a conference of the Contracting States for revising or amending it.

2. Any instrument of ratification, acceptance, approval or accession deposited after the entry into force of an amendment to this Convention, is deemed to apply to the Convention as amended.

Article 33—Revision of the limitation amounts and unit of account or monetary unit

1. Notwithstanding the provisions of Article 32, a conference only for the purpose of altering the amount specified in Article 6 and paragraph 2 of Article 26, or of substituting either or both of the units defined in paragraphs 1 and 3 of Article 26 by other units is to be convened by the depositary in accordance with paragraph 2 of this article. An alteration of the amounts shall be made only because of a significant change in their real value.

2. A revision conference is to be convened by the depositary when not less than one-fourth of the Contracting States so request.

3. Any decision by the conference must be taken by a two-thirds majority of the participating States. The amendment is communicated by the depositary to all the Contracting States for acceptance and to all the States signatories of the Convention for information.

4. Any amendment adopted enters into force on the first day of the month following one year after its acceptance by two-thirds of the Contracting States. Acceptance is to be effected by the deposit of a formal instrument to that effect, with the depositary.

5. After entry into force of an amendment a Contracting State which has accepted the amendment is entitled to apply the Convention as amended in its relations with Contracting States which have not within six months after the adoption of the amendment notified the depositary that they are not bound by the amendment.

6. Any instrument of ratification, acceptance, approval or accession deposited after the entry into force of an amendment to this Convention, is deemed to apply to the Convention as amended.

Article 34—Denunciation

1. A Contracting State may denounce this Convention at any time by means of a notification in writing addressed to the depositary.

2. The denunciation takes effect on the first day of the month following the expiration of one year after the notification is received by the depositary. Where a longer period is specified in the notification, the denunciation takes effect upon the expiration of such longer period after the notification is received by the depositary.

Done at Hamburg, this thirty-first day of March one thousand nine hundred and seventy-eight, in a single original, of which the Arabic, Chinese, English, French, Russian and Spanish texts are equally authentic.

In witness whereof the undersigned plenipotentiaries, being duly authorised by their respective Governments, have signed the present Convention.

Common understanding adopted by the United Nations Conference on the Carriage of Goods by Sea (A/CONF.89/13, annex II)

It is the common understanding that the liability of the carrier under this Convention is based on the principle of presumed fault or neglect. This means that, as a rule, the burden of proof rests on the carrier but, with respect to certain cases, the provisions of the Convention modify this rule.

Resolution adopted by the United Nations Conference on the Carriage of Goods by Sea (A/CONF.89/13, annex III)

The United Nations Conference on the Carriage of Goods by Sea,

Noting with appreciation the kind invitation of the Federal Republic of Germany to hold the Conference in Hamburg,

Being aware that the facilities placed at the disposal of the Conference and the generous hospitality bestowed on the participants by the Government of the Federal Republic of Germany and by the Free and Hanseatic City of Hamburg, have in no small measure contributed to the success of the Conference.

Expresses its gratitude to the Government and people of the Federal Republic of Germany, and

Having adopted the Convention on the Carriage of Goods by Sea on the basis of a draft Convention prepared by the United Nations Commission on International Trade Law at the request of the United Nations Conference on Trade and Development,

Expresses its gratitude to the United Nations Commission on International Trade Law and to the United Nations Conference on Trade and Development for their outstanding contribution to the simplification and harmonisation of the law of the carriage of goods by sea, and

Decides to designate the Convention adopted by the Conference as the: "UNITED NATIONS CONVENTION ON THE CARRIAGE OF GOODS BY SEA, 1978", and

Recommends that the rules embodied therein be known as the "HAMBURG RULES".

ELECTROCHEMICAL SERIES

Oxidation Corrosion Scale

Do not corrode: Copper, Mercury, Silver, Platinum, and Gold.

Oxidation + Corrosion at ordinary temperatures: Potassium, Sodium, Calcium, Magnesium, Aluminium, Manganese, Zinc, Iron, Nickel, Lead, and Tin.

(*The above are in order of their importance, with gold being the most noble metal at the top end of the scale in the first part and tin in the second part.*)

PH VALUES

Ph is used as a measure of acidity of a solution on a scale of 1 to 13 (pure water being 7 and neutral). Concentrations below 7 indicate acidity and above that alkalinity. On the scale sea water (NaCl) is indicated by 7.8–8.2, hydrochloric acid is 2, sulphuric acid 1.7 and caustic soda 12.

SPECIFIC GRAVITY OF VARIOUS ELEMENTS

Element	**Symbol**	**Melting Point**		**SG**
		F°	**C°**	
Aluminium	Al	1215	657.2	2.7
Calcium	Ca	1490	810	1.55
Chromium	Cr	2750	1510	6.92
Copper	Cu	1981	1082	8.96
Hydrogen	H	−485		0.695
Iron	Fe	2795	1536	7.85
Lead	Pb	621	327.2	11.3
Manganese	Mn	2268	1242	7.2
Nickel	Ni	2646	1452	8.9
Potassium	K	150	65.5	0.859
Sodium	Na	208	97.7	0.972
Tin	Sn	449	231.7	7.3
Zinc	Zn	787	419.4	7.2

STOWAGE FACTOR CONVERSION TABLE

FT^3/LT	FT^3/MT	M^3/LT	M^3/MT	LT/M^3	MT/M^3
29	28.5433	0.8212	0.8083	1.2177	1.2372
30	29.5276	0.8495	0.8361	1.1172	1.1960
32	31.4961	0.9061	0.8919	1.1036	1.1212
34	33.4646	0.9628	0.9476	1.0387	1.0553
36	35.4331	1.0194	1.0034	0.9810	0.9967
38	37.4016	1.0760	1.0591	0.9293	0.9442
40	39.3701	1.1327	1.1148	0.8829	0.8970
41	40.3543	1.1610	1.1427	0.8613	0.8751
42	41.3386	1.1893	1.1706	0.8408	0.8543
43	42.3228	1.2176	1.1984	0.8213	0.8344
44	43.3071	1.2459	1.2263	0.8026	0.8154
45	44.2913	1.2743	1.2542	0.7848	0.7973
46	45.2757	1.3026	1.2821	0.7677	0.7800
47	46.2598	1.3309	1.3099	0.7514	0.7634
48	47.2441	1.3592	1.3378	0.7357	0 7475
49	48.2283	1.3875	1.3657	0.7207	0.7322
50	49.2126	1.4158	1.3935	0.7063	0.7176
52	51.1811	1.4725	1.4493	0.6791	0.6900
54	53.1496	1.5291	1.5050	0.6540	0.6644
56	55.1181	1.5857	1.5608	0.6306	0.6407
58	57.0866	1.6424	1.6165	0.6089	0.6186
60	59.0551	1.6990	1.6723	0.5886	0.5980
62	61.0236	1.7556	1.7280	0.5696	0.5787
64	62.9921	1.8123	1.7837	0.5518	0.5606
66	64.9606	1.8689	1.8395	0.5351	0.5436
68	66.9291	1.9255	1.8952	0.5193	0.5276
70	68.8976	1.9822	1.9510	0.5045	0.5126

TEMPERATURE CONVERSION TABLE

°F	°C	°F	°C	°F	°C	°F	°C	°F	°C	°F	°C
14	−10	33.5	0.8	52.7	11.5	73	22.8	93	33.9	113	45
14.5	−9.7	33.8	1	53	11.7	73.4	23	93.2	34	113.5	45.3
14.9	−9.5	34	1.1	53.5	11.9	73.5	23.1	93.5	34.2	113.9	45.5
15	−9.4	34.5	1.4	53.6	12	74	23.3	94	34.4	114	45.6
15.5	−9.2	34.7	1.5	54	12.2	74.3	23.5	94.1	34.5	114.5	45.8
15.8	−9	35	1.7	55	12.8	74.5	23.6	94.5	34.7	114.8	46
16	−8.9	35.5	1.9	55.4	13	75	23.9	95	35	115	46.1
16.5	−8.6	35.6	2	55.5	13.1	75.2	24	95.5	35.3	115.5	46.4
16.7	−8.5	36	2.2	56	13.3	75.5	24.2	95.9	35.5	115.7	46.5
17	−8.3	36.5	2.5	56.1	13.4	76	24.4	96	35.6	116	46.7
17.5	−8.1	37	2.8	56.3	13.5	76.1	24.5	96.5	35.8	116.5	46.9
17.6	−8	37.4	3	56.5	13.6	76.5	24.7	96.8	36	116.6	47
18	−7.8	37.5	3.1	57	13.9	77	25	97	36.1	117	47.2
18.5	−7.5	38	3.3	57.2	14	77.5	25.3	97.5	36.4	117.5	47.5
19	−7.2	38.3	3.5	57.5	14.2	77.9	25.5	97.7	36.5	118	47.8
19.4	−7	38.5	3.6	58	14.4	78	25.6	98	36.7	118.4	48
19.5	−6.9	39	3.9	58.1	14.5	78.5	25.8	98.5	36.9	118.5	48.1
20	−6.7	39.2	4	58.5	14.7	78.8	26	98.6	37	119	48.3
20.3	−6.5	39.5	4.2	59	15	79	26.1	99	37.2	119.3	48.5
20.5	−6.4	40	4.4	59.5	15.3	79.5	26.4	99.5	37.5	119.5	48.6
21	−6.1	40.1	4.5	59.9	15.5	79.7	26.5	100	37.8	120	48.9
21.2	−6	40.5	4.7	60	15.6	80	26.7	100.4	38	120.2	49
21.5	−5.8	41	5	60.5	15.8	80.5	26.9	100.5	38.1	120.5	49.2
22	−5.6	41.5	5.3	60.8	16	80.6	27	101	38.3	121	49.4
22.1	−5.5	41.9	5.5	61	16.1	81	27.2	101.3	38.5	121.1	49.5
22.5	−5.3	42	5.6	61.5	16.4	81.5	27.5	101.5	38.6	121.5	49.7
23	−5	42.1	5.6	61.7	16.5	82	27.8	102	38.9	122	50
23.5	−4.7	42.5	5.8	62	16.7	82.4	28	102.2	39	122.5	50.3
23.9	−4.5	42.8	6	62.5	16.9	82.5	28.1	102.5	39.2	122.9	50.5
24	−4.4	43	6.1	62.6	17	83	28.3	103	39.4	123	50.6
24.1	−4.4	43.1	6.2	63	17.2	83.3	28.5	103.1	39.5	123.5	50.8
24.5	−4.2	43.5	6.4	63.5	17.5	83.5	28.6	103.5	39.7	123.8	51
24.8	−4	43.7	6.5	64	17.8	84	28.9	104	40	124	51.1
25	−3.9	44	6.7	64.4	18	84.2	29	104.5	40.3	124.5	51.4
25.5	−3.6	44.5	6.9	64.5	18.1	84.5	29.2	104.9	40.5	124.7	51.5
25.7	−3.5	44.6	7	65	18.3	85	29.4	105	40.6	125	51.7
26	−3.3	45	7.2	65.3	18.5	85.1	29.5	105.5	40.8	125.5	51.9
26.5	−3.1	45.5	7.5	65.5	18.6	85.5	29.7	105.8	41	125.6	52
26.6	−3	46	7.8	66	18.9	86	30	106	41.1	126	52.2
27	−2.8	46.4	8	66.2	19	86.5	30.3	106.5	41.4	126.5	52.5
27.5	−2.5	46.5	8.1	66.5	19.2	86.9	30.5	106.7	41.5	127	52.8
28	−2.2	47	8.3	67	19.4	87	30.6	107	41.7	127.4	53
28.4	−2	47.3	8.5	67.1	19.5	87.5	30.8	107.5	41.9	127.5	53.1
28.5	−1.9	47.5	8.6	67.5	19.7	87.8	31	107.6	42	128	53.3
29	−1.7	48	8.9	68	20	88	31.1	108	42.2	128.3	53.5
29.3	−1.5	48.2	9	68.5	20.3	88.5	31.4	108.5	42.5	128.5	53.6
29.5	−1.4	48.5	9.2	68.9	20.5	88.7	31.5	109	42.8	129	53.9
30	−1.1	49	9.4	69	20.6	89	31.7	109.4	43	129.2	54
30.2	−1	49.1	9.5	69.5	20.8	89.5	31.9	109.5	43.1	129.5	54.2
30.5	−0.8	49.5	9.7	69.8	21	89.6	32	110	43.3	130	54.4
31	−0.6	50	10	70	21.1	90	32.2	110.3	43.5		
31.1	−0.5	50.5	10.3	70.5	21.4	90.5	32.5	110.5	43.6		
31.5	−0.3	50.9	10.5	70.7	21.5	91	32.8	111	43.9		
32	0	51	10.6	71	21.7	91.4	33	111.2	44		
32.1	0.1	51.5	10.8	71.5	21.9	91.5	33.1	111.5	44.2		
32.5	0.3	51.8	11	71.6	22	92	33.3	112	44.4		
32.9	0.5	52	11.1	72	22.2	92.3	33.5	112.1	44.5		
33	0.6	52.5	11.4	72.5	22.5	92.5	33.6	112.5	44.7		

COIL DIAMETERS (IN METRES)

Width (h) / Tonnes	0.900	0.940	1.000	1.100	1.200	1.250	1.300	1.340	1.400	1.500	1.530	1.560
8.0	1.430	1.408	1.377	1.333	1.295	1.278	1.262	1.250	1.233	1.207	1.200	1.193
9.0	1.492	1.469	1.436	1.388	1.347	1.329	1.311	1.298	1.280	1.252	1.244	1.237
10.0	1.552	1.527	1.492	1.441	1.397	1.377	1.359	1.345	1.325	1.295	1.287	1.279
11.0	1.610	1.584	1.547	1.492	1.445	1.424	1.405	1.390	1.369	1.337	1.328	1.319
12.0	1.666	1.638	1.599	1.542	1.492	1.470	1.449	1.433	1.411	1.377	1.368	1.359
13.0	1.720	1.691	1.650	1.590	1.538	1.514	1.492	1.476	1.452	1.417	1.407	1.397
14.0	1.773	1.742	1.699	1.636	1.582	1.557	1.534	1.517	1.492	1.455	1.445	1.434
15.0	1.824	1.791	1.747	1.681	1.625	1.599	1.575	1.557	1.531	1.492	1.481	1.471
16.0	1.873	1.840	1.793	1.725	1.666	1.640	1.615	1.596	1.569	1.529	1.517	1.506
17.0	1.922	1.887	1.839	1.768	1.707	1.679	1.654	1.634	1.606	1.564	1.552	1.541
18.0	1.969	1.933	1.883	1.810	1.747	1.718	1.692	1.671	1.643	1.599	1.587	1.575
19.0	2.015	1.978	1.926	1.851	1.786	1.756	1.729	1.708	1.678	1.633	1.620	1.608
20.0	2.060	2.022	1.969	1.891	1.824	1.793	1.765	1.743	1.713	1.666	1.653	1.641
21.0	2.104	2.064	2.010	1.930	1.861	1.830	1.801	1.778	1.747	1.699	1.686	1.673
22.0	2.147	2.107	2.051	1.969	1.898	1.866	1.835	1.813	1.780	1.731	1.717	1.704
23.0	2.189	2.148	2.091	2.007	1.934	1.901	1.870	1.846	1.813	1.763	1.748	1.735
24.0	2.231	2.188	2.130	2.044	1.969	1.935	1.903	1.879	1.845	1.793	1.779	1.765
25.0	2.271	2.228	2.168	2.080	2.003	1.969	1.936	1.912	1.877	1.824	1.809	1.795
26.0	2.311	2.267	2.206	2.116	2.037	2.002	1.969	1.944	1.908	1.854	1.839	1.824
27.0	2.351	2.305	2.243	2.151	2.071	2.035	2.001	1.975	1.939	1.883	1.868	1.853

VOYAGE TEMPERATURE RECORDS

DATE	TIME	OUTSIDE AIR			HOLD No....			HOLD No....			HOLD No....			HOLD No....			HOLD No....			VENTING (Yes or No)	SEA TEMP	GENERAL REMARKS (If not possible to ventilate, state reasons)
		Wet Bulb	Dry Bulb	Dew Point	Wet Bulb	Dry Bulb	Dew Point	Wet Bulb	Dry Bulb	Dew Point	Wet Bulb	Dry Bulb	Dew Point	Wet Bulb	Dry Bulb	Dew Point	Wet Bulb	Dry Bulb	Dew Point			
	0000																					
	0400																					
	0800																					
	1200																					
	1600																					
	2000																					
	0000																					
	0400																					
	0800																					
	1200																					
	1600																					
	2000																					
	0000																					
	0400																					
	0800																					
	1200																					
	1600																					
	2000																					
	0000																					
	0400																					
	0800																					
	1200																					
	1600																					
	2000																					

Average cargo temperature at loading ________ °C

Master ____________________ Chief Officer ____________________

For steel cargoes, the cargo temperature per cargo hold must be included.

HUMIDITY CHART FOR WET AND DRY BULB THERMOMETERS

Based on Pernter's Formula for air speeds past the wet bulb of about 1 to 1.5 m/s

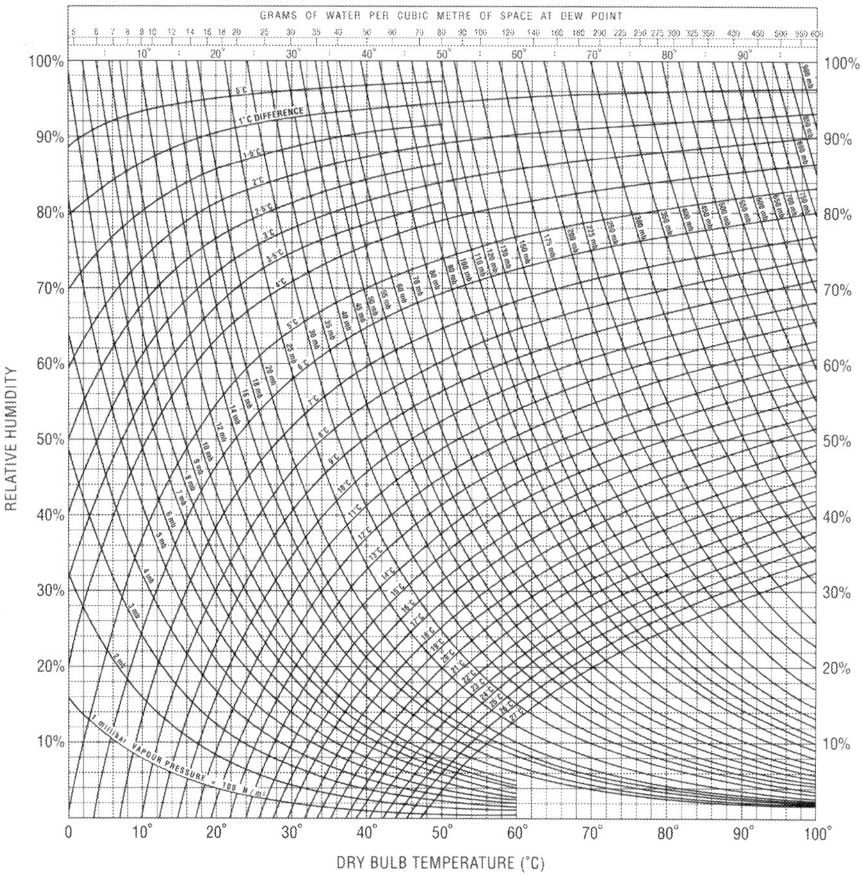

Reproduced with the kind permission of TRADA Technology Limited.

EQUILIBRIUM MOISTURE CONTENT CURVES FOR WOOD

Based on average values obtained during drying from green condition

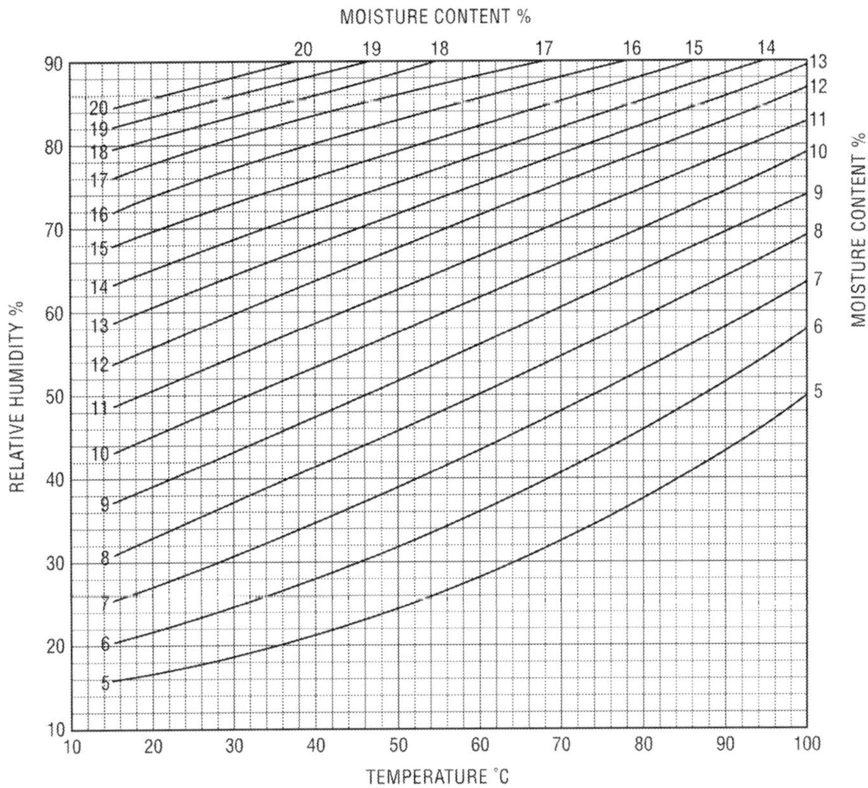

Reproduced with the kind permission of TRADA Technology Limited.

CONVERSION FACTOR TABLE

Centimetres	× 0.3937	= Inches
Inches	× 2.54	= Centimeters
Circumference	× 0.31831	= Diameter
Diameter	× 3.1416	= Circumference
Cu. Centimeters	× 0.061024	= Cu. Inches
Cu. Inches	× 16.387	= Cu. Centimeters
Cu. Meters	× 35.3148	= Cu. Feet
Cu. Feet	× 0.0283167	= Cu. Meters
Grams	× 0.083527369	= Ounces
Ounces	× 28.349527	= Grams
Metric Tons	× 0.98421	= Long Tons
Long Tons	× 1.01605	= Metric Tons
Feet	× 0.3048	= Meters
Kilograms	× 2.20462	= Pounds
Long Tons	× 1016.047	= Kilograms
Long Tons	× 2240	= Pounds
MetricTons	× 2204.6223	= Pounds
Pounds	× 0.4536	= Kilograms
Metric Tons	× 1000	= Kilograms
One Hundredweight (CWT)		= 112 Pounds
		= 50.8 Kilograms

°F = (°C × 9/5) + 32   °C = (°F − 32) × 5/9

LIST OF USEFUL ADDRESSES

Associated British Ports
150 Holborn, London, EC1N 2LR, United Kingdom
Tel: 020 7430 1177
Fax: 020 7430 1384
E-mail: pr@abports.co.uk
Web: *www.abports.co.uk*

BCA (British Cement Association)
Century House, Telford Avenue, Crowthorne, Berkshire RG45 6YS, United Kingdom
Tel: 01344 762676
Fax: 01344 761214
E-mail: info@bca.org.uk
Web: *www.bca.org.uk*
Information on re-bars, pre-stressing wire rod, for reinforcing concrete.

British Metals Recycling Association
(formerly British Scrap Federation)
16 High Street, Brampton, Huntingdon, Cambs PE18 8TU, United Kingdom
Tel: 01480 455249
Fax: 01480 453680
E-mail: admin@britmetrec.org.uk
Web: *www.britmetrec.org.uk*
Contact: Mr Marshall

British Standards Institute
389 Chiswick High Road, London W4 4AL, United Kingdom
Tel: 020 8996 9000
Fax: 020 8996 9001
E-mail: cservices@bsi-global.com
Web: *www.bsi-global.com*

Cares (UK Certification Authority for Reinforcing Steels)
Pembroke House, 21 Pembroke Road, Sevenoaks, Kent, TN13 1XR, United Kingdom
Tel: 01732 450000
Fax: 01732 455917
E-mail: general@ukcares.com
Web: *www.ukcares.com*

Infospectrum Reporting Services (Infospectrum.net Limited)
60 St Aldates, Oxford, Oxon OX1 1ST, United Kingdom
Tel: 01865 420400
Fax: 01865 420401
E-mail: info@infospectrum.net
Web: *www.infospectrum.net*
Infospectrum is the leading specialist in marine, aviation and commodity Credit and Due Diligence Reports looking at companies on a worldwide scale.

Intercargo (International Association of Dry Cargo Shipowners)
2nd Floor, 4 London Wall Buildings, Blomfield Street, London EC2M 59T, United Kingdom
Tel: 020 7638 3989
Fax: 020 7638 3943
E-mail: info@intercargo.org
Web: *www.intercargo.org*

IACS (International Association of Classification Societies)
5 Old Street, London SW1H 9JA, United Kingdom
Tel: 020 7976 0660
Fax: 020 7976 0440
E-mail: permsec@iacs.org.uk
Web: *www.iacs.org.uk*

IMO (International Maritime Organisation)
4 Albert Embankment, London SE1 7SR, United Kingdom
IMO Publishing
Tel: 020 7463 4137
Fax: 020 7587 3210
E-mail: Publications-sales@.imo.org
Web: *www.imo.org*

International Shipping Federation
12 Carthusian Street, London EC1M 6EZ, United Kingdom
Tel: 020 7417 8844
Fax: 020 7417 8877
E-mail: isf@marisec.org
Web: *www.marisec.org/isf/index.htm*

Iron & Steel Society
186 Thorn Hill Road, Warrendale, PA 15086–7528, USA
Tel: 001 724 776 1535
Fax: 001 724 776 0430
E-mail: custerv@iss.org
Web: *www.issource.org*

ISSB (Iron and Steel Statistics Bureau Limited)
1 Carlton House Terrace, London SW1Y 5DB, United Kingdom
Tel: +44 (0) 20 7343 3900
Fax: +44 (0) 20 7343 3902
E-mail: information@issb.co.uk
Web: *www.issb.co.uk*
ISSB is a publisher of steel information/statistics and is not a steel trader. It claims to be the leading producer of steel industry statistics in the UK with information covering both the national and international steel industries.

MEPS (Europe) (Management Engineering & Production Services Limited)
263 Glossop Road, Sheffield S10 2GZ, United Kingdom
Tel: 0114 275 0570
Fax: 0114 275 9808
e-mail: pmfish@meps.co.uk
Web: *www.meps.co.uk*
Managing Director: M. Fish
This company monitors steel prices around the world and issues a newsletter.

Metal Bulletin plc
Park House, 3 Park Terrace, Worcester Park, Surrey KT4 7HY, United Kingdom
Tel: 020 7827 9977
Fax: 020 7827 5290
E-mail: enquiries@metalbulletin.com
Web: *www.metalbulletin.co.uk*

Meteorological Office
London Road, Bracknell, Berkshire RG12 2SZ, United Kingdom
Tel: 0845 300 0300
Fax: 0845 300 1300
E-Mail: enquiries@metoffice.com
Web: *www.met-office.gov.uk*
(NB: The Meteorological Office is moving to Exeter in October 2003.)

Ocean Routes (WNI Ocean Routes UK Limited)
Weathernews House, Hareness Circle, Altens, Aberdeen AB18 3LY, Scotland
Tel: 01224 248080
Fax: 01224 248250
E-mail: webmaster@oceanroutes.co.uk
Web: *www.wni.co.uk*

Stainless Steel Advisory Service
Broomgrove, 59 Clarkehouse Road, Sheffield S10 2LE, United Kingdom
Tel: 0114 267 1260
Fax: 0114 266 1252
E-mail: enquiry@bssa.org.uk or ssas@materials.org.uk
Web: *www.bssa.orq.uk*

Steel Construction Institute
Silwood Park, Ascot, Berkshire SL5 7QN, United Kingdom
Tel: 01344 623345
Fax: 01344 622944
E-mail: reception@steel-sci.com
Web: *www.steel-sci.org*

Steel Times International
Queensbury House, 2 Queensway, Redhill, Surrey RH1 1QS, United Kingdom
Tel: 01737 768611
Fax: 01737 855469
E-mail: timsmith@uk.dmgworldmedia.com
Web: *www.steel-times.com*

TSL (Technical Standards Services Limited)
3 Bury Mead Road, Hitchin, Hertfordshire SG5 1RT, United Kingdom
Tel: 01462 453211
Fax: 01462 457714
E-mail: info@techstandards.co.uk
Web: *www.tssonline.net*

United Kingdom Hydrographic Office
Admiralty Way, Taunton, Somerset TA1 2DN, United Kingdom
Tel: 01823 337900
Fax: 01823 284077
E-mail: general.enquiries@ukho.gov.uk
Web: *www.ukho.gov.uk*

UK Steel Association (part of the Enginering Employers' Federation)
Broadway House, Tothill Street, London SW1H 9NQ, United Kingdom
Tel: 020 7222 7777
Fax: 020 7222 3531
E-mail: enquiries@uksteel.org.uk
Web: *www.uksteel.org.uk*

ABBREVIATIONS

AICS	Association of International Classification Societies
AP	After Perpendicular (after-edge of vessel's stern post)
AK	Aluminium Killed (a method of removing oxygen in steel making)
MPA	American Association of Port Authorities
AC	Alternating Current
ABS	American Bureau of Shipping
API	American Petroleum Institute (sets standards for pipes in the industry)
AISI	American Iron & Steel Industry
ASTM	American Standards for Testing Metals
AFNOR	French Standards
ADR	Alternative Dispute Resolution (alternative method of settling claims)
APT	After Peak Tank (Ship's Plans)
AISE	Association of Iron and Steel Engineers
ABP	Associated British Ports
AGOC	Apparent Good Order and Condition
AIMS	Australasian Institute of Marine Surveyors

AIIS	American Institute for Imported Steel
BSI	British Standards Institute (see also **List of Useful Addresses**)
BV	Bureau Veritas, Classification Society
BC Code	Code of Safe Practice for Solid Bulk Cargoes (IMO Publication)
BL	Bill of Lading
BL	Base Line (Ship's Plans)
BS	British Standards
BIMCO	Baltic & International Maritime Conference Organisation
BMLD	Moulded Breadth (Ship's Plans)
BMRA	British Metals Recycling Association (see also **List of Useful Addresses**)
BSSA	British Stainless Steel Association (see also "Stainless Steel Advisory Service" in **List of Useful Addresses**)
BSPA	The British Iron & Steel Producers' Association
BW	Bulk Loading and Unloading Code (IMO)
BS	British Corporation Classification Society
BSF	British Scrap Federation (see BMRA)
BCA	British Cement Association*
COGSA	Carriage of Goods by Sea Act
CIF	Cost Insurance and Freight. An INCOTRANS term for invoice values and conditions of financing a commercial enterprise. Used extensively in the shipping industry. —Seller: arranges for the contract, pays the freight to the named port, delivers the goods on board and completely arranges all details of the shipment, carriage and discharge. —Buyer: Accepts delivery of the goods. Pays costs of unloading if not covered by the freight.
CFR	Code of Federal regulations Shredded Scrap ISRI 211
C&F	Cost & Freight. An INCOTRANS term for invoice values and conditions for financing a commercial enterprise. —Seller: Arranges for the contract of carriage and the payment of freight, delivers the goods on board. Arranges export licence, Invoice and clean bill of lading. Pays loading costs and unloading costs. —Buyer: Arranges reception of the goods and pays discharge if not included in the freight.
COFF	Cofferdam (Ship's Plans)
CMI	Comité Maritime International
CLC	Convention on Civil Liability for Oil Pollution Damage, CLC protocol (IMO)
CEDR	Center for Dispute Resolution (London)
CSC	Convention for Safe Containers. In force from 6/9/77 (Container Safety Committee)
CRSS	Cold Rolled Steel Sheeting (coils)
CET	Central European Time
CSI	California Steel Industries
CP	Capacity Plan of a sea-going vessel
C/P	Charter Party
CQD	Customary Quick Despatch
CAL	Continuous Annealing Lines
CGL	Continuous Galvanising Lines (steel)
Cbf	Cubic feet (ft^3)
Cbm	Cubic metres (m^3)

CARES	A UK system of controlling the quality of steel reinforcing bars
CQ	Commercial Quality (steel)
CSS (Code)	Code of Safe Practice for Stowage and Securing (an IMO publication appertaining to ships' cargoes)
CSS	Californian Slab Stowage (steel)
CAPL	Continuous Annealing & Pickling Line (steel)
CDW	Cold Drawn Welded Tube (steel)
CHS	Circular Hollow Section (steel)
CR	Cold Rolled (steel)
CRTK	Centre Tank
DC	Direct Current
DNV	Den Norske Veritas (Norwegian classification society)
DRI	Direct Reduced Iron (Pellets—Cold Moulded Briquettes)
DPT	Dew Point Temperature
DIN	Deutsches Institut fur Noemung (German standards)
DOT	Diesel Oil Tank (Ship's Plans)
D.Mld	Depth Moulded (Ship's Plans)
DW	Deadweight (Ship's Plans)
DWTS	Deadweight Tonnage Summer Draft
DWTW	Deadweight Tonnage Winter Draft
DWAT	Deadweight All Told (Charter party term)
DOC	Document of Competence Compliance (ISM)
DBT	Double Bottom Tank (Ship's Plans)
DB	Double Bottom (Ship's Plans)
DSA+	Dead Slow Ahead
DSA−	Dead Slow Astern
ERW	Electric Resistive Welded (steel pipes)
EN	Enhanced Strength (New regulations for Bulk Carriers) IACS Enhanced Survey program 1993
ERH	Environmental Relative Humidity
ELEC	Electrician (Ship's Plans)
STEW	Steward (Ship's Plans)
ENG	Engineer (Ship's Plans)
BOS'N	Boson (Ship's Plans)
E/R	Engine Room (Ship's Plans)
ESP	Enhanced Survey Program (Bulk carriers)
ECSC	Europeon Coal & Steel Community
ECISS	European Committee on Iron & Steel Standardisation
EMS Guide	Emergency Procedure for Ships Carrying Dangerous Goods (IMO)
ESS	Electrical Steel Sheeting (steel)
ETA	Expected Time of Arrival
EU	European Union
ES	European Standards
EDD	Extra Deep Drawing (steel wire)
EGL	Electro Galvanising Line (steel)
FOB	Free On Board. Incotrans term —Seller: Delivers the goods on board. Provides export licence and a clean Mates Receipt. Pays loading costs if these are not included in the freight. —Buyer: Nominates the carrier and arranges contract of carriage also pays freight. Pays the costs of discharge.
FIOSLSD	Free In and Out Stowed, Lashed, Secured, Dunnaged (charter party clause).

FIOS Free In and Out Stowed (charter party clause).
FOT Fuel Oil Tank (Ship's Plans)
FPT Fore Peak Tank (Ship's Plans)
FL Fairlead (ship's plans)
FP Forward Perpendicular (ship's plans)
FCL Full Container Load. One bill of lading involved
FWE Finished With Engines
FEMAS Federation of Marine Engineers (European)
FEU Forty foot empty sea container units
FAS Free Alongside Ship (Incoterms). The sellers obligations are fulfilled when the goods are placed alongside the ship ready for loading.
FIO Free In and Out (charter party clause)
FW Fresh Water
GCEI Guide for Container Equipment Inspection
GRP Glass Reinforced Plastic
GT Gross Tonnage (Ship's Plans)
GA General Average
General arrangement plan (Ship's Plans)
GOST Russian Standards
GI Galvanized Iron (steel)
GP Galvanized Iron General Purpose (steel)
GSI Global Steel Industry
GL Germanisher Lloyd (classification society)
HRP+O Hot Rolled Pickled and Oiled (steel)
HRSS Hot Rolled Steel Sheet (steel coils)
HMB Hot Moulded Briquette
HBI Hot Briquetted Iron
HDG Hot Dip Galvanizing (steel)
HSS High Strength Steel
HR Hellenic Register of Shipping
HMS Heavymetal Scrap (steel)
HFO Heavy Fuel Oil
HR Hot Rolled (steel)
HSLA High Strength Low Alloy (steel)
HSS High Speed Steel (steel)
ISRI Institute of Steel Recycling Industries (see CFR)
ISWG International Standard Wire Gauge
ICHCA International Cargo Handling Coordination Association
ICS International Chamber of Shipping
ICS International Chamber of Shipbrokers
IOPPC International Oil Pollution Protection Conference
ITF International Transport Federation
ITM Inch Trim Moment
ISBA Club International Ship Brokers and Agents Protection and Indemnity Club Ltd
ID Inner Diameter Edge Protectors (steel coils)
IUA International Underwriting Association
IIMS International Institute of Marine Surveyors (London)
ILC International Loadline Convention
IMO International Maritime Organisation (see also **List of Useful Addresses**)
ISSB Iron & Steel Statistics Bureau (UK) (London)

ISS	Iron & Steel Society (American) (see also List of Useful Addresses)
JIS	Japanese Industrial Standards
JCCC	Joint Customs Consultative Committee
JCS	Joint Container Service
JR	Yugoslavia Register of Shipping (classification society)
JISF	Japan Iron & Steel Federation
KND	Completely Knocked Down
KR	Korean Register of Shipping (classification «07»society)
Kg	Kilogram
LR	Lloyd's Register of Shipping (classification society)
LFQ	Lock Forming Quality (galvanized sheeting)
LOI	Letter of Indemnity
LT	Long Tons
LO	Lubricating Oil
LOA	Length Over All
LMld	Length Moulded
LCL	Less than Container Load (more than one bill of lading in container)
LMAA	London Maritime Arbitrators Association.
LPG	Liquid Petroleum Gas
LC	Letter of Credit.
LME	London Metal Exchange
MARPOL	Marine Pollution Convention
MT	Metric Tons
MWG	Marine Weather Group
MLA	Maritime Law Association (American)
MSC	Marine Safety Committee (IMO)
MEPC	Marine Environment Protection Committee (IMO)
MRCCs	Maritime Rescue and Co-ordination Centres
MOUs	Mobile Offshore Units
MSC	Maritime Safety Committee (IMO)
MOWCA	Maritime Organization of West & Central Africa
MCA	Maritime & Coastguard Agency (UK)
MEPS	Management Engineering & Production Services Limited (steel prices around the world) (see also **List of Useful Addresses**)
NKK	Nipon Kaiji Kyokai (NK) (Japanese classification society)
NYPE	New York Produce Exchange, a form of charter party used extensively with steel cargoes
NAMS	National Association of Marine Surveyors Inc. (USA)
NOR	Notice of Readiness
NT	Net Tonnage
NAFTA	North America Free Trade Area
OBO	Ore. Bulk. Oil. Carrier (interchangeable)
OHV	Open HatchValue (ultrasonic, testing of steel hatches)
OSV	Off Shore Supply Vessel (OSV Code (IMO))
OS&D	Overside and Discharge Report (American Stevedores)
ODP	Outer Diameter Edge Protectors
PRS	Partly Rust Stained
PRS	Polish Register of Shipping
P&I	Protection and Indemnity Association (P&I Club)
PSSA	Particularly Sensitive Sea Area
PSC	Port State Control

POB	Pilot on Board
PPM	Parts Per Million
(P)	Prime Condition
QDA	Quick Dispatch Arrangement
RSJ	Rolled Steel Joist
RH	Relative Humidity
RP	Rust Protective
RS	Russian Register of Shipping
RS	Rust Stained
RNIA	Registro Italiano Navale
SHS	Square Hollow Section
SOLAS	Safety of Life at Sea (IMO)
SCI	Steel Construction Institute
SAE	Society of Automotive Engineers (an American standard used with steel)
SIS	Swedish Industrial Standards
SQ	Special Quality
SMS	Safety Management System (IMO)
SWL	Safe Working Load
SWBT	Salt Water Ballast Tanks (Ship's Plans)
STCWT	Stern Tube Cooling Water Tank (Ship's Plans)
STCW	Standard of Training and Watchkeeping – for seafarers (IMO)
SQE	Safety, Quality & Environmental Management
SA	Salvage Association (London)
SOF	Statement of Facts (cargo)
SO	Shipping Order (cargo)
SBE	Stand By Engines
SMV	Sound Market Value (cargo)
SW	Salt Water
Swg	Standard Wire Gauge (Imperial)
SSAS	Standard Steel Advisory Service (see also **List of Useful Addresses**)
TST	Top Side Tank (upper wing tank in a bulk carrier vessel)
TEU	Twenty Foot Empty Units (sea containers)
TF	Tropical Fresh Load Line
TLL	Timber Load Line
TPI	Tons Per Inch immersion
TCI	Tons Per Centimeter immersion
TSS	Technical Standards Services Ltd (see also **List of Useful Addresses**)
USDV	Ultra Shallow Draft Vessel
UMS	Universal Measuring System (for measuring ships' tonnages)
UTS	Ultra Tensile Strength of Steel
ULSAB	Ultra Light Steel Auto Body
USCG	United States Coast Guard
UKC	Under Keel Clearence
UBS	Universal Bulk Ship
Unclos	United Nations Convention on the Law of the Sea
ULBC	Ultra Large Bulk Carrier
UN	United Nations
UAP	Channel with parallel flanges
VDR	Voyage Data Recorders
VP	Vapour Pressure

VCI	Volatile Corrosion Inhibitors
VCG	Vertical Centre of Gravity
VLBC	Very Large Bulk Carrier
VPP	Variable Pitch Propellor
VRP	Vessels Response Plan
VTIS	Vessels Traffic Information System
VTSS	Vessels Traffic Separation Scheme
WBT	Water Ballast Tank
WB	Water Ballast
WMO	World Meteorological Organisation (Geneva)
WMU	World Maritime University (Sweden)
WRIC	Wire Rod in Coil
WFB	Wide Flanged Beams (steel)
WBMS	World Bureau of Metal Statistics
WBS	Wet Before Shipment

Index